How to Solve Applied
MATHEMATICS
PROBLEMS

B. L. Moiseiwitsch

Department of Applied Mathematics and Theoretical Physics
The Queen's University of Belfast

DOVER PUBLICATIONS, INC.
Mineola, New York

Bibliographical Note

How to Solve Applied Mathematics Problems is a new work, first published
by Dover Publications, Inc., in 2011.

International Standard Book Number

ISBN-13: 978-0-486-47927-9
ISBN-10: 0-486-47927-7

Manufactured in the United States by Courier Corporation
47927701
www.doverpublications.com

CONTENTS

PREFACE

It is only possible to have a good understanding of applied mathematics and theoretical physics by spending a long time solving a wide variety of problems. However, most students find that after attending lectures or reading books on applied mathematics, they discover to their dismay that they are simply unable to solve many of the problems. This book attempts to show students of applied mathematics how problems can be solved. The subject of applied mathematics is extremely wide ranging, as can be seen from the contents list of this book which is far from being comprehensive in scope since otherwise the book would be enormously long.

One can try to formulate basic principles for solving problems in applied mathematics, and to a certain extent this can be done, but the best way is to study the solutions of a large representative selection of problems.

It is important to realize that solving problems in applied mathematics is strongly dependent on *understanding*, and is not just a matter of memory, although a knowledge of the relevant formulae plays a significant role.

Most of the selection of problems in this book belong to the standard repertoire and all have been given to mathematics and physics undergraduate students at the Queen's University of Belfast as examples or exercises at one time or another. As far as possible lengthy and very complicated problems have been avoided and the problems have been chosen to illustrate general principles. All the problems can be solved in closed analytical forms in terms of elementary functions or simple integrals.

The book starts with an *introductory* chapter in which there is a survey of the range of subjects which are covered in the subsequent chapters by means of a small representative selection of problems, together with some general principles for solving problems in applied mathematics.

The domain of applied mathematics and theoretical physics, or mathematical physics as it is often called, may be viewed as a set of nested Chinese boxes in which it is not possible to get at the inner boxes without first opening the outer ones.

Each chapter in this book opens a new box. The outermost box which we open in this book of elementary solved problems is labelled *vector algebra*. This subject is basic to much of applied mathematics since it is a three dimensional algebra which provides the formal basis for dealing with problems in space, for example geometry.

From here we proceed to *kinematics* or the study of the motion of points in space. Next we examine the *dynamics of a particle* which provides basic examples of the use of Newton's laws of motion, including projectiles, vibrations, orbits, the motion of a charged particle in crossed electric and magnetic fields, and rotating frames of reference.

We now come to *vector field theory* which is concerned with vector functions of position in regions of space and the important theorems first proved by Gauss, Green and Stokes. This subject enables us to solve problems in *Newtonian gravitation, electricity and magnetism,* and *fluid dynamics.*

We then return to the solution of problems in *classical dynamics* through the use of Lagrange's equations and Hamilton's equations, as well as Hamilton-Jacobi theory which is fundamental to *quantum theory* discussed towards the end of the book.

Although the subjects of *Fourier series, Fourier and Laplace transforms,* and *integral equations,* are not strictly applied mathematics, they are essential for the study of *wave motions,* including *vibrating strings, sound waves* and *water waves,* and for the study of *heat conduction.* Thus the boxes labelled *wave motion* and *heat conduction* cannot be opened before the box containing *Fourier series* has been examined in some detail.

Next we turn to *tensor analysis,* which enables us to express Maxwell's electromagnetic equations in a relativistically invariant form, and to develop the *special theory* and *general theory of relativity.*

At this stage of the book we show how to treat problems in elementary *quantum theory* and finally we look at the solution of problems by using *variational principles* and *variational methods* which are extremely important for many aspects of applied mathematics and theoretical physics. This last chapter contains problems associated with many of the preceding chapters and relates them by means of the calculus of variations.

There are many applied mathematics boxes which we have not explored in this book. Several contain more than one box immediately within, some of which we have not opened, such as for the subjects of elasticity and statistical mechanics, for example.

There is a short *bibliography* at the end of this book. Many of the older books are the best and the ones I have chosen include some of the classic volumes in applied mathematics and theoretical physics. These cover the wide range of subjects which are encompassed in this book and should be referred to for details of the relevant subject matter.

To summarize, my aim has been to attempt to bridge the gap between a course of lectures in which concepts are developed and the basic applied mathematics analyzed, and the problems which are provided for students to test their understanding of the subject being studied.

INTRODUCTION

In this introductory chapter we shall try to discover how to solve problems in applied mathematics by considering a representative selection of examples from the branches of the subject to be examined in much greater detail in the subsequent chapters. This chapter will also provide a overall view of the subject of applied mathematics without going into too much depth.

We start with an elementary problem in *geometry* which we shall solve by using *vector algebra,* the subject matter of Chapter 1.

0.01 What is the angle subtended by a diameter of a circle at the circumference of the circle?

- *Solution*

In the solution of any problem the choice of *notation* is important.

We let O be the centre of the circle and let the diameter be AB where A and B are points on the circumference of the circle at opposite ends of the diameter.

Let P be any other point on the circumference of the circle.

Then we have to *find* the angle between the straight lines AP and BP.

We shall do this by using vector algebra in which we shall denote vectors by an arrow, for example $\overrightarrow{AB}$, or by bold type, for example $\mathbf{a}$, and scalar and vector products by . and × respectively.

The *scalar product* between two vectors $\mathbf{a}$ and $\mathbf{b}$ is given by $\mathbf{a}.\mathbf{b} = ab\cos\theta$ where a and b are the magnitudes of the vectors and θ is the angle between the vectors. Thus we can find the angle between AP and BP by evaluating the scalar product of the vectors $\overrightarrow{AP}$ and $\overrightarrow{BP}$.

Now by the *triangle law* $\overrightarrow{AP} = \overrightarrow{AO} + \overrightarrow{OP}$ and $\overrightarrow{BP} = \overrightarrow{BO} + \overrightarrow{OP}$. Hence

$$\overrightarrow{AP}.\overrightarrow{BP} = \left(\overrightarrow{AO} + \overrightarrow{OP}\right).\left(\overrightarrow{BO} + \overrightarrow{OP}\right)$$

Since AB is a diameter we have $\overrightarrow{BO} = -\overrightarrow{AO}$ and so we get

$$\overrightarrow{AP}.\overrightarrow{BP} = \left(\overrightarrow{AO} + \overrightarrow{OP}\right).\left(-\overrightarrow{AO} + \overrightarrow{OP}\right) = \overrightarrow{OP}.\overrightarrow{OP} - \overrightarrow{OA}.\overrightarrow{OA} = (OP)^2 - (OA)^2$$

where we have expanded the scalar product by using the *distributive law* $\mathbf{a}.(\mathbf{b} + \mathbf{c}) = \mathbf{a}.\mathbf{b} + \mathbf{a}.\mathbf{c}$ and cancelled the cross terms. Now $(OP)^2 = (OA)^2$ since OP and OA are both equal to the radius of the circle and so

$$\overrightarrow{AP}.\overrightarrow{BP} = 0.$$

Neither $\overrightarrow{AP}$ nor $\overrightarrow{BP}$ are zero vectors and so the cosine of the angle between them must be zero. It follows that this angle is a right angle or $\pi/2$.

The next problem we shall solve is on the change of position of a point with respect to time which is called *kinematics* and is the subject matter of Chapter 2.

0.02 A point is moving in a plane with constant angular velocity in such a way that its rate of change of acceleration is purely in the radial direction. Find the equation of the path of the point in polar coordinates r, θ given that $dr/dt = 0$ and $r = a$ when $\theta = \alpha$.

- Solution

It is convenient to denote differentiation with respect to time t by a dot so that $\dot{r} = dr/dt$.

Also we shall denote the position vector $\overrightarrow{OP}$ of the point P referred to an origin O by $\mathbf{r}$.

The problem refers to the radial direction and so it is clear that we need to use the *radial* and *transverse resolutes* formula for the acceleration

$$\ddot{\mathbf{r}} = (\ddot{r} - r\,\dot{\theta}^2)\hat{\mathbf{r}} + (2\,\dot{r}\dot{\theta} + r\,\ddot{\theta})\hat{\mathbf{s}} \tag{0.1}$$

where $\hat{\mathbf{r}}$ and $\hat{\mathbf{s}}$ are unit vectors in the radial and transverse directions which are related by

$$\dot{\hat{\mathbf{r}}} = \dot{\theta}\,\hat{\mathbf{s}}, \quad \dot{\hat{\mathbf{s}}} = -\dot{\theta}\,\hat{\mathbf{r}}. \tag{0.2}$$

We are given that the angular velocity $\dot{\theta}$ is a constant, which we shall denote by ω, and so we have $\ddot{\theta} = 0$. It follows that $\ddot{\mathbf{r}} = (\ddot{r} - r\omega^2)\hat{\mathbf{r}} + 2\,\dot{r}\,\omega\hat{\mathbf{s}}$ and so using (0.2) we get, on differentiating with respect to t, that

$$\dddot{\mathbf{r}} = (\dddot{r} - \dot{r}\,\omega^2 - 2\,\dot{r}\,\omega^2)\hat{\mathbf{r}} + (3\,\ddot{r} - r\omega^2)\omega\hat{\mathbf{s}}.$$

Now we are also given that the rate of change of acceleration is purely in the radial direction and so we have $3\,\ddot{r} - r\omega^2 = 0$, that is

$$\frac{d^2r}{dt^2} = \frac{1}{3}\omega^2 r.$$

The general solution of this differential equation is $r = A\cosh(\omega t/\sqrt{3} + \epsilon)$ where A and ϵ are arbitrary constants.

We can take $\theta = 0$ when $t = 0$ without any loss of generality, and then we have $\omega t = \theta$ so that the solution may be written in the form

$$r = A\cosh(\theta/\sqrt{3} + \epsilon).$$

Finally we use $dr/dt = 0$ and $r = a$ when $\theta = \alpha$. Then we get $\sinh(\alpha/\sqrt{3} + \epsilon) = 0$ so that $\epsilon = -\alpha/\sqrt{3}$, and $A = a$ giving

$$r = a\cosh\left[\frac{1}{\sqrt{3}}(\theta - \alpha)\right]$$

for the path of the moving point.

We note that in deriving this solution we have had to take account of *all* of the conditions determining the behaviour of the moving point including the *initial conditions*.

The problem could not have been solved completely if any of these conditions had not been used. This, of course, demands that the given conditions are *consistent*.

The previous problem did not involve any dynamical considerations. However our next problem is on the *dynamics of a particle* to be discussed more fully in Chapter 3, and requires the use of *Newton's second law of motion*.

0.03 A ball of mass m is projected with velocity $\mathbf{u}$ at time $t = 0$. If the force of gravity is given by $m\mathbf{g}$ and the resistance of the air is given by $\mathbf{R} = -mk\mathbf{v}$, where $\mathbf{v}$ is the velocity and k is a positive constant, show that the equation of motion can be written $d(\mathbf{v}e^{kt})/dt = \mathbf{g}e^{kt}$ where t is the time. Hence show that $\mathbf{v} = \mathbf{g}/k + (\mathbf{u} - \mathbf{g}/k)e^{-kt}$ and that the position vector of the ball referred to the point of projection as origin is $\mathbf{r} = \mathbf{g}t/k + (\mathbf{u} - \mathbf{g}/k)(1 - e^{-kt})/k$.

- *Solution*

Newton's second law of motion is

$$m\frac{d\mathbf{v}}{dt} = m\mathbf{g} + \mathbf{R} \tag{0.3}$$

and so we have

$$\frac{d\mathbf{v}}{dt} + k\mathbf{v} = \mathbf{g}. \tag{0.4}$$

Hence

$$\frac{d(\mathbf{v}e^{kt})}{dt} = \mathbf{g}e^{kt}.$$

Integrating with respect to t gives

$$\mathbf{v}e^{kt} = \frac{1}{k}\mathbf{g}e^{kt} + \mathbf{c}$$

where $\mathbf{c}$ is a constant vector, and since $\mathbf{v} = \mathbf{u}$ when $t = 0$ we obtain

$$\mathbf{v} = \frac{d\mathbf{r}}{dt} = \frac{\mathbf{g}}{k} + \left(\mathbf{u} - \frac{\mathbf{g}}{k}\right)e^{-kt}. \tag{0.5}$$

Now integrating again with respect to t we find

$$\mathbf{r} = \frac{\mathbf{g}t}{k} + \frac{1}{k}(\mathbf{u} - \frac{\mathbf{g}}{k})(1 - e^{-kt}) \tag{0.6}$$

since $\mathbf{r} = \mathbf{0}$ when $t = 0$.

Our next step is to solve a problem in *vector field theory* which is the subject of Chapter 4.

It provides an elementary exercise on the use of the *del* or *nabla operator* ∇ in div and curl, and *Laplace's operator* ∇^2. The name nabla for ∇ comes from the Greek word $\nu\alpha\beta\lambda\alpha$ for a harp and was introduced by Hamilton.

0.04 If $\mathbf{k}$ is a unit vector in the direction of the z-axis and

$$\mathbf{B} = \frac{\mathbf{k} \times \mathbf{r}}{r^6}$$

is a vector field, show that div $\mathbf{B} = 0$.

Further show that $\mathbf{B} = \operatorname{curl}\mathbf{A}$ where

$$\mathbf{A} = \frac{\mathbf{k}}{4r^4}$$

so that $\mathbf{A}$ is a vector potential for $\mathbf{B}$.

If
$$\mathbf{F} = \mathbf{A} + \nabla f$$

and div $\mathbf{F} = 0$ find the equation satisfied by f.

Show that
$$f = \frac{\cos \theta}{4r^3}$$

is a solution of this equation.

- *Solution*

We have
$$\operatorname{div} \mathbf{B} = \nabla.\mathbf{B} = \nabla.\left(\frac{\mathbf{k} \times \mathbf{r}}{r^6}\right) = -\mathbf{k}.\left(\nabla \times \frac{\mathbf{r}}{r^6}\right)$$

since $\mathbf{k}$ is a constant vector and making a cyclic interchange in the scalar triple product.

Now
$$\nabla \times \frac{\mathbf{r}}{r^6} = \frac{1}{r^6}\nabla \times \mathbf{r} + \left(\nabla \frac{1}{r^6}\right) \times \mathbf{r} = 0$$

since $\nabla \times \mathbf{r} = 0$ and $(\nabla r^{-6}) \times \mathbf{r} = -6r^{-8}\mathbf{r} \times \mathbf{r} = 0$, and so div $\mathbf{B} = 0$.

Also
$$\operatorname{curl} \mathbf{A} = \nabla \times \mathbf{A} = \nabla \times \frac{\mathbf{k}}{4r^4} = \left(\nabla \frac{1}{4r^4}\right) \times \mathbf{k} = \mathbf{k} \times \frac{\mathbf{r}}{r^6} = \mathbf{B}$$

and so we have shown that $\mathbf{A}$ is a *vector potential* for $\mathbf{B}$ as required.

Next we have that
$$\operatorname{div} \mathbf{F} = \operatorname{div} \mathbf{A} + \operatorname{div} \operatorname{grad} f = \operatorname{div} \mathbf{A} + \nabla^2 f$$

where div $\mathbf{A} = \frac{1}{4}\nabla.(r^{-4}\mathbf{k}) = \frac{1}{4}(\nabla r^{-4}).\mathbf{k} = -r^{-6}\mathbf{r}.\mathbf{k} = -r^{-5}\cos\theta$ since $\mathbf{r}.\mathbf{k} = r\cos\theta$ in spherical polar coordinates. But we are given that div $\mathbf{F} = 0$ and so f satisfies
$$\nabla^2 f = \frac{\cos \theta}{r^5}. \tag{0.7}$$

Laplace's operator
$$\nabla^2 = \frac{\partial^2}{\partial x^2} + \frac{\partial^2}{\partial y^2} + \frac{\partial^2}{\partial z^2} \tag{0.8}$$

where x, y, z are rectangular Cartesian coordinates, has the form in spherical polar coordinates r, θ, ϕ
$$\nabla^2 = \frac{1}{r^2}\frac{\partial}{\partial r}\left(r^2\frac{\partial}{\partial r}\right) + \frac{1}{r^2\sin\theta}\frac{\partial}{\partial \theta}\left(\sin\theta\frac{\partial}{\partial \theta}\right) + \frac{1}{r^2\sin^2\theta}\frac{\partial^2}{\partial \phi^2}. \tag{0.9}$$

To obtain a solution of (0.7) we put $f(r, \theta) = g(r)\cos\theta$ and then using (0.9) we get
$$\frac{d}{dr}\left(r^2\frac{dg}{dr}\right) - 2g = \frac{1}{r^3}$$

and it can be easily verified that this has the solution $g(r) = \frac{1}{4}r^{-3}$ as required.

To solve the above problem we have had to use standard formulae involving the nabla operator, for example div $\mathbf{B} = \nabla.\mathbf{B}$, curl $\mathbf{A} = \nabla \times \mathbf{A}$, $\nabla g(r) = (dg/dr)\hat{\mathbf{r}} = (r^{-1}dg/dr)\mathbf{r}$, as well as the expression (0.9) for Laplace's operator ∇^2.

An important theorem in *Newtonian gravitation* and *electricity and magnetism*, the subjects of Chapters 5 and 6, is *Gauss's theorem of total normal intensity*. We shall use this theorem in the following elementary problem in *electrostatics*.

0.05 Show that the electrostatic field arising from a uniform distribution of charge in a sphere of radius a is

$$\mathbf{E} = \begin{cases} r^{-2}Q\,\hat{\mathbf{r}} & (r > a) \\ a^{-3}Qr\,\hat{\mathbf{r}} & (r < a) \end{cases}$$

where Q is the total charge.

Further show that the formula in Gaussian units for the electrostatic energy

$$U = \frac{1}{8\pi} \int \mathbf{E}^2 \, d\tau, \tag{0.10}$$

where the integration is over all space, gives

$$U = \frac{3}{5}\frac{Q^2}{a} \tag{0.11}$$

for the electrostatic energy of the spherical distribution of charge.

- *Solution*

Gauss's theorem of total normal intensity states that

$$\int_S \mathbf{E}.\hat{\mathbf{n}}\, dS = 4\pi Q \tag{0.12}$$

where Q is the charge contained within the closed surface S and $\mathbf{E}.\hat{\mathbf{n}}$ is the normal component of the electrostatic field at the closed surface taken in the *outward direction* given by the unit vector $\hat{\mathbf{n}}$.

In the present problem for a spherical surface S with radius $r > a$ and concentric with the sphere of charge, we get $4\pi r^2 E_r = 4\pi Q$ using the spherical symmetry of the distribution, that is $E_r = r^{-2}Q$ for the radial component of the electric field strength, as required.

For a spherical surface S with radius $r < a$ we obtain $4\pi r^2 E_r = 4\pi(r^3/a^3)Q$ which gives $E_r = a^{-3}rQ$ as required.

Then the formula (0.10) for the electrostatic energy yields

$$U = \frac{1}{8\pi} \int_0^a \left(\frac{rQ}{a^3}\right)^2 4\pi r^2 dr + \frac{1}{8\pi} \int_a^\infty \left(\frac{Q}{r^2}\right)^2 4\pi r^2 dr$$

so that

$$U = \frac{Q^2}{2a^6} \int_0^a r^4\, dr + \frac{Q^2}{2} \int_a^\infty \frac{1}{r^2}\, dr = \frac{3}{5}\frac{Q^2}{a}$$

as required.

We now turn to *fluid dynamics* which is the subject of Chapter 7. We shall consider a problem on a non-viscous fluid which will involve the use of the *continuity equation*

$$\frac{\partial \rho}{\partial t} + \text{div}(\rho\mathbf{v}) = 0 \tag{0.13}$$

which corresponds to *conservation of mass*, and *Euler's equation of motion* for an inviscid fluid

$$\frac{d\mathbf{v}}{dt} = \mathbf{F} - \frac{1}{\rho}\nabla p \tag{0.14}$$

where ρ is the density of the fluid, $\mathbf{v}$ is the velocity of the fluid, p is the pressure in the fluid, and $\mathbf{F}$ is the external force acting on the fluid per unit mass. Euler's equation is based on Newton's second law of motion.

0.06 Show that for an incompressible fluid subject to external forces which form a conservative system

$$\frac{\partial \omega}{\partial t} = \omega.\nabla\mathbf{v} - \mathbf{v}.\nabla\omega \tag{0.15}$$

where $\omega = \text{curl } \mathbf{v}$ is the *vorticity*.

• *Solution*

For an incompressible fluid, or liquid, having constant density, the continuity equation becomes

$$\text{div } \mathbf{v} = 0. \tag{0.16}$$

Now

$$\frac{d\mathbf{v}}{dt} = \frac{\partial \mathbf{v}}{\partial t} + \mathbf{v}.\nabla\mathbf{v}$$

and since $\mathbf{v} \times (\nabla \times \mathbf{v}) = \nabla(\frac{1}{2}v^2) - \mathbf{v}.\nabla\mathbf{v}$ we may write Euler's equation (0.14) in the form

$$\frac{\partial \mathbf{v}}{\partial t} + \omega \times \mathbf{v} + \nabla(\frac{1}{2}v^2) = \mathbf{F} - \rho^{-1}\nabla p.$$

Since the external forces form a conservative system we may set $\mathbf{F} = -\nabla\Omega$ and so Euler's equation becomes

$$\frac{\partial \mathbf{v}}{\partial t} + \omega \times \mathbf{v} = -\nabla U \tag{0.17}$$

where $U = \Omega + \frac{1}{2}v^2 + \rho^{-1}p$ since ρ is constant for an incompressible fluid.

We now take the cross product with ∇ and this gives

$$\frac{\partial \omega}{\partial t} + \nabla \times (\omega \times \mathbf{v}) = \mathbf{0} \tag{0.18}$$

since $\nabla \times \nabla U = \mathbf{0}$.

Expanding the vector triple product gives

$$\nabla \times (\omega \times \mathbf{v}) = \mathbf{v}.\nabla\omega + \omega\nabla.\mathbf{v} - \mathbf{v}\nabla.\omega - \omega.\nabla\mathbf{v} = \mathbf{v}.\nabla\omega - \omega.\nabla\mathbf{v} \tag{0.19}$$

since the continuity equation is $\nabla.\mathbf{v} = 0$ and $\nabla.\omega = \nabla.(\nabla \times \mathbf{v}) = 0$. Substituting (0.19) into (0.18) then yields the required result (0.15).

Chapter 8 is on *classical dynamics* in which we shall use more advanced methods based on *Lagrange's equations*, *Hamilton's equations* and also the *Hamilton-Jacobi equation*.

The *Lagrangian function* is defined to be

$$L(q_r, \dot{q}_r) = T - V \tag{0.20}$$

where $T(q_r, \dot{q}_r)$ is the kinetic energy and $V(q_r)$ is the potential energy. For a dynamical system which is completely specified by n generalized coordinates $q_1, q_2, ..., q_n$, *Lagrange's equations* are

$$\frac{d}{dt}\left(\frac{\partial L}{\partial \dot{q}_r}\right) = \frac{\partial L}{\partial q_r} \quad (r = 1, ..., n). \tag{0.21}$$

The *Hamiltonian function* is defined in terms of the generalized momenta $p_r = \partial L / \partial \dot{q}_r$ $(r = 1, 2, ..., n)$ to be

$$H(q_r, p_r) = \sum_{r=1}^{n} \dot{q}_r \, \frac{\partial L}{\partial \dot{q}_r} - L \tag{0.22}$$

which for a conservative system is given by

$$H = T + V. \tag{0.23}$$

Then *Hamilton's equations* are

$$\dot{q}_r = \frac{\partial H}{\partial p_r}, \quad \dot{p}_r = -\frac{\partial H}{\partial q_r} \quad (r = 1, 2, ..., n). \tag{0.24}$$

Note that Lagrange's equations are a set of n second-order differential equations while Hamilton's equations are a set of $2n$ first-order equations.

We shall apply these equations to the following problem.

0.07 Use (i) Lagrange's equations and (ii) Hamilton's equations to obtain the equations of motion in circular polar coordinates for a particle of mass m moving in a plane under the action of a central force with potential energy $V(r)$.

• Solution

(i) The kinetic energy of a particle of mass m having velocity **v** is

$$T = \frac{1}{2} m \mathbf{v}^2 = \frac{1}{2} m (\dot{x}^2 + \dot{y}^2)$$

where $x = r \cos \theta$ and $y = r \sin \theta$ in circular polar coordinates r, θ so that in these coordinates the kinetic energy is given by

$$T = \frac{1}{2} m (\dot{r}^2 + r^2 \dot{\theta}^2). \tag{0.25}$$

The Lagrange equation for the coordinate r is

$$\frac{d}{dt} \left(\frac{\partial T}{\partial \dot{r}} \right) - \frac{\partial T}{\partial r} = -\frac{\partial V}{\partial r} \tag{0.26}$$

while the Lagrange equation for the coordinate θ is

$$\frac{d}{dt} \left(\frac{\partial T}{\partial \dot{\theta}} \right) - \frac{\partial T}{\partial \theta} = -\frac{\partial V}{\partial \theta} \tag{0.27}$$

which give

$$m (\ddot{r} - r \dot{\theta}^2) = -\frac{dV}{dr} \tag{0.28}$$

and

$$\frac{d}{dt} (mr^2 \dot{\theta}) = 0 \tag{0.29}$$

respectively.

From equation (0.29) we get $r^2 \dot{\theta} = h$ where h is the areal constant, and so equation (0.28) may be written as

$$m (\ddot{r} - h^2/r^3) = -\frac{dV}{dr}. \tag{0.30}$$

Using $\ddot{r}= d\left(\frac{1}{2}\dot{r}^2\right)/dr$ and integrating with respect to r then yields

$$m\left(\dot{r}^2 +\frac{h^2}{r^2}\right) = C - 2V \tag{0.31}$$

where C is a constant. This is equivalent to conservation of energy $T + V = \frac{1}{2}C$.

(ii) Since
$$p_r = \partial L/\partial \dot{r}= m\,\dot{r}, \quad p_\theta = \partial L/\partial \dot{\theta}= mr^2\,\dot{\theta} \tag{0.32}$$
we may write the Hamiltonian $H = T + V$ in the form

$$H = \frac{1}{2m}\left(p_r^2 +\frac{p_\theta^2}{r^2}\right) + V(r).$$

Then the Hamilton equations (0.24) give

$$\dot{p}_r= -\frac{\partial H}{\partial r} = \frac{1}{mr^3}p_\theta^2 - \frac{dV}{dr}$$

which is the same as (0.28), and

$$\dot{p}_\theta= -\frac{\partial H}{\partial \theta} = 0$$

which is the same as (0.29).

The other Hamilton's equations

$$\dot{r}= \frac{\partial H}{\partial p_r} = \frac{p_r}{m}, \quad \dot{\theta}= \frac{\partial H}{\partial p_\theta} = \frac{p_\theta}{mr^2}$$

just yield the definitions of the generalized momenta given in (0.32).

Of course, these equations could have been obtained directly by using Newton's second law of motion. Nevertheless Lagrange's equations and Hamilton's equations are often convenient to use when the system is complicated and characterized by a set of curvilinear coordinates as, for example, the theory of a spinning top to be considered in Chapter 8, since only the kinetic energy T and the potential energy V need to be known.

However the great importance of Lagrange's equations and Hamilton's equations arises more from their fundamental role in the general theory of classical dynamics than in applications.

The next topic we shall look at is *wave motion* which is described by a *partial differential equation* known as the *wave equation*. This is the subject of Chapter 11.

For a *vibrating string* the wave equation takes the form

$$\frac{\partial^2 y}{\partial x^2} = \frac{1}{c^2}\frac{\partial^2 y}{\partial t^2} \tag{0.33}$$

where $y(x,t)$ is the displacement of the string at the point x at time t.

Problems on wave motion are often solved by using *Fourier series*, to be discussed in Chapter 9. Thus consider the following problem.

0.08 A uniform string which is stretched at tension P and has length l, is fixed at its two ends. If the mid-point of the string is plucked through a small distance d and then released from rest, find the total energy of the vibrating string and show that it is the same as the work done against the tension in plucking the string at the start.

• *Solution*

We separate the x and t variables by writing the displacement of the string in the form

$$y(x,t) = (a\cos kx + b\sin kx)(A\cos ckt + B\sin ckt) \qquad (0.34)$$

which is clearly a solution of the partial differential equation (0.33).

We let the ends of the string be at $x = 0$ and $x = l$. Since the string is fixed at $x = 0$ we have $a = 0$, and since the string is also fixed at $x = l$ we have $\sin kl = 0$ so that $k = n\pi/l$ where $n = 1, 2, \ldots$. Taking $b = 1$ we now have, summing over all values of n,

$$y(x,t) = \sum_{n=1}^{\infty} \sin\frac{n\pi x}{l}\left(A_n\cos\frac{n\pi ct}{l} + B_n\sin\frac{n\pi ct}{l}\right). \qquad (0.35)$$

At time $t = 0$ we have,

$$y = \sum_{n=1}^{\infty} A_n\sin\frac{n\pi x}{l} = \begin{cases} 2dx/l & (0 \le x \le l/2) \\ 2d(l-x)/l & (l/2 \le x \le l) \end{cases}$$

and

$$\frac{\partial y}{\partial t} = \frac{\pi c}{l}\sum_{n=1}^{\infty} nB_n\sin\frac{n\pi x}{l} = 0.$$

Hence $B_n = 0$ and

$$A_n = \frac{2}{l}\int_0^{l/2}\frac{2d}{l}x\sin\frac{n\pi x}{l}\,dx + \frac{2}{l}\int_{l/2}^{l}\frac{2d}{l}(l-x)\sin\frac{n\pi x}{l}\,dx$$

since

$$\frac{2}{l}\int_0^l \sin\frac{n\pi x}{l}\sin\frac{m\pi x}{l}\,dx = \delta_{nm} \qquad (0.36)$$

where

$$\delta_{nm} = \begin{cases} 1 & (n = m) \\ 0 & (n \ne m) \end{cases} \qquad (0.37)$$

is the *Kronecker delta*.

Now $\int \sin kx\,dx = k^{-1}\cos kx$ and $\int x\sin kx\,dx = -k^{-1}x\cos kx + k^{-2}\sin kx$ which give, on putting in the limits,

$$A_n = \frac{8d}{n^2\pi^2}\sin\frac{n\pi}{2}$$

so that

$$y(x,t) = \frac{8d}{\pi^2}\sum_{n=1}^{\infty}\frac{1}{n^2}\sin\frac{n\pi}{2}\sin\frac{n\pi x}{l}\cos\frac{n\pi ct}{l}. \qquad (0.38)$$

Hence the kinetic energy of the vibrating string is

$$T = \frac{1}{2}\int_0^l \rho\left(\frac{\partial y}{\partial t}\right)^2 dx = \frac{16c^2d^2\rho}{l\pi^2}\sum_{n=1}^{\infty}\frac{1}{n^2}\left(\sin\frac{n\pi}{2}\right)^2\left(\sin\frac{n\pi ct}{l}\right)^2$$

where ρ is the mass per unit length of the string.

Since the extension of an element δx of the string is given by $\sqrt{\delta x^2 + \delta y^2} - \delta x \simeq \frac{1}{2}(\partial y/\partial x)^2\delta x$, the potential energy of the vibrating string is

$$V = \frac{1}{2}\int_0^l P\left(\frac{\partial y}{\partial x}\right)^2 dx = \frac{16d^2 P}{l\pi^2}\sum_{n=1}^{\infty}\frac{1}{n^2}\left(\sin\frac{n\pi}{2}\right)^2\left(\cos\frac{n\pi ct}{l}\right)^2.$$

Now $c^2 = P/\rho$ for a string having line density ρ which is stretched to tension P, and so the total energy of the vibrating string is

$$E = T + V = \frac{16d^2 P}{l\pi^2} \sum_{n=1}^{\infty} \frac{1}{n^2} \left(\sin \frac{n\pi}{2} \right)^2 = \frac{16d^2 P}{l\pi^2} \sum_{n=0}^{\infty} \frac{1}{(2n+1)^2}.$$

It is shown in problem 9.04 to be found in Chapter 9 on Fourier series etc., that

$$\sum_{n=0}^{\infty} \frac{1}{(2n+1)^2} = \frac{\pi^2}{8}$$

and so we get $E = 2d^2 P/l$ for the total energy of the vibrating string.

The work done against the tension P in plucking the string through a small distance d is $W = P \times$ extension where the extension is given by $2\sqrt{d^2 + l^2/4} - l \simeq 2d^2/l$ and so $W = 2d^2 P/l = E$ as required.

In solving this problem we have:

(i) used the *wave equation* of a vibrating string derived from Newton's second law of motion,

(ii) *expanded* the solution of the wave equation in the form of a *Fourier series*,

(iii) used the conditions which apply at the ends of the string, that is the *boundary conditions*,

(iv) used the *initial conditions* applying to the string, namely the initial shape of the string and the initial transverse velocity of the string,

(v) used the *formulae* for the *kinetic energy* and the *potential energy* of the string as well as other basic formulae,

(vi) shown that the energy given to the string by plucking it is equal to its energy at any later time.

The next problem we shall study is on *heat conduction* which is the subject of Chapter 12.

The partial differential equation for the conduction of heat is characteristic of diffusion and has the form

$$\frac{\partial^2 V}{\partial x^2} = \kappa \frac{\partial V}{\partial t} \tag{0.39}$$

where $V(x, t)$ is the temperature at the point x at time t and κ is called the *diffusivity*.

The solution of problems on the conduction of heat in slabs of material can be solved by using Fourier series.

However for infinite and semi-infinite solids the problems are solved by using *Fourier transforms* which are discussed in Chapter 9, together with Fourier series.

On the right hand side of the heat conduction equation (0.39) there is a first-order derivative with respect to time t and so the solution is not oscillatory as in the case of wave motion which involves a second-order time derivative.

Let us consider the following problem.

0.09 The temperature throughout a semi-infinite homogeneous solid occupying the region $x > 0$ is initially zero while that of the face $x = 0$ is maintained at temperature A. Obtain a definite integral for the temperature distribution $V(x, t)$ at time t.

• *Solution*

We use the *Fourier sine transform* by putting

$$\overline{V}(\xi,t) = \sqrt{\frac{2}{\pi}} \int_0^\infty \sin\xi x\, V(x,t)\,dx \qquad (0.40)$$

and then we see that

$$\frac{\partial \overline{V}}{\partial t} = \sqrt{\frac{2}{\pi}}\kappa \int_0^\infty \sin\xi x \frac{\partial^2 V}{\partial x^2}\,dx$$

using the heat conduction equation (0.39). Now, integrating by parts, we get

$$\int_0^\infty \sin\xi x \frac{\partial^2 V}{\partial x^2}\,dx = \left[\sin\xi x \frac{\partial V}{\partial x} - \xi\cos\xi x\, V\right]_0^\infty -\xi^2\int_0^\infty \sin\xi x\, V\,dx = A\xi-\xi^2\int_0^\infty \sin\xi x\, V\,dx$$

assuming that V and $\partial V/\partial x \to 0$ as $x \to \infty$, and putting $V = A$ at $x = 0$. Then we obtain

$$\frac{\partial \overline{V}}{\partial t} = \sqrt{\frac{2}{\pi}}\kappa A\xi - \kappa\xi^2\overline{V}$$

which has the solution

$$\overline{V}(\xi,t) = \sqrt{\frac{2}{\pi}}\frac{A}{\xi}\left[1 - \exp(-\kappa\xi^2 t)\right]$$

since $\overline{V}(\xi,0) = 0$ because the temperature of the solid is zero initially.

Hence using the *reciprocal Fourier sine transform* we get

$$V(x,t) = \sqrt{\frac{2}{\pi}}\int_0^\infty \sin\xi x\,\overline{V}(\xi,t)\,d\xi = \frac{2}{\pi}A\int_0^\infty \frac{\sin\xi x}{\xi}\left[1 - \exp(-\kappa\xi^2 t)\right]\,d\xi.$$

We note that as $t \to \infty$

$$V(x,t) \to \frac{2}{\pi}A\int_0^\infty \frac{\sin\xi x}{\xi}\,d\xi = A$$

and so the temperature becomes A throughout the semi-infinite solid after a long period of time.

We now turn our attention to the *special theory of relativity*, studied in Chapter 14 and consider the following dynamical problem.

0.10 Two particles having rest masses m_1, m_2 are moving with momenta $\mathbf{p}_1, \mathbf{p}_2$ and energies E_1, E_2 respectively relative to an inertial frame of reference S. If the particles collide and coalesce show that the rest mass m of the combined particle is given by

$$m^2 = m_1^2 + m_2^2 + 2c^{-4}(E_1 E_2 - c^2\mathbf{p}_1{\cdot}\mathbf{p}_2). \qquad (0.41)$$

Rewrite the formula for m^2 in terms of the velocities $\mathbf{v}_1, \mathbf{v}_2$ of the incident particles and hence prove that $m \geq m_1 + m_2$.

• *Solution*

The momentum of the combined particle is given by

$$\mathbf{p} = \mathbf{p}_1 + \mathbf{p}_2$$

using the principle of conservation of linear momentum, and the energy of the combined particle is given by

$$E = E_1 + E_2$$

using the principle of conservation of energy.

In the special theory of relativity the energy E and the momentum $\mathbf{p}$ of a particle having rest mass m are related by

$$E = c(m^2c^2 + \mathbf{p}^2)^{1/2} \tag{0.42}$$

where c is the speed of light, and so we have

$$E^2 = E_1^2 + E_2^2 + 2E_1E_2$$

where $E^2 = c^2(m^2c^2 + \mathbf{p}^2)$, $E_1^2 = c^2(m_1^2c^2 + \mathbf{p}_1^2)$, $E_2^2 = c^2(m_2^2c^2 + \mathbf{p}_2^2)$.

Hence

$$m^2c^2 + \mathbf{p}^2 = m_1^2c^2 + \mathbf{p}_1^2 + m_2^2c^2 + \mathbf{p}_2^2 + 2c^{-2}E_1E_2$$

which gives

$$m^2 = m_1^2 + m_2^2 + c^{-2}(\mathbf{p}_1^2 + \mathbf{p}_2^2 - \mathbf{p}^2) + 2c^{-4}E_1E_2$$

from which follows the required result (0.41).

In the special theory of relativity, the energy and momentum of a particle moving with velocity $\mathbf{v}$ referred to an inertial frame of reference S are given by $E = m\gamma c^2$ and $\mathbf{p} = m\gamma\mathbf{v}$ where $\gamma = (1 - v^2/c^2)^{-1/2}$ is the Lorentz factor.

Hence we have $E_1 = m_1\gamma_1 c^2$, $E_2 = m_2\gamma_2 c^2$, $\mathbf{p}_1 = m_1\gamma_1\mathbf{v}_1$, $\mathbf{p}_2 = m_2\gamma_2\mathbf{v}_2$ and so

$$c^{-4}(E_1E_2 - c^2\mathbf{p}_1.\mathbf{p}_2) = m_1 m_2 \gamma_1\gamma_2(1 - c^{-2}\mathbf{v}_1.\mathbf{v}_2)$$

from which it follows that

$$m^2 = m_1^2 + m_2^2 + 2m_1 m_2 \gamma_1\gamma_2\left(1 - \frac{\mathbf{v}_1.\mathbf{v}_2}{c^2}\right).$$

Now

$$(1 - c^{-2}\mathbf{v}_1.\mathbf{v}_2)^2 \geq (1 - c^{-2}v_1v_2)^2 = 1 + c^{-4}v_1^2v_2^2 - 2c^{-2}v_1v_2 \geq 1 + c^{-4}v_1^2v_2^2 - c^{-2}(v_1^2 + v_2^2)$$

since $(v_1 - v_2)^2 \geq 0$. Hence $(1 - c^{-2}\mathbf{v}_1.\mathbf{v}_2)^2 \geq (1 - v_1^2/c^2)(1 - v_2^2/c^2)$ so that $\gamma_1\gamma_2(1 - c^{-2}\mathbf{v}_1.\mathbf{v}_2) \geq 1$.

Thus we see that $m^2 \geq m_1^2 + m_2^2 + 2m_1m_2 = (m_1 + m_2)^2$ giving $m \geq m_1 + m_2$ as required.

We continue this introductory chapter by considering the following problem in *Quantum theory* which is the subject matter of Chapter 15.

It is analogous to the problem 0.08 on a plucked string.

0.11 A particle is confined within the one-dimensional square well potential with rigid walls

$$V(x) = \begin{cases} 0 & (0 \leq x \leq a) \\ \infty & (x < 0, x > a) \end{cases} \tag{0.43}$$

and has the wave function at time $t = 0$ given by the V-shape

$$\psi(x, t = 0) = u(x) = \begin{cases} 2Cx/a & (0 \leq x \leq a/2) \\ 2C(a - x)/a & (a/2 \leq x \leq a) \end{cases} \tag{0.44}$$

where C is a normalization constant.

Find the probability amplitudes for the different energy eigenvalues at time $t = 0$ and hence find an expansion for the wave function $\psi(x, t)$ at any later time t.

Further find the expectation value of the energy.

• *Solution*

In quantum theory the motion of a particle is determined by the *time-dependent Schrödinger equation*

$$H\psi = i\hbar \frac{\partial \psi}{\partial t} \tag{0.45}$$

where H is the *Hamiltonian operator* given by

$$H = \frac{\mathbf{p}^2}{2m} + V(x) \tag{0.46}$$

and $\mathbf{p} = -i\hbar\nabla$ is the *momentum operator*, where $\hbar = h/2\pi$ and h is *Planck's constant*. Then the one-dimensional time-dependent Schrödinger wave equation has the form

$$\left[-\frac{\hbar^2}{2m} \frac{d^2}{dx^2} + V(x) \right] \psi(x, t) = i\hbar \frac{\partial}{\partial t} \psi(x, t) \tag{0.47}$$

where we have taken the momentum operator to be $p = -i\hbar\partial/\partial x$.

We note that the derivative with respect to time in the Schrödinger equation is first-order although with a complex coefficient $i\hbar$, whereas the wave equation for a vibrating string has a second-order time derivative.

We now separate the variables by setting $\psi(x, t) = u(x)\exp(-iEt/\hbar)$ where E is the energy. Then we get the *time-independent Schrödinger equation*

$$\left[-\frac{\hbar^2}{2m} \frac{d^2}{dx^2} + V(x) \right] u(x) = Eu(x). \tag{0.48}$$

For $0 \leq x \leq a$ we have

$$\frac{d^2u}{dx^2} + k^2 u = 0$$

where $k = (2mE/\hbar^2)^{1/2}$, whose general solution is $u(x) = A\cos kx + B\sin kx$. Now $u(0) = u(a) = 0$ since the potential jumps to ∞ at $x = 0$ and $x = a$, and so $A = 0$ and $\sin ka = 0$ which gives $ka = n\pi$ for $n = 1, 2, \ldots$. Hence the *eigenfunctions* are $u_n(x) = B\sin(n\pi x/a)$ and the associated *energy eigenvalues* are

$$E_n = \frac{\pi^2 \hbar^2}{2ma^2} n^2 \quad (n = 1, 2, \ldots).$$

We *normalize* the eigenfunctions to unity by setting

$$\int_0^a |u_n(x)|^2 \, dx = 1$$

which gives $B = \sqrt{2/a}$. Thus the normalized eigenfunctions are

$$u_n(x) = \sqrt{\frac{2}{a}} \sin \frac{n\pi x}{a} \quad (n = 1, 2, \ldots). \tag{0.49}$$

We now normalize the wave function at time $t = 0$ given by (0.44) by setting $\int_0^a |\psi(x, t = 0)|^2 \, dx = 1$. Then we get $2(2C/a)^2 \int_0^{a/2} x^2 \, dx = 1$ which gives $C = \sqrt{3/a}$.

We note that $\psi(x, t = 0)$ vanishes at $x = 0$ and $x = a$ and so it satisfies the required boundary conditions.

Next we expand $\psi(x, t)$ in terms of the orthogonal and normalized set of functions $u_n(x) \exp(-iE_n t/\hbar)$ given by (0.49). Thus we write

$$\psi(x, t) = \sum_{n=1}^{\infty} c_n u_n(x) \exp(-iE_n t/\hbar) \tag{0.50}$$

where, according to the fourth postulate of quantum theory given at the beginning of Chapter 15, the coefficient $c_n = (u_n, u) = \int_0^a u_n^*(x) u(x) \, dx$ is the *probability amplitude* for finding the particle in the nth state of the system.

We have

$$c_n = \frac{2}{a} \sqrt{\frac{3}{a}} \sqrt{\frac{2}{a}} \left[\int_0^{a/2} x \sin \frac{n\pi x}{a} \, dx + \int_{a/2}^a (a - x) \sin \frac{n\pi x}{a} \, dx \right] = \frac{4\sqrt{6}}{n^2 \pi^2} \sin \frac{n\pi}{2}.$$

We see that $c_n = 0$ if n is even and that if n is odd and we put $n = 2s + 1$ we get

$$c_s = \frac{4\sqrt{6}}{(2s + 1)^2 \pi^2} \cos s\pi$$

and so we obtain

$$\psi(x, t) = \frac{4\sqrt{6}}{\pi^2} \sum_{s=0}^{\infty} \frac{(-1)^s}{(2s + 1)^2} \sqrt{\frac{2}{a}} \sin \frac{(2s + 1)\pi x}{a} \exp \left[-\frac{i\pi^2 \hbar}{2ma^2} (2s + 1)^2 t \right].$$

Since $|c_n|^2$ is the probability of finding the particle in the state with energy E_n the *expectation value* of the energy of the particle is given by

$$< E >= \sum_{n=1}^{\infty} |c_n|^2 E_n = \frac{48\hbar^2}{ma^2 \pi^2} \sum_{s=0}^{\infty} \frac{1}{(2s + 1)^2}.$$

But $\sum_{s=0}^{\infty} (2s + 1)^{-2} = \pi^2/8$ by the problem 9.04 and so we get

$$< E >= \frac{6\hbar^2}{ma^2}.$$

Since the wave function $u(x)$ was normalized to unity we must have $\sum_{n=0}^{\infty} |c_n|^2 = 1$. This is indeed satisfied since

$$\sum_{n=0}^{\infty} |c_n|^2 = \frac{96}{\pi^4} \sum_{s=0}^{\infty} \frac{1}{(2s + 1)^4} = 1$$

where we have used the result $\sum_{s=0}^{\infty} (2s + 1)^{-4} = \pi^4/96$ derived in problem 9.05.

It is always instructive to try to solve a problem by using a different approach.

In the present problem we may use the definition of the expectation value of an operator Ω for a wave function ψ given by $< \Omega >= \int \psi^* \Omega \psi \, d\tau$.

The expectation value of the energy is then given by $< E >= \int \psi^* H \psi \, d\tau$ where H is the Hamiltonian operator.

Since $H = -(\hbar^2/2m) d^2/dx^2$ for $0 \leq x \leq a$ in the present one-dimensional case, we have

$$< E >= -\frac{\hbar^2}{2m} \int_0^a \psi(x, t = 0) \frac{d^2}{dx^2} \psi(x, t = 0) \, dx.$$

Now $d^2\psi(x,t=0)/dx^2 = 0$ except at $x = a/2$ and hence

$$< E >= -\frac{\hbar^2}{2m}C \lim_{\epsilon \to +0} \left[\frac{d}{dx}\psi(x,t=0)\right]_{x=a/2-\epsilon}^{x=a/2+\epsilon}.$$

But the first derivative of $\psi(x,t=0)$ has a discontinuity of $4C/a$ at $x = a/2$ and so $< E >= 2\hbar^2 C^2/ma = 6\hbar^2/ma^2$.

Thus we have obtained the same result as before.

Remember that we also used two different methods to obtain the energy for the case of the plucked string discussed in problem 0.08.

The last problem we shall consider in this introductory chapter involves the use of a *variational method* which will be discussed more fully in Chapter 15 on *Quantum theory* and the final chapter.

The variational method is a very powerful and useful procedure for obtaining upper bounds to eigenvalues.

Consider the wave equation

$$\nabla^2\Psi = \frac{1}{c^2}\frac{\partial^2\Psi}{\partial t^2} \tag{0.51}$$

and let us separate the spatial variables $\mathbf{r}$ from the time t by setting
$\Psi(\mathbf{r},t) = \psi(\mathbf{r})\sin(\omega t + \epsilon)$ where ω is the angular frequency of the wave motion. Then we get the Helmholtz equation

$$\nabla^2\psi + \frac{\omega^2}{c^2}\psi = 0. \tag{0.52}$$

It will be shown in the section on eigenvalue problems in Chapter 16 on *Variational principles* that the functional

$$I[\chi] = \frac{\int (\nabla\chi)^2\, d\tau}{\int \chi^2\, d\tau} \tag{0.53}$$

gives an *upper bound* to ω_1^2/c^2 where ω_1 is the least angular frequency of vibration.

The trial function should be chosen so that it satisfies the boundary conditions as in the following problem.

0.12 A semi-circular membrane of radius a is clamped at its boundary composed of the semi-circular arc $r = a$ and the straight edges given by $\theta = 0$ and $\theta = \pi$ where r, θ are circular polar coordinates. Using the trial function

$$\chi(r,\theta) = r(a^2 - r^2)\sin\theta \tag{0.54}$$

find an upper bound to the least angular frequency of vibration of the membrane.

• *Solution*

We are solving a two-dimensional problem here and so (0.53) becomes

$$I[\chi] = \frac{\int (\nabla\chi)^2\, rdrd\theta}{\int \chi^2\, rdrd\theta} \tag{0.55}$$

where

$$\int \chi^2\, rdrd\theta = \int_0^\pi \sin^2\theta\, d\theta \int_0^a r^3(a^2 - r^2)^2\, dr = \frac{\pi a^8}{48}$$

and

$$(\nabla\chi)^2 = \left(\frac{\partial\chi}{\partial r}\right)^2 + \frac{1}{r^2}\left(\frac{\partial\chi}{\partial\theta}\right)^2 = (a^2 - 3r^2)^2 \sin^2\theta + (a^2 - r^2)^2 \cos^2\theta$$

giving, after some simple integrations,

$$\int (\nabla\chi)^2 \, r dr d\theta = \frac{\pi a^6}{3}.$$

Hence $I[\chi] = 16/a^2$ and so $\omega_1^2/c^2 \leq 16/a^2$ from which we get

$$\omega_1 \leq \frac{4c}{a}.$$

The exact value of the least angular frequency is $\omega_1 = 3.832c/a$ and so the upper bound to ω_1 obtained above using the variational method is reasonably good.

Extremely accurate results can be obtained by using sufficiently flexible trial functions and the method has been widely used to solve energy eigenvalue problems in quantum theory.

We conclude this chapter with some straightforward general principles .

In order to obtain the solution to a problem in applied mathematics you should remember to:

(i) introduce a good *notation*,

(ii) try to *understand* what you want to prove,

(iii) try to *remember* the fundamental *mathematical formulae* and *physical laws* appropriate to the problem,

(iv) put in all the *constraints* which are given,

(v) take account of the *boundary conditions* to the problem when these occur,

(vi) make use of the *initial conditions* of the problem when they are relevant.

In addition,

(vii) it is always instructive to use *more than one approach* in solving a problem whenever this is possible.

These considerations are well illustrated by the solutions to problem 0.08 on a vibrating string and problem 0.11 on a quantum mechanical particle in a box.

1
VECTOR ALGEBRA

One of the fundamental objectives of Applied Mathematics is to use mathematics to aid the understanding of the nature of the physical world. To this end it is necessary to develop a mathematical structure which is capable of dealing with events in three-dimensional space.

The problem of constructing an algebra for a triplet of numbers, representing a point in space, was resolved by the American Josiah Willard Gibbs (1839-1903). Further developments were made by the English mathematical physicist Oliver Heaviside (1850-1925) in 1891.

A controversy arose over the relative merits of vectors and quaternions, 2×2 matrices, introduced by the Irish mathematician William Rowen Hamilton (1805-1865), which satisfy the rules of an associative algebra. A strong supporter of quaternions was the Scotsman Peter Guthrie Tait (1831-1901) who was professor of mathematics at Queen's College Belfast from 1854 to 1860.

However the non-associative, non-commutative vector algebra developed by Gibbs and Heaviside proved to be the most practical system.

We shall denote vectors by using an arrow, for example $\overrightarrow{AB}$, or by using bold type, for example **a**, and scalar and vector products by . and $\times$ respectively.

We start this chapter with the application of vector methods to problems in *geometry*.

1.01 Prove the theorem of Apollonius, that if D is the mid-point of the side BC of a triangle ABC then $AB^2 + AC^2 = 2 \left(AD^2 + BD^2 \right).$

- *Solution*

The solution of any applied mathematics problem depends on knowing which are the relevant fundamental formulae.

In this problem we use the *triangle law*. For the triangle ABC we have $\overrightarrow{AB} = \overrightarrow{AD} + \overrightarrow{DB}$ and $\overrightarrow{AC} = \overrightarrow{AD} + \overrightarrow{DC}$. Hence

$$
\begin{aligned}
AB^2 + AC^2 &= (\overrightarrow{AD} + \overrightarrow{DB})^2 + (\overrightarrow{AD} + \overrightarrow{DC})^2 \\
&= 2AD^2 + 2DB^2 + 2\overrightarrow{AD}.\overrightarrow{DB} + 2\overrightarrow{AD}.\overrightarrow{DC} \\
&= 2(AD^2 + BD^2)
\end{aligned}
$$

since $\overrightarrow{DC} = -\overrightarrow{DB}$.

In the next three problems we use the following important result:

- *Theorem.*

If $\overrightarrow{OA}$ and $\overrightarrow{OB}$ are two vectors and X is a point in the line AB such that

$$
\lambda \overrightarrow{AX} = \mu \overrightarrow{XB} \tag{1.1}
$$

then

$$
\lambda \overrightarrow{OA} + \mu \overrightarrow{OB} = (\lambda + \mu) \overrightarrow{OX} \tag{1.2}
$$

1.02 Show that the position vector of the *centroid G* of a triangle ABC is given by

$$\overrightarrow{OG} = \frac{1}{3}(\overrightarrow{OA} + \overrightarrow{OB} + \overrightarrow{OC}) \qquad (1.3)$$

where O is an arbitrary origin.

• *Solution*

Let D, E, F be the mid-points of the sides BC, CA, AB respectively of a triangle ABC.

Setting $\lambda = 1, \mu = 1$ in (1.2) we have $\overrightarrow{OB} + \overrightarrow{OC} = 2\overrightarrow{OD}$. Further setting $\lambda = 1, \mu = 2$ in (1.2) we have $\overrightarrow{OA} + 2\overrightarrow{OD} = 3\overrightarrow{OX_1}$ where X_1 divides AD in the ratio 2:1. Hence

$$\overrightarrow{OA} + \overrightarrow{OB} + \overrightarrow{OC} = 3\overrightarrow{OX_1}.$$

Similarly

$$\overrightarrow{OA} + \overrightarrow{OB} + \overrightarrow{OC} = 3\overrightarrow{OX_2}$$

where X_2 divides BE in the ratio 2:1, and

$$\overrightarrow{OA} + \overrightarrow{OB} + \overrightarrow{OC} = 3\overrightarrow{OX_3}$$

where X_3 divides CF in the ratio 2:1. Hence $\overrightarrow{OX_1} = \overrightarrow{OX_2} = \overrightarrow{OX_3}$ and so the points X_1, X_2, X_3 coincide. The point of coincidence is called the centroid G and

$$\overrightarrow{OA} + \overrightarrow{OB} + \overrightarrow{OC} = 3\overrightarrow{OG}.$$

1.03 Show that the four lines joining the centroids of the faces of a tetrahedron $ABCD$ to the opposite vertices are concurrent.

• *Solution*

If G is the centroid of the face ABC we have from the previous problem that

$$\overrightarrow{OA} + \overrightarrow{OB} + \overrightarrow{OC} + \overrightarrow{OD} = 3\overrightarrow{OG} + \overrightarrow{OD} = 4\overrightarrow{OX}$$

where X divides DG in the ratio 3:1. Similarly X lies on the other lines joining the centroids to the opposite vertices and divides them in the same ratio. The point of concurrence is given by

$$\overrightarrow{OX} = \frac{1}{4}\left(\overrightarrow{OA} + \overrightarrow{OB} + \overrightarrow{OC} + \overrightarrow{OD}\right)$$

1.04 If E, F, G, H are the mid-points of the edges AB, BC, CD, DA respectively of a tetrahedron $ABCD$, show that EG and FH meet at their mid-points and that this is the same as the point of concurrence of the lines joining the vertices to the centroids of the opposite faces given in the previous problem.

Further show that $EFGH$ is a parallelogram.

• *Solution*

We have $\overrightarrow{OA} + \overrightarrow{OB} = 2\overrightarrow{OE}$ and $\overrightarrow{OC} + \overrightarrow{OD} = 2\overrightarrow{OG}$. Hence

$$\overrightarrow{OA} + \overrightarrow{OB} + \overrightarrow{OC} + \overrightarrow{OD} = 2(\overrightarrow{OE} + \overrightarrow{OG}) = 4\overrightarrow{OX_1}$$

where X_1 is the mid-point of EG. Similarly

$$\overrightarrow{OA} + \overrightarrow{OB} + \overrightarrow{OC} + \overrightarrow{OD} = 4\overrightarrow{OX_2}$$

where X_2 is the mid-point of FH. Hence EG and FH meet at their mid-points X where

$$\overrightarrow{OX} = \frac{1}{4}(\overrightarrow{OA} + \overrightarrow{OB} + \overrightarrow{OC} + \overrightarrow{OD}).$$

Further $\overrightarrow{EF} = \overrightarrow{OF} - \overrightarrow{OE} = \frac{1}{2}(\overrightarrow{OB} + \overrightarrow{OC}) - \frac{1}{2}(\overrightarrow{OA} + \overrightarrow{OB}) = \frac{1}{2}(\overrightarrow{OC} - \overrightarrow{OA}) = \frac{1}{2}\overrightarrow{AC}$ and
$\overrightarrow{HG} = \overrightarrow{OG} - \overrightarrow{OH} = \frac{1}{2}(\overrightarrow{OC} + \overrightarrow{OD}) - \frac{1}{2}(\overrightarrow{OA} + \overrightarrow{OD}) = \frac{1}{2}(\overrightarrow{OC} - \overrightarrow{OA}) = \frac{1}{2}\overrightarrow{AC}$.

Hence $\overrightarrow{EF} = \overrightarrow{HG} = \frac{1}{2}\overrightarrow{AC}$. Similarly $\overrightarrow{EH} = \overrightarrow{FG} = \frac{1}{2}\overrightarrow{BD}$. It follows that $EFGH$ is a parallelogram since its opposite sides are equal in length and are parallel.

1.05 Any five points may be taken in 10 ways, as one pair of points and a set of three points. H is the mid-point of the line joining the pair and K is the centroid of the triangle formed by the other three points. Show that the ten lines HK are concurrent.

• *Solution*

This is a problem which would require some considerable skill to establish using ordinary geometrical methods but which comes out easily using (1.2).

Let the five points be A_1, A_2, A_3, A_4, A_5. Further let H_{12} be the mid-point of the line A_1A_2 and K_{345} be the centroid of the triangle $A_3A_4A_5$. Take any origin O. Then $\overrightarrow{OA_1} + \overrightarrow{OA_2} = 2\overrightarrow{OH_{12}}$ and $\overrightarrow{OA_3} + \overrightarrow{OA_4} + \overrightarrow{OA_5} = 3\overrightarrow{OK_{345}}$.

Hence

$$\overrightarrow{OA_1} + \overrightarrow{OA_2} + \overrightarrow{OA_3} + \overrightarrow{OA_4} + \overrightarrow{OA_5} = 2\overrightarrow{OH_{12}} + 3\overrightarrow{OK_{345}} = 5\overrightarrow{OG_{12,345}}.$$

Therefore $\frac{1}{5}\sum_{i=1}^{5} \overrightarrow{OA_i}$ is the position vector of a point $G_{12,345}$ on the line $H_{12}K_{345}$ with $2H_{12}G_{12,345} = 3K_{345}G_{12,345}$. Similarly it is the position vector of a point G on each of the other lines HK. Hence the lines are concurrent at G.

Next we consider the application of vector methods to some *trigonometry* problems.

1.06 If $\mathbf{a}, \mathbf{b}, \mathbf{c}$ are vectors determining the sides BC, CA, AB respectively of a triangle ABC, show that $\mathbf{a} + \mathbf{b} + \mathbf{c} = \mathbf{0}$. Hence prove the cosine law $a^2 = b^2 + c^2 - 2bc \cos A$ where A is the angle opposite to the side BC of length a.

• *Solution*

Since $\overrightarrow{BC} + \overrightarrow{CA} + \overrightarrow{AB} = \mathbf{0}$ it follows at once that $\mathbf{a} + \mathbf{b} + \mathbf{c} = \mathbf{0}$. Hence $\mathbf{a} = -(\mathbf{b} + \mathbf{c})$ and so

$$\begin{aligned}
a^2 &= (\mathbf{b} + \mathbf{c})^2 \\
&= b^2 + c^2 + 2\mathbf{b}.\mathbf{c} \\
&= b^2 + c^2 - 2bc \cos A
\end{aligned}$$

using the definition of the *scalar product* $\mathbf{b.c} = bc \cos A$.

1.07 If $\mathbf{a} + \mathbf{b} + \mathbf{c} = 0$ show that $\mathbf{a} \times \mathbf{b} = \mathbf{b} \times \mathbf{c} = \mathbf{c} \times \mathbf{a}$ and hence prove the sine law

$$\frac{\sin A}{a} = \frac{\sin B}{b} = \frac{\sin C}{c}$$

where A, B, C are the internal angles of a triangle opposite to the sides of lengths a, b, c respectively.

- *Solution*

Since $\mathbf{a} = -(\mathbf{b} + \mathbf{c})$ we have that $\mathbf{a} \times \mathbf{b} = -(\mathbf{b} + \mathbf{c}) \times \mathbf{b} = -\mathbf{c} \times \mathbf{b} = \mathbf{b} \times \mathbf{c}$ using $\mathbf{b} \times \mathbf{b} = 0$ and $\mathbf{b} \times \mathbf{c} = -\mathbf{c} \times \mathbf{b}$. Similarly $\mathbf{a} \times \mathbf{b} = \mathbf{c} \times \mathbf{a}$ and thus we get $\mathbf{a} \times \mathbf{b} = \mathbf{b} \times \mathbf{c} = \mathbf{c} \times \mathbf{a}$. Now, using the definition of the *vector product* $\mathbf{a} \times \mathbf{b} = ab \sin C \, \hat{\mathbf{n}}$ where $\hat{\mathbf{n}}$ is a unit vector perpendicular to $\mathbf{a}$ and $\mathbf{b}$, we obtain $ab \sin C = bc \sin A = ca \sin B$ and the sine law follows.

1.08 Show that

$$(\mathbf{a} \times \mathbf{b}).(\mathbf{b} \times \mathbf{c}) = \mathbf{a.b} \ \mathbf{b.c} - b^2 \ \mathbf{a.c} \tag{1.4}$$

and hence show that

$$\cos(A + B) = \cos A \cos B - \sin A \sin B$$

- *Solution*

Using the *cyclic interchange property* of the *scalar triple product*

$$\mathbf{p}.(\mathbf{q} \times \mathbf{r}) = \mathbf{q}.(\mathbf{r} \times \mathbf{p}) = \mathbf{r}.(\mathbf{p} \times \mathbf{q}) \tag{1.5}$$

and the formula for the *vector triple product*

$$\mathbf{p} \times (\mathbf{q} \times \mathbf{r}) = \mathbf{p.r}\,\mathbf{q} - \mathbf{p.q}\,\mathbf{r} \tag{1.6}$$

we see that

$$(\mathbf{a} \times \mathbf{b}).(\mathbf{b} \times \mathbf{c}) = [\mathbf{b} \times (\mathbf{b} \times \mathbf{c})].\mathbf{a} = (\mathbf{b.c}\,\mathbf{b} - b^2\,\mathbf{c}).\mathbf{a} = \mathbf{a.b}\,\mathbf{b.c} - b^2\,\mathbf{a.c}$$

which proves the vector formula (1.4). Now taking A to be the angle between $\mathbf{a}$ and $\mathbf{b}$, and taking B to be the angle between $\mathbf{b}$ and $\mathbf{c}$, and further assuming that $\mathbf{a}$, $\mathbf{b}$, $\mathbf{c}$ are coplanar, we obtain $\sin A \sin B = \cos A \cos B - \cos(A + B)$ which yields the trigonometric formula for the cosine of the sum of two angles.

1.09 Solve the simultaneous vector equations $\mathbf{r} + \mathbf{c} \times \mathbf{s} = \mathbf{a}$, $\mathbf{s} + \mathbf{c} \times \mathbf{r} = \mathbf{b}$.

- *Solution*

Since a scalar triple product vanishes if two of its vectors are the same, we see that $\mathbf{c.r} = \mathbf{c.a}$ and $\mathbf{c.s} = \mathbf{c.b}$. Now from the first vector equation $\mathbf{c} \times \mathbf{r} + \mathbf{c} \times (\mathbf{c} \times \mathbf{s}) = \mathbf{c} \times \mathbf{a}$ so that $\mathbf{c} \times \mathbf{r} + \mathbf{c.s}\,\mathbf{c} - c^2\,\mathbf{s} = \mathbf{c} \times \mathbf{a}$. It follows that $\mathbf{b} - \mathbf{s} + \mathbf{c.b}\,\mathbf{c} - c^2\,\mathbf{s} = \mathbf{c} \times \mathbf{a}$ and so we get

$$\mathbf{s} = \frac{\mathbf{b} + \mathbf{b.c}\,\mathbf{c} - \mathbf{c} \times \mathbf{a}}{1 + c^2}$$

Hence

$$r = a - \frac{c \times b - c \times (c \times a)}{1 + c^2} = a + \frac{b \times c + c.a\,c - c^2\,a}{1 + c^2} = \frac{a + c.a\,c + b \times c}{1 + c^2}.$$

1.10 Show that the volume of a tetrahedron whose vertices have position vectors a, b, c, d is given by $\frac{1}{6}[a.(b \times c) - a.(b \times d) + a.(c \times d) - b.(c \times d)]$.

• *Solution*

The volume of a tetrahedron $ABCD$ is given by

$$V = \frac{1}{3}\text{height} \times \text{area of base} = \frac{1}{6}\overrightarrow{DA}.(\overrightarrow{DB} \times \overrightarrow{DC}) \qquad (1.7)$$

Hence

$$V = \frac{1}{6}(a - d).[(b - d) \times (c - d)] = \frac{1}{6}(a - d).(b \times c - d \times c - b \times d)$$

and the result follows.

1.11 If a, b, c, d are four vectors such that $a + b + c + d = 0$ show that the magnitude of each vector is proportional to the volume of the parallelepiped determined by the unit vectors of the other three vectors.

• *Solution*

Let $a = a\hat{a}, b = b\hat{b}, c = c\hat{c}, d = d\hat{d}$ so that we have

$$a\hat{a} + b\hat{b} + c\hat{c} + d\hat{d} = 0.$$

Let us take the scalar product with $\hat{c} \times \hat{d}$. Then we get $a\hat{a}.(\hat{c} \times \hat{d}) + b\hat{b}.(\hat{c} \times \hat{d}) = 0$ since $\hat{c}.(\hat{c} \times \hat{d}) = 0$ and $\hat{d}.(\hat{c} \times \hat{d}) = 0$. It follows that

$$\frac{a}{\hat{b}.(\hat{c} \times \hat{d})} = -\frac{b}{\hat{a}.(\hat{c} \times \hat{d})}.$$

Similarly

$$\frac{a}{\hat{c}.(\hat{b} \times \hat{d})} = -\frac{c}{\hat{a}.(\hat{b} \times \hat{d})}$$

and

$$\frac{a}{\hat{d}.(\hat{b} \times \hat{c})} = -\frac{d}{\hat{a}.(\hat{b} \times \hat{c})}.$$

Hence

$$\frac{a}{\hat{b}.(\hat{c} \times \hat{d})} = -\frac{b}{\hat{a}.(\hat{c} \times \hat{d})} = \frac{c}{\hat{a}.(\hat{b} \times \hat{d})} = -\frac{d}{\hat{a}.(\hat{b} \times \hat{c})}$$

which proves the required result since, for example, $\hat{a}.(\hat{b} \times \hat{c})$ is the volume of the parallelepiped determined by the unit vectors $\hat{a}, \hat{b}, \hat{c}$.

1.12 If p, q, r are three linearly independent vectors and

$$a = p + \alpha q, \quad b = q + \beta r, \quad c = r + \gamma p$$

where α, β, γ are scalars, show that $\alpha\beta\gamma = -1$ if and only if a, b, c are parallel to the same plane and that in this case

$$a = \alpha b + \gamma^{-1} c.$$

• *Solution*

If $\mathbf{a}, \mathbf{b}, \mathbf{c}$ are parallel to the same plane we may write $\mathbf{a} = \lambda\mathbf{b} + \mu\mathbf{c}$ where λ and μ are scalars. Then

$$\mathbf{p} + \alpha\mathbf{q} = \lambda(\mathbf{q} + \beta\mathbf{r}) + \mu(\mathbf{r} + \gamma\mathbf{p})$$

so that

$$(1 - \mu\gamma)\mathbf{p} + (\alpha - \lambda)\mathbf{q} - (\lambda\beta + \mu)\mathbf{r} = 0$$

Since $\mathbf{p}, \mathbf{q}$ and $\mathbf{r}$ are linearly independent their coefficients in the above equation must vanish and so we have $\mu\gamma = 1, \lambda = \alpha, \lambda\beta + \mu = 0$ giving $\alpha\beta\gamma = -1$.

Conversely if $\alpha\beta\gamma = -1$ then

$$\mathbf{a} = \mathbf{p} + \alpha\mathbf{q} = \mathbf{p} + \alpha(\mathbf{b} - \beta\mathbf{r}) = \alpha\mathbf{b} + \mathbf{p} - \alpha\beta\mathbf{r} = \alpha\mathbf{b} + \mathbf{p} + \gamma^{-1}\mathbf{r} = \alpha\mathbf{b} + \gamma^{-1}\mathbf{c}.$$

1.13 Two straight lines are represented by $\mathbf{r} = \mathbf{a} + \lambda\hat{\mathbf{u}}$ and $\mathbf{r} = \mathbf{b} + \mu\hat{\mathbf{v}}$ where $\hat{\mathbf{u}}$ and $\hat{\mathbf{v}}$ are unit vectors in the directions of the lines. Show that if they intersect then $\hat{\mathbf{v}}.(\mathbf{a} \times \hat{\mathbf{u}}) = \hat{\mathbf{v}}.(\mathbf{b} \times \hat{\mathbf{u}})$ and find the point of intersection.

• *Solution*

The lines intersect if we have $\mathbf{a} + \lambda\hat{\mathbf{u}} = \mathbf{b} + \mu\hat{\mathbf{v}}$ for some λ and μ, that is if $\mathbf{a}.(\hat{\mathbf{u}} \times \hat{\mathbf{v}}) = \mathbf{b}.(\hat{\mathbf{u}} \times \hat{\mathbf{v}})$ or, using the property of the invariance of the scalar triple product under cyclic interchange, if $\hat{\mathbf{v}}.(\mathbf{a} \times \hat{\mathbf{u}}) = \hat{\mathbf{v}}.(\mathbf{b} \times \hat{\mathbf{u}})$. At the point of intersection $\mathbf{a}.(\mathbf{b} \times \hat{\mathbf{v}}) + \lambda\hat{\mathbf{u}}.(\mathbf{b} \times \hat{\mathbf{v}}) = 0$ so that

$$\lambda = \frac{\mathbf{a}.(\mathbf{b} \times \hat{\mathbf{v}})}{\hat{\mathbf{v}}.(\mathbf{b} \times \hat{\mathbf{u}})} = \frac{\mathbf{a}.(\mathbf{b} \times \hat{\mathbf{v}})}{\hat{\mathbf{v}}.(\mathbf{a} \times \hat{\mathbf{u}})}$$

Hence the point of intersection is given by the position vector

$$\mathbf{a} + \frac{\mathbf{a}.(\mathbf{b} \times \hat{\mathbf{v}})}{\mathbf{a}.(\hat{\mathbf{u}} \times \hat{\mathbf{v}})} \, \hat{\mathbf{u}}$$

1.14 Show that the perpendicular distance of the origin from a plane passing through three non-collinear points A, B, C with position vectors $\mathbf{a}, \mathbf{b}, \mathbf{c}$ is given by

$$\frac{|\mathbf{a}.(\mathbf{b} \times \mathbf{c})|}{|\mathbf{a} \times \mathbf{b} + \mathbf{b} \times \mathbf{c} + \mathbf{c} \times \mathbf{a}|}$$

• *Solution*

A vector which is perpendicular to the plane is $(\mathbf{a} - \mathbf{b}) \times (\mathbf{a} - \mathbf{c}) = \mathbf{a} \times \mathbf{b} + \mathbf{b} \times \mathbf{c} + \mathbf{c} \times \mathbf{a}$ and so a unit vector in this direction is

$$\hat{\mathbf{n}} = \frac{\mathbf{a} \times \mathbf{b} + \mathbf{b} \times \mathbf{c} + \mathbf{c} \times \mathbf{a}}{|\mathbf{a} \times \mathbf{b} + \mathbf{b} \times \mathbf{c} + \mathbf{c} \times \mathbf{a}|}$$

The perpendicular distance of the origin from the plane is therefore

$$p = |\mathbf{a}.\hat{\mathbf{n}}| = \frac{|\mathbf{a}.(\mathbf{a} \times \mathbf{b} + \mathbf{b} \times \mathbf{c} + \mathbf{c} \times \mathbf{a})|}{|\mathbf{a} \times \mathbf{b} + \mathbf{b} \times \mathbf{c} + \mathbf{c} \times \mathbf{a}|} = \frac{|\mathbf{a}.(\mathbf{b} \times \mathbf{c})|}{|\mathbf{a} \times \mathbf{b} + \mathbf{b} \times \mathbf{c} + \mathbf{c} \times \mathbf{a}|}$$

We conclude this chapter on vectors with the following interesting result on vector triple products.

1.15 Show that $\mathbf{a} \times (\mathbf{b} \times \mathbf{c}) + \mathbf{b} \times (\mathbf{c} \times \mathbf{a}) + \mathbf{c} \times (\mathbf{a} \times \mathbf{b}) = 0$.

• *Solution*

Using the formula for the vector triple product $\mathbf{a} \times (\mathbf{b} \times \mathbf{c}) = \mathbf{a}.\mathbf{c} \, \mathbf{b} - \mathbf{a}.\mathbf{b} \, \mathbf{c}$ we get

$$\mathbf{a} \times (\mathbf{b} \times \mathbf{c}) + \mathbf{b} \times (\mathbf{c} \times \mathbf{a}) + \mathbf{c} \times (\mathbf{a} \times \mathbf{b}) = \mathbf{a}.\mathbf{c} \, \mathbf{b} - \mathbf{a}.\mathbf{b} \, \mathbf{c} + \mathbf{b}.\mathbf{a} \, \mathbf{c} - \mathbf{b}.\mathbf{c} \, \mathbf{a} + \mathbf{c}.\mathbf{b} \, \mathbf{a} - \mathbf{c}.\mathbf{a} \, \mathbf{b} = 0$$

2
KINEMATICS

This chapter is concerned with problems on the *motion of points or particles in space,* without taking account of the dynamics of the motion.

The first pair of problems we shall discuss are three-dimensional.

2.01 The position vectors of the vertices A_1, A_2, A_3 of a triangle $A_1 A_2 A_3$ are $\mathbf{r_1}, \mathbf{r_2}, \mathbf{r_3}$ respectively, and σ is the vector area of the triangle. If the time rates of change of the position vectors are given by

$$\frac{d\mathbf{r_1}}{dt} = -n\mathbf{r_1}, \quad \frac{d\mathbf{r_2}}{dt} = -n\mathbf{r_2}, \quad \frac{d\mathbf{r_3}}{dt} = -n\mathbf{r_3},$$

find $d\sigma/dt$ in terms of σ and the constant n.

If the position vectors of A_1, A_2, A_3 are $\mathbf{a_1}, \mathbf{a_2}, \mathbf{a_3}$ respectively at time $t = 0$, find σ as a function of the time t.

- *Solution*

The *vector area* of a triangle with sides determined by the vectors $\mathbf{a}$ and $\mathbf{b}$ is given by $\frac{1}{2}\mathbf{a} \times \mathbf{b}$. Hence we have

$$\sigma = \frac{1}{2}(\mathbf{r_2} - \mathbf{r_1}) \times (\mathbf{r_3} - \mathbf{r_1}) = \frac{1}{2}(\mathbf{r_2} \times \mathbf{r_3} + \mathbf{r_3} \times \mathbf{r_1} + \mathbf{r_1} \times \mathbf{r_2})$$

and so

$$\frac{d\sigma}{dt} = \frac{1}{2}\left(\frac{d\mathbf{r_2}}{dt} \times \mathbf{r_3} + \mathbf{r_2} \times \frac{d\mathbf{r_3}}{dt} + \frac{d\mathbf{r_3}}{dt} \times \mathbf{r_1} + \mathbf{r_3} \times \frac{d\mathbf{r_1}}{dt} + \frac{d\mathbf{r_1}}{dt} \times \mathbf{r_2} + \mathbf{r_1} \times \frac{d\mathbf{r_2}}{dt} \right)$$

giving

$$\frac{d\sigma}{dt} = -n(\mathbf{r_2} \times \mathbf{r_3} + \mathbf{r_3} \times \mathbf{r_1} + \mathbf{r_1} \times \mathbf{r_2}) = -2n\sigma.$$

Therefore

$$\sigma = \frac{1}{2}(\mathbf{a_2} \times \mathbf{a_3} + \mathbf{a_3} \times \mathbf{a_1} + \mathbf{a_1} \times \mathbf{a_2})e^{-2nt}.$$

2.02 Four points A_1, A_2, A_3, A_4 with position vectors $\mathbf{r_1}, \mathbf{r_2}, \mathbf{r_3}, \mathbf{r_4}$ are moving with velocities $\mathbf{v_1}, \mathbf{v_2}, \mathbf{v_3}, \mathbf{v_4}$ respectively. Find the time rate of change of the volume of the tetrahedron $A_1 A_2 A_3 A_4$.

- *Solution*

The volume of a tetrahedron having three concurrent edges determined by the vectors $\mathbf{a}, \mathbf{b}, \mathbf{c}$ is given by the scalar triple product $V = \frac{1}{6}\mathbf{a}.(\mathbf{b} \times \mathbf{c})$. Hence the volume of the tetrahedron $A_1 A_2 A_3 A_4$ is

$$V = \frac{1}{6}(\mathbf{r_2} - \mathbf{r_1}).[(\mathbf{r_3} - \mathbf{r_1}) \times (\mathbf{r_4} - \mathbf{r_1})].$$

It follows that

$$
\begin{aligned}
\frac{dV}{dt} &= \tfrac{1}{6}\{(\dot{\mathbf{r}}_2 - \dot{\mathbf{r}}_1)\cdot[(\mathbf{r}_3 - \mathbf{r}_1) \times (\mathbf{r}_4 - \mathbf{r}_1)] + (\mathbf{r}_2 - \mathbf{r}_1)\cdot[(\dot{\mathbf{r}}_3 - \dot{\mathbf{r}}_1) \times (\mathbf{r}_4 - \mathbf{r}_1)] \\
&\quad +(\mathbf{r}_2 - \mathbf{r}_1)\cdot[(\mathbf{r}_3 - \mathbf{r}_1) \times (\dot{\mathbf{r}}_4 - \dot{\mathbf{r}}_1)]\} \\
&= \tfrac{1}{6}\{(\mathbf{v}_2 - \mathbf{v}_1)\cdot[(\mathbf{r}_3 - \mathbf{r}_1) \times (\mathbf{r}_4 - \mathbf{r}_1)] + (\mathbf{r}_2 - \mathbf{r}_1)\cdot[(\mathbf{v}_3 - \mathbf{v}_1) \times (\mathbf{r}_4 - \mathbf{r}_1)] \\
&\quad +(\mathbf{r}_2 - \mathbf{r}_1)\cdot[(\mathbf{r}_3 - \mathbf{r}_1) \times (\mathbf{v}_4 - \mathbf{v}_1)]\}
\end{aligned}
$$

where we have denoted differentiation with respect to time t by a dot so that $\dot{\mathbf{r}}_1 = \mathbf{v}_1, \dot{\mathbf{r}}_2 = \mathbf{v}_2, \dot{\mathbf{r}}_3 = \mathbf{v}_3$. Hence

$$
\begin{aligned}
\frac{dV}{dt} &= \tfrac{1}{6}\{(\mathbf{v}_2 - \mathbf{v}_1)\cdot[(\mathbf{r}_3 - \mathbf{r}_1) \times (\mathbf{r}_4 - \mathbf{r}_1)] + (\mathbf{v}_3 - \mathbf{v}_1)\cdot[(\mathbf{r}_4 - \mathbf{r}_1) \times (\mathbf{r}_2 - \mathbf{r}_1)] \\
&\quad +(\mathbf{v}_4 - \mathbf{v}_1)\cdot[(\mathbf{r}_2 - \mathbf{r}_1) \times (\mathbf{r}_3 - \mathbf{r}_1)]\}
\end{aligned}
$$

which may be rewritten, on expanding the cross products, in the simpler form

$$
\begin{aligned}
\frac{dV}{dt} &= \tfrac{1}{6}[\mathbf{v}_1\cdot(\mathbf{r}_3 \times \mathbf{r}_2 + \mathbf{r}_4 \times \mathbf{r}_3 + \mathbf{r}_2 \times \mathbf{r}_4) + \mathbf{v}_2\cdot(\mathbf{r}_1 \times \mathbf{r}_3 + \mathbf{r}_3 \times \mathbf{r}_4 + \mathbf{r}_4 \times \mathbf{r}_1) \\
&\quad +\mathbf{v}_3\cdot(\mathbf{r}_1 \times \mathbf{r}_4 + \mathbf{r}_2 \times \mathbf{r}_1 + \mathbf{r}_4 \times \mathbf{r}_2) + \mathbf{v}_4\cdot(\mathbf{r}_1 \times \mathbf{r}_2 + \mathbf{r}_2 \times \mathbf{r}_3 + \mathbf{r}_3 \times \mathbf{r}_1)].
\end{aligned}
$$

2-1 Radial and transverse resolutes of velocity and acceleration

The next few problems are two-dimensional and use the *radial and transverse resolutes* formulae for the velocity $\mathbf{v}$ and the acceleration $\mathbf{a}$:

$$\mathbf{v} = \dot{\mathbf{r}} = \dot{r}\,\hat{\mathbf{r}} + r\,\dot{\theta}\,\hat{\mathbf{s}} \tag{2.1}$$

$$\mathbf{a} = \ddot{\mathbf{r}} = (\ddot{r} - r\,\dot{\theta}^2)\hat{\mathbf{r}} + (2\,\dot{r}\dot{\theta} + r\,\ddot{\theta})\hat{\mathbf{s}} \tag{2.2}$$

where $\hat{\mathbf{r}}$ and $\hat{\mathbf{s}}$ are unit vectors in the radial and transverse directions respectively and r, θ are circular polar coordinates. Here we have denoted differentiation with respect to time t by a dot so that $\dot{\mathbf{r}} = d\mathbf{r}/dt$ and $\ddot{\mathbf{r}} = d^2\mathbf{r}/dt^2$.

2.03 If

$$\hat{\mathbf{r}} = \cos\theta\,\mathbf{i} + \sin\theta\,\mathbf{j}, \quad \hat{\mathbf{s}} = -\sin\theta\,\mathbf{i} + \cos\theta\,\mathbf{j}$$

where $\mathbf{i}$ and $\mathbf{j}$ are mutually perpendicular unit vectors, show that $\hat{\mathbf{s}}.\hat{\mathbf{r}} = 0$ and

$$\dot{\hat{\mathbf{r}}} = \dot{\theta}\,\hat{\mathbf{s}}, \quad \dot{\hat{\mathbf{s}}} = -\dot{\theta}\,\hat{\mathbf{r}}. \tag{2.3}$$

- *Solution*

We have that $\hat{\mathbf{s}}.\hat{\mathbf{r}} = -\cos\theta\,\sin\theta + \sin\theta\,\cos\theta = 0$ and

$$\dot{\hat{\mathbf{r}}} = -\sin\theta\,\dot{\theta}\,\mathbf{i} + \cos\theta\,\dot{\theta}\,\mathbf{j} = \dot{\theta}\,\hat{\mathbf{s}}, \quad \dot{\hat{\mathbf{s}}} = -\cos\theta\,\dot{\theta}\,\mathbf{i} - \sin\theta\,\dot{\theta}\,\mathbf{j} = -\dot{\theta}\,\hat{\mathbf{r}}.$$

2.04 A point P moves along a circle of radius a so that its acceleration is directed towards a point O of its circumference. Show that the acceleration towards O is $8a^2h^2/r^5$ where h is the areal constant.

- *Solution*

The equation of a circle of radius a is $r = 2a \cos\theta$ where the polar angle θ is referred to a diameter and the origin is on the circumference of the circle. Hence $\dot{r} = -2a \sin\theta\,\dot\theta$ and so $\ddot{r} = -2a \cos\theta\,\dot\theta^2 - 2a \sin\theta\,\ddot\theta$. Since the acceleration is purely radial, the transverse component of acceleration must vanish. Now

$$2\,\dot{r}\dot\theta + r\,\ddot\theta = \frac{1}{r}\frac{d}{dt}(r^2\,\dot\theta) \tag{2.4}$$

and so $h = r^2\,\dot\theta$ must be a constant, called the *areal constant*. Using $\ddot\theta = -2r^{-1}\,\dot{r}\dot\theta$ we see that

$$\ddot{r} - r\,\dot\theta^2 = -2a \cos\theta\,\dot\theta^2 - 8r^{-1}a^2 \sin^2\theta\,\dot\theta^2 - r\,\dot\theta^2$$
$$= -h^2r^{-3} - 2h^2(4a^2 - 4a^2 \cos^2\theta)r^{-5} - h^2r^{-3}$$

giving for the radial component of acceleration

$$\ddot{r} - r\,\dot\theta^2 = -8h^2a^2r^{-5}.$$

2.05 A point describes a curve whose polar equation is $r^2 = \sin 2\theta$ in such a way that the time t is connected with θ by the formula $ht = \sin^2\theta$ where h is a constant. Prove that the acceleration is radial and express it as a function of r only.

• *Solution*

We see that $r^2\,\dot\theta = \sin 2\theta\,\dot\theta = d(\sin^2\theta)/dt = d(ht)/dt = h$ since h is a constant and so the acceleration is purely radial since the transverse component vanishes.

Now $r^2 = \sin 2\theta$ and so $r\,\dot{r} = \cos 2\theta\,\dot\theta$ giving $\dot{r} = hr^{-3}\cos 2\theta$ since $\dot\theta = hr^{-2}$. Hence

$$\ddot{r} = -2hr^{-3}\sin 2\theta\,\dot\theta - 3hr^{-4}\cos 2\theta\,\dot{r}$$
$$= -2h^2r^{-5}\sin 2\theta - 3h^2r^{-7}\cos^2 2\theta$$
$$= -2h^2r^{-3} - 3h^2(1 - r^4)r^{-7}$$
$$= h^2r^{-3} - 3h^2r^{-7}$$

and so we get

$$\ddot{\mathbf{r}} = (\ddot{r} - r\,\dot\theta^2)\hat{\mathbf{r}} = (h^2r^{-3} - 3h^2r^{-7} - h^2r^{-3})\hat{\mathbf{r}} = -3h^2r^{-7}\hat{\mathbf{r}}.$$

2.06 A point describes a curve whose polar equation is

$$\frac{1}{r} = \lambda + \mu \sin p\theta.$$

If the acceleration is purely in the radial direction show that it is given by

$$h^2\left(\frac{p^2\lambda}{r^2} + \frac{1 - p^2}{r^3}\right)$$

towards the origin, where h is the areal constant.

• *Solution*

We see that $\dot{r}\,r^{-2} = -p\mu \cos p\theta\,\dot\theta$ and since $r^2\,\dot\theta = h$ we obtain $\dot{r} = -hp\mu \cos p\theta$ giving $\ddot{r} = hp^2\mu \sin p\theta\,\dot\theta$. Hence the acceleration in the radial direction away from the origin is

$$\ddot{r} - r\,\dot\theta^2 = \frac{h^2p^2}{r^2}\left(\frac{1}{r} - \lambda\right) - \frac{h^2}{r^3}$$

and so the formula for the acceleration towards the origin is verified.

2.07 A point describes the spiral $r = ae^{-\alpha\theta}$ with constant speed u, where a and α are constants and r, θ are polar coordinates referred to an origin O. Show that the radial acceleration of P towards O is given by $[u^2/(\alpha^2 + 1)]r^{-1}$.

- *Solution*

Since $r = ae^{-\alpha\theta}$ we have $\dot{r} = -a\alpha e^{-\alpha\theta}\,\dot\theta$. Since

$$\mathbf{v}^2 = \dot{r}^2 + r^2\,\dot\theta^2 \tag{2.5}$$

we have $u^2 = a^2(\alpha^2 + 1)e^{-2\alpha\theta}\,\dot\theta^2$ and so $\dot\theta = ue^{\alpha\theta}/(a\sqrt{\alpha^2+1})$ giving $\ddot\theta = \alpha\,\dot\theta^2$. Also, differentiating $\dot{r}$ with respect to t we get $\ddot{r} = a\alpha^2 e^{-\alpha\theta}\,\dot\theta^2 - a\alpha e^{-\alpha\theta}\,\ddot\theta = 0$ and hence it follows that the acceleration towards O is

$$r\,\dot\theta^2 = \frac{u^2 r e^{2\alpha\theta}}{a^2(\alpha^2+1)} = \frac{u^2}{\alpha^2+1}\frac{1}{r}.$$

We note that the transverse component of acceleration does not vanish since $r^2\,\dot\theta = aue^{-\alpha\theta}/\sqrt{\alpha^2+1} = ur/\sqrt{\alpha^2+1}$ which is not a constant.

2-2 Tangential and normal resolutes of velocity and acceleration

To solve the remaining problems we use the *tangential and normal resolutes* formulae for velocity and acceleration:

$$\mathbf{v} = \dot{s}\,\hat{\mathbf{t}} \tag{2.6}$$

$$\mathbf{a} = \ddot{s}\,\hat{\mathbf{t}} + \dot{s}\,\omega\hat{\mathbf{n}} \tag{2.7}$$

where $\hat{\mathbf{t}}$ and $\hat{\mathbf{n}}$ are unit vectors in the tangential and normal directions, s is the arc length and $\omega = \dot\psi$ is the rate of rotation of the tangent, ψ being the angle which the tangent at a point P makes with the tangent at a fixed point P_0 of the path.

An alternative formula for the acceleration is

$$\mathbf{a} = \ddot{s}\,\hat{\mathbf{t}} + \frac{\dot{s}^2}{\rho}\hat{\mathbf{n}} \tag{2.8}$$

where $\rho = ds/d\psi$ is the radius of curvature.

2.08 A point P moves along a circle of radius a so that the ratio of the tangential to the normal resolutes of acceleration of P is a constant α. If the initial speed of P is v_0 show that the time taken by P to describe the circumference of the circle is given by $a(1 - e^{-2\pi\alpha})/v_0\alpha$.

- *Solution*

Since the radius of curvature for a circle of radius a is just $\rho = a$ we have $\ddot{s} = \alpha\,\dot{s}^2/a$. Now

$$\ddot{s} = \frac{d}{ds}\left(\frac{1}{2}\dot{s}^2\right) \tag{2.9}$$

and so

$$\frac{d}{ds}\left(\frac{1}{2}\dot{s}^2\,e^{-2\alpha s/a}\right) = 0$$

which gives

$$\dot{s}^2\,e^{-2\alpha s/a} = v_0^2$$

taking $s = 0$ when $t = 0$.

Hence the time taken by the point to describe the circumference of the circle of radius a is given by

$$T = \int_0^{2\pi a} \frac{ds}{\dot{s}} = \frac{1}{v_0} \int_0^{2\pi a} e^{-\alpha s/a} \, ds = \frac{1}{v_0} \left[-\frac{a}{\alpha} e^{-\alpha s/a} \right]_0^{2\pi a} = \frac{a}{v_0 \alpha} (1 - e^{-2\pi \alpha}).$$

2.09 If the normal and tangential resolutes of acceleration of a point are λ and $\lambda \tan \psi$ respectively, where ψ is the angle between the tangent at P and the tangent at P_0, show that the speed v of the point is given by $v = v_0 \cos \psi$ where v_0 is the speed of the point at P_0.

If λ is a constant show that the path of the point is the catenary $s = (v_0^2/\lambda) \tan \psi$ where s is the arc distance of P from P_0.

- *Solution*

Putting $v = \dot{s}$ we have $\dot{v} = \lambda \tan \psi$ and $v\omega = \lambda$. Hence

$$\frac{\dot{v}}{v} = \tan \psi \frac{d\psi}{dt}$$

and so $\ln v = \int \tan \psi \, d\psi + \text{constant} = \ln \sec \psi + \text{constant}$. Since $v = v_0$ when $\psi = 0$ we have $v = v_0 \sec \psi$.

Now $\lambda \dot{s} = v^2 \omega = v_0^2 \sec^2 \psi \, d\psi/dt$ and hence

$$\lambda s = v_0^2 \int \sec^2 \psi \, d\psi + \text{constant} = v_0^2 \tan \psi + \text{constant}.$$

Taking $s = 0$ when $\psi = 0$ yields

$$s = \frac{v_0^2}{\lambda} \tan \psi.$$

2.10 A particle describes a plane curve so that its tangential resolute of acceleration at a point P is a continuous function $f(s)$ of the distance s along the arc of P from a fixed point P_0. Show that the normal resolute of acceleration at P is

$$\kappa \left[v_0^2 + 2 \int_0^s f(s) \, ds \right]$$

where v_0 is the speed of the particle at P_0 and κ is the curvature at P.

If the normal resolute of acceleration of the particle is constant and the motion along the arc is simple harmonic about P_0 as centre with time period $2\pi/\omega$ show that the equation of the curve is

$$s = \frac{v_0}{\omega} \tanh \frac{\omega \psi}{v_0 \kappa_0}$$

where κ_0 is the curvature at P_0 and ψ is the angle between the tangents at P and P_0.

- *Solution*

The acceleration vector can be written in the alternative form

$$\mathbf{a} = \ddot{s} \,\widehat{\mathbf{t}} + \dot{s}^2 \,\kappa \widehat{\mathbf{n}} \tag{2.10}$$

where the curvature κ and the radius of curvature are related by $\kappa = d\psi/ds = 1/\rho$.

We have $\ddot{s} = f(s)$, that is $d(\frac{1}{2}\dot{s}^2)/ds = f(s)$ and hence

$$\dot{s}^2 = v_0^2 + 2\int_0^s f(s)\, ds$$

which yields the required formula for the normal resolute of acceleration.

For simple harmonic motion with period $2\pi/\omega$ we have $\ddot{s} = -\omega^2 s$ and so $f(s) = -\omega^2 s$. Since the normal resolute of acceleration is constant we have $v^2\kappa = v_0^2\kappa_0$, so that $\kappa[v_0^2 + 2\int_0^s f(s)\, ds] = v_0^2\kappa_0$. It follows that

$$\frac{d\psi}{ds}\left(v_0^2 - \omega^2 s^2\right) = v_0^2\kappa_0$$

giving

$$\psi = \int_0^s \frac{v_0^2\kappa_0}{v_0^2 - \omega^2 s^2}\, ds = \frac{v_0\kappa_0}{\omega}\tanh^{-1}\frac{\omega s}{v_0}$$

which leads to the required formula for the path of the moving point.

2.11 Show that the radius of curvature of the path of a point moving with velocity **v** and acceleration **a** is given by

$$\rho = \frac{v^3}{|\mathbf{v}\times\mathbf{a}|}.$$

- *Solution*

Since $\mathbf{v} = v\hat{\mathbf{t}}$ and $\mathbf{a} = \dot{v}\,\hat{\mathbf{t}} + (v^2/\rho)\hat{\mathbf{n}}$ we see that $\mathbf{v}\times\mathbf{a} = (v^3/\rho)\hat{\mathbf{t}}\times\hat{\mathbf{n}} = (v^3/\rho)\hat{\mathbf{b}}$ where $\hat{\mathbf{b}} = \hat{\mathbf{t}}\times\hat{\mathbf{n}}$ is called the *unit binormal*, and $\hat{\mathbf{t}}, \hat{\mathbf{n}}, \hat{\mathbf{b}}$ form a right-handed system of mutually perpendicular unit vectors. Hence

$$|\mathbf{v}\times\mathbf{a}| = \frac{v^3}{\rho}$$

and the required result follows.

The plane containing $\hat{\mathbf{t}}$ and the *principal normal* unit vector $\hat{\mathbf{n}}$ is called the *osculating plane* or the *plane of curvature*.

3
DYNAMICS OF A PARTICLE

In this chapter we shall use Newton's *second law of motion* to solve dynamics problems involving particles.

The science of dynamics originated with the investigations of Galilei Galileo (1564-1642) who studied the motion under gravity of a rolling bronze ball down an inclined plane. He also considered the motion of a projectile and showed that the path was a parabola. He discovered the *principle of inertia* which states that a particle moves in a straight line with constant speed unless influenced by external forces. Isaac Newton (1642-1727) introduced the concept of mass and generalized the idea of force which he defined in his second law of motion to be the change of *momentum*, that is *mass* × *velocity*, generated per unit time. Thus for a particle of mass m moving with velocity $\mathbf{v}$ Newton's second law states that

$$\frac{d\mathbf{p}}{dt} = \mathbf{F} \tag{3.1}$$

where $\mathbf{p} = m\mathbf{v}$ is the linear momentum and $\mathbf{F}$ is the total force acting on the particle.

We shall suppose that there exist frames of reference relative to which a particle that is not under the influence of external forces moves in a straight line with constant speed. Such a frame of reference is known as an *inertial frame* since the principle of inertia applies to it.

3-1 One-dimensional motion

The first few problems are one-dimensional and their solutions are all based on the correct formulation of Newton's second law of motion in one dimension.

3.01 A particle of mass m is projected with speed u along a horizontal plane. If the magnitude of the resistive force acting on the particle is mkv^{n+1} where v is the speed of the particle and k, $n(>1)$ are constants, show that after time t has elapsed

$$v = \frac{u}{(1 + nku^n t)^{1/n}}$$

and the distance the particle has moved is given by $(v^{1-n} - u^{1-n})/k(n-1)$.

- *Solution*

In this problem Newton's second law of motion takes the form

$$m\frac{dv}{dt} = -mkv^{n+1}.$$

Integrating with respect to time we get

$$k \int_0^t dt = -\int_u^v \frac{dv}{v^{n+1}} = \left[\frac{1}{n}\frac{1}{v^n}\right]_u^v$$

and so

$$t = \frac{1}{nk}\left(\frac{1}{v^n} - \frac{1}{u^n}\right)$$

giving $v = u/(1 + nku^n t)^{1/n}$ as required.

Since

$$\frac{dv}{dt} = v\frac{dv}{dx} \tag{3.2}$$

we may rewrite Newton's second law of motion for the particle as

$$\frac{dv}{dx} = -kv^n.$$

Now integrating with respect to the distance x gives

$$k\int_0^x dx = -\int_u^v \frac{dv}{v^n} = \left[\frac{1}{n-1}\frac{1}{v^{n-1}}\right]_u^v$$

and so

$$x = \frac{1}{(n-1)k}\left(\frac{1}{v^{n-1}} - \frac{1}{u^{n-1}}\right).$$

3.02 Show that the kinetic energy of two particles of masses m_1 and m_2 moving with velocities $\mathbf{v}_1$ and $\mathbf{v}_2$ relative to a Newtonian inertial frame of reference is $T = \frac{1}{2}M\bar{\mathbf{v}}^2 + \frac{1}{2}\mu\mathbf{v}^2$ where $M = m_1 + m_2$ and $\mu = m_1 m_2/(m_1 + m_2)$ is the *reduced mass*, $\mathbf{v} = \mathbf{v}_1 - \mathbf{v}_2$ is the velocity of relative motion and $\bar{\mathbf{v}} = (m_1\mathbf{v}_1 + m_2\mathbf{v}_2)/M$ is the velocity of the centre of mass of the two particles.

A nucleus of mass M, originally at rest, splits up into two nuclei as the result of an internal disintegration which generates kinetic energy T. Show that the two nuclei move off in opposite directions, and prove that the least value of their speed of separation is $2(2T/M)^{\frac{1}{2}}$.

- *Solution*

The first part of this problem is an important general result for two particles which holds in three-dimensions.

We have

$$\frac{1}{2}M\bar{\mathbf{v}}^2 = \frac{1}{2(m_1 + m_2)}(m_1^2 v_1^2 + m_2^2 v_2^2 + 2m_1 m_2\mathbf{v}_1.\mathbf{v}_2)$$

and hence

$$\frac{1}{2}M\bar{\mathbf{v}}^2 + \frac{1}{2}\mu\mathbf{v}^2 = \frac{1}{2(m_1 + m_2)}\left[m_1^2 v_1^2 + m_2^2 v_2^2 + 2m_1 m_2\mathbf{v}_1.\mathbf{v}_2 + m_1 m_2(v_1^2 + v_2^2 - 2\mathbf{v}_1.\mathbf{v}_2)\right].$$

It follows that

$$\frac{1}{2}M\bar{\mathbf{v}}^2 + \frac{1}{2}\mu\mathbf{v}^2 = \frac{1}{2}m_1 v_1^2 + \frac{1}{2}m_2 v_2^2 \tag{3.3}$$

which is the kinetic energy of the two particles.

In the second part of this problem, since the nucleus is originally at rest we have, by conservation of linear momentum, that $\bar{\mathbf{v}} = \mathbf{0}$ so that $m_1\mathbf{v}_1 + m_2\mathbf{v}_2 = \mathbf{0}$. Hence the two nuclei move off in a straight line but in opposite directions.

Also $T = \mu\mathbf{v}^2/2$ and so $\mathbf{v}^2 = (\mathbf{v}_1 - \mathbf{v}_2)^2 = 2MT/m_1 m_2$. Since $m_1 m_2 = m_1(M - m_1)$ where $M = m_1 + m_2$, we see that $m_1 m_2$ has its maximum value when $m_1 = m_2 = M/2$ and so $(\mathbf{v}_1 - \mathbf{v}_2)^2$ attains its minimum value $8T/M$ when the two nuclei have equal masses $M/2$.

Hence the smallest value of the velocity of separation is $(8T/M)^{\frac{1}{2}}$.

Our next two problems are on *vertical motion under gravity*.

3.03 A raindrop having initial mass m falls from rest under the action of gravity through a cloud of water vapour which is at rest. If the mass of the raindrop increases owing to the accretion of water vapour so that its mass after time t is $m(1 + \lambda t)^3$, find its speed at time t. Hence show that when the raindrop has fallen through a distance h, its mass M satisfies $h = (g/8\lambda^2)[(M/m)^{\frac{1}{3}} - (m/M)^{\frac{1}{3}}]^2$.

- *Solution*

An important aspect of this problem is that the mass changes with time. Then from Newton's second law of motion we have

$$\frac{d(M\,\dot{z})}{dt} = Mg$$

where $M = m(1 + \lambda t)^3$, g is the acceleration of gravity, and $\dot{z}$ is the speed of the raindrop in the downward direction. Hence

$$\frac{d}{dt}[m(1 + \lambda t)^3\,\dot{z}] = m(1 + \lambda t)^3 g$$

and so integrating with respect to time t we get $(1 + \lambda t)^3\,\dot{z} = g[(1 + \lambda t)^4 - 1]/4\lambda$ since $\dot{z} = 0$ when $t = 0$. Thus the downward speed is

$$\frac{dz}{dt} = \frac{g}{4\lambda}\left[1 + \lambda t - \frac{1}{(1 + \lambda t)^3}\right]$$

and the distance the raindrop falls is given by

$$z = \frac{g}{8\lambda^2}\left[(1 + \lambda t)^2 + \frac{1}{(1 + \lambda t)^2} - 2\right] = \frac{g}{8\lambda^2}\left(1 + \lambda t - \frac{1}{1 + \lambda t}\right)^2$$

so that

$$h = \frac{g}{8\lambda^2}\left[\left(\frac{M}{m}\right)^{1/3} - \left(\frac{m}{M}\right)^{1/3}\right]^2.$$

3.04 If a particle is projected vertically upwards with initial speed u in a medium whose resistance varies as the square of the speed, show that the time of ascent is $(V/g)\tan^{-1}(u/V)$ and the distance ascended is $(V^2/2g)\ln(1 + u^2/V^2)$ where $V = \sqrt{g/k}$ is the terminal speed of the particle. Further show that the particle returns to the point of projection with speed $uV/(u^2 + V^2)^{\frac{1}{2}}$.

- *Solution*

It is important to realize that Newton's second law of motion is different for the case of upward motion from the case of downward motion when the resistance is an even function of the speed v.

For *upward motion* in the present problem Newton's second law is

$$\frac{dv}{dt} = -g - kv^2 = -k(V^2 + v^2)$$

since $V^2 = g/k$ where k is the coefficient of resistance. Hence the time of ascent is

$$T = -\frac{1}{k}\int_u^0 \frac{dv}{V^2 + v^2} = -\frac{1}{kV}\left[\tan^{-1}\frac{v}{V}\right]_u^0 = \frac{V}{g}\tan^{-1}\frac{u}{V}.$$

If z is the upward vertical displacement we have

$$\frac{dv}{dt} = v\frac{dv}{dz} = -k(V^2 + v^2)$$

and so the distance ascended is

$$h = -\frac{1}{k}\int_u^0 \frac{vdv}{V^2 + v^2} = -\frac{1}{2k}\left[\ln(V^2 + v^2)\right]_u^0 = \frac{V^2}{2g}\ln(1 + \frac{u^2}{V^2}).$$

For *downward motion* Newton's second law becomes

$$\frac{dv}{dt} = g - kv^2$$

where v is the speed in the downward direction. Hence we have

$$v\frac{dv}{dz} = -k(v^2 - V^2)$$

and so the downward displacement is

$$z = -\frac{1}{k}\int_0^v \frac{vdv}{v^2 - V^2}.$$

Hence if w is the speed of the particle when it returns to the point of projection, we have

$$h = -\frac{1}{k}\int_0^w \frac{vdv}{v^2 - V^2} = -\frac{V^2}{2g}\left[\ln(v^2 - V^2)\right]_0^w = \frac{V^2}{2g}\ln\frac{V^2}{V^2 - w^2}$$

Hence $V^2/(V^2 - w^2) = 1 + u^2/V^2$ from which it follows by simple algebra that $w^2 = u^2V^2/(u^2 + V^2)$ which gives $w = uV/\sqrt{u^2 + V^2}$ as required.

3-2 Projectile motion

Our next few problems are concerned with projectile motion and are two-dimensional.

3.05 A particle of mass m is projected with speed u at an inclination α to the horizontal in a medium whose resistance is mkv for particle speed v. Show that the direction of motion of the particle will again make an angle α with the horizontal after a time $k^{-1}\ln[1 + (2kv/g)\sin\alpha]$.

- *Solution*

If v_x and v_y are the horizontal and vertical components of the velocity $\mathbf{v}$, then Newton's second law of motion gives, taking horizontal and vertical components respectively:

$$\frac{dv_x}{dt} = -kv_x, \quad \frac{dv_y}{dt} = -g - kv_y. \tag{3.4}$$

Integrating with respect to time leads to

$$v_x = u\cos\alpha\, e^{-kt}, \quad v_y = \left(\frac{g}{k} + u\sin\alpha\right)e^{-kt} - \frac{g}{k}. \tag{3.5}$$

Hence the direction of motion of the particle will again make an angle α with the horizontal when

$$-\frac{v_y}{v_x} = \tan\alpha,$$

that is when

$$\frac{\frac{g}{k} - \left(\frac{g}{k} + u\sin\alpha\right)e^{-kt}}{u\cos\alpha\ e^{-kt}} = \tan\alpha$$

giving

$$\frac{g}{k} = \left(\frac{g}{k} + 2u\sin\alpha\right)e^{-kt}$$

so that the time taken is given by

$$t = k^{-1}\ln\left(1 + \frac{2ku}{g}\sin\alpha\right).$$

3.06 A particle of mass m moving under gravity is projected at an angle α with the horizontal in a medium whose resistance is mkv. It arrives at the horizontal plane through the point of projection at an angle β and the time of flight is T. Show that

$$\frac{\tan\alpha}{\tan\beta} = \frac{e^{kT} - 1 - kT}{e^{-kT} - 1 + kT}.$$

• *Solution*

From the second equation of (3.5) in problem 3.05 we have

$$\frac{dy}{dt} = \left(\frac{g}{k} + u\sin\alpha\right)e^{-kt} - \frac{g}{k}$$

and so integrating with respect to time we obtain for the vertical height of the particle above the point of projection

$$y = \frac{1}{k}\left(\frac{g}{k} + u\sin\alpha\right)\left(1 - e^{-kt}\right) - \frac{g}{k}t. \qquad (3.6)$$

Since the particle reaches the horizontal through the point of projection again when $y = 0$, the time of flight T satisfies

$$gT = \left(\frac{g}{k} + u\sin\alpha\right)\left(1 - e^{-kT}\right).$$

Now from (3.5) of the previous problem, we have

$$\tan\beta = -\frac{v_y}{v_x} = \frac{\frac{g}{k} - \left(\frac{g}{k} + u\sin\alpha\right)e^{-kT}}{u\cos\alpha\ e^{-kT}} = \frac{gT - u\sin\alpha}{u\cos\alpha\ e^{-kT}}$$

so that

$$\frac{\tan\beta}{\tan\alpha} = \frac{gT - u\sin\alpha}{u\sin\alpha\ e^{-kT}}$$

which produces the required result on performing some elementary algebra.

3.07 A particle of mass m is projected with initial speed u at an angle α with the horizontal in a medium whose resistance is mkv. Show that the range of the particle is a maximum if $\ln(1 + \lambda\operatorname{cosec}\alpha) = \lambda(1 + \lambda\sin\alpha)/(\lambda + \sin\alpha)$ where $\lambda = uk/g$.

• *Solution*

We have from the first equation of (3.5) in problem 3.05 that

$$\frac{dx}{dt} = u \cos \alpha \; e^{-kt}$$

and so the distance travelled horizontally is

$$x = \frac{u \cos \alpha}{k} \left(1 - e^{-kt}\right).$$

Hence from equation (3.6) of problem 3.06 we get for the equation of the path of the particle

$$y = (V + u \sin \alpha) \frac{x}{u \cos \alpha} + \frac{V^2}{g} \ln \left(1 - \frac{gx}{V u \cos \alpha}\right)$$

where $V = g/k$ is the terminal velocity. Now $y = 0$ when $x = R$, the range of the projectile, and so

$$0 = (1 + \lambda \sin \alpha) \frac{R}{\cos \alpha} + \frac{V\lambda}{k} \ln \left(1 - \frac{kR}{V \lambda \cos \alpha}\right).$$

The range is a maximum when $dR/d\alpha = 0$. Some analysis then shows that the maximum range is given by

$$R = \frac{V\lambda^2 \cos \alpha}{k(\lambda + \sin \alpha)}$$

from which the required result follows.

3.08 A package of mass m is dropped from an airplane flying horizontally with speed u at a height h above level ground. The resisting force is mkv. Show that the horizontal distance R travelled by the package, from the time $t = 0$ when it leaves the airplane to the time T when it reaches the ground, is given by $R = u(1 - e^{-kT})/k$ where $h = gT/k - g(1 - e^{-kT})/k^2$.

● *Solution*

The equations of motion are

$$\frac{dv_x}{dt} = -kv_x, \quad \frac{dv_y}{dt} = g - kv_y$$

where v_x is the horizontal speed and v_y is the downward vertical speed.

Since $v_x = dx/dt$ we have $d^2x/dt^2 = -kdx/dt$ giving $dx/dt = -kx + u$ because $dx/dt = u$ when $x = 0$. If R is the horizontal range and T is the time of descent we see that

$$-k \int_0^T dt = \int_0^R \frac{dx}{x - u/k}$$

giving $-kT = [\ln(x - u/k)]_0^R = \ln(1 - kR/u)$ and so $R = u(1 - e^{-kT})/k$.

Since $v_y = dy/dt$ we see that $d^2y/dt^2 = g - kdy/dt$ giving $dy/dt = gt - ky$ because $dy/dt = 0$ when $y = 0$ and $t = 0$.

Hence

$$\frac{d(ye^{kt})}{dt} = gte^{kt}$$

and so

$$he^{kT} = g \int_0^T te^{kt}dt = \frac{gT}{k}e^{kT} - \frac{g}{k^2}(e^{kT} - 1)$$

giving

$$h = \frac{gT}{k} - \frac{g}{k^2}(1 - e^{-kT}).$$

Hence $R = u(T - kh/g)$.

3.09 A particle of mass m moves under gravity in a medium whose resistance is
mR. If u is the horizontal resolute of the velocity $\mathbf{v}$ show that $du/dt = -R\cos\psi$ and
$u d\psi/dt = -g\cos^2\psi$ where ψ is the angle which the velocity makes with the horizontal.

If $R = kv^n$ where k and n are constants, and u_1, u_0, u_2 are the values of u for $\psi = \alpha$,
$\psi = 0$, $\psi = -\alpha$ respectively, show that

$$(i)\ u_1^{-n} + u_2^{-n} = 2u_0^{-n}, \quad (ii)\ u_2^{-n} - u_1^{-n} = (2kn/g)\int_0^\alpha \sec^{n+1}\psi d\psi.$$

- *Solution*

The equation of motion is

$$\frac{d\mathbf{v}}{dt} = \mathbf{g} - R\hat{\mathbf{v}}$$

where $\mathbf{g}$ is the vector acceleration of gravity, and so resolving horizontally we have

$$\frac{du}{dt} = -R\cos\psi$$

and resolving vertically we have

$$\frac{d(u\tan\psi)}{dt} = -g - R\sin\psi.$$

It follows that $\dot{u} = du/dt = -R\cos\psi$,which is the first result, and
$\dot{u}\tan\psi + u\sec^2\psi\,\dot{\psi} = -g - R\sin\psi$. Hence $u\,\dot{\psi} = -g\cos^2\psi$ which is the second result.
We now see that

$$\frac{du}{d\psi} = \frac{uR}{g\cos\psi}$$

and taking $R = kv^n = ku^n\sec^n\psi$ we obtain

$$\frac{du}{d\psi} = \frac{k}{g}u^{n+1}\sec^{n+1}\psi$$

so that, on integrating, we find

$$\int_{u_0}^{u_1}\frac{du}{u^{n+1}} = \frac{k}{g}\int_0^\alpha \sec^{n+1}\psi\,d\psi$$

and

$$\int_{u_0}^{u_2}\frac{du}{u^{n+1}} = \frac{k}{g}\int_0^{-\alpha}\sec^{n+1}\psi\,d\psi = -\frac{k}{g}\int_0^\alpha\sec^{n+1}\psi\,d\psi.$$

Hence we get

$$\frac{1}{u_0^n} - \frac{1}{u_1^n} = \frac{nk}{g}\int_0^\alpha\sec^{n+1}\psi\,d\psi$$

and

$$\frac{1}{u_0^n} - \frac{1}{u_2^n} = -\frac{nk}{g}\int_0^\alpha\sec^{n+1}\psi\,d\psi$$

from which the results (i) and (ii) follow at once.

3-3 Vibrational motion

3.10 A particle oscillates with small amplitude β about the vertex O of the catenary with equation $s = a \tan \psi$ where s is the arc distance from O and ψ is the angle which the tangent makes with the horizontal, O being chosen as the lowest point. Show that the period of the motion is, to second order in β/a,

$$T = 2\pi \sqrt{\frac{a}{g}} \left(1 + \frac{3\beta^2}{16a^2} \right).$$

- *Solution*

The equation of motion is $\ddot{s} = -g \sin \psi$ where $s = a \tan \psi$. Since $\sin \psi = s/\sqrt{a^2 + s^2}$ we have

$$\ddot{s} = -\frac{gs}{\sqrt{a^2 + s^2}}.$$

Now $\ddot{s} = d(\frac{1}{2} \dot{s}^2)/ds$ and so integrating with respect to s gives $\frac{1}{2} \dot{s}^2 = -g\sqrt{a^2 + s^2} + $ constant. But $\dot{s} = 0$ when $s = \beta$ so that we have

$$\dot{s}^2 = 2g \left(\sqrt{a^2 + \beta^2} - \sqrt{a^2 + s^2} \right)$$

which gives for the periodic time

$$T = \frac{4}{\sqrt{2g}} \int_0^\beta \frac{ds}{\left(\sqrt{a^2 + \beta^2} - \sqrt{a^2 + s^2} \right)^{1/2}}.$$

Putting $s = \beta \sin \phi$ we obtain

$$T = \frac{4}{\sqrt{2g}} \int_0^{\pi/2} \frac{\beta \cos \phi \, d\phi}{\left(\sqrt{a^2 + \beta^2} - \sqrt{a^2 + \beta^2 \sin^2 \phi} \right)^{1/2}}$$

and expanding in powers of β^2/a^2 using the binomial theorem we find that to second order in β/a:

$$T \simeq 4 \sqrt{\frac{a}{g}} \int_0^{\pi/2} \left[1 + \frac{1}{8} \frac{\beta^2}{a^2} (1 + \sin^2 \phi) \right] d\phi.$$

Now $\int_0^{\pi/2} \sin^2 \phi \, d\phi = \pi/4$ and so the result follows.

3.11 A particle of mass m moves on a cycloid with equation $s = 4a \sin \psi$ whose axis is vertical and vertex downward. In addition to the force of gravity mg, the particle is subject to a small resistance $mk \, ds/dt$. Show that the particle oscillates with period

$$T = \frac{4\pi}{\sqrt{g/a - k^2}}$$

which is independent of amplitude and thus *isochronous*.

- *Solution*

The equation of motion is

$$m\,\ddot{s}= -mg\sin\psi - mk\,\dot{s}\,.$$

For a cycloid $s = 4a\sin\psi$ this becomes

$$\ddot{s} + k\,\dot{s} + gs/4a = 0$$

whose solution can be written in the form

$$s = Ce^{-kt/2}\sin(\omega t + \epsilon)$$

where C and ϵ are constants and

$$\omega^2 = \frac{1}{4}\left(\frac{g}{a} - k^2\right).$$

Hence the period is $2\pi/\omega = 4\pi/\sqrt{g/a - k^2}$.

3.12 A pendulum is held making a small angle α with the downward vertical and then released. If the mass of the pendulum bob is m and the air resistance is k times the speed of the bob, show that the angle θ which the pendulum makes with the vertical at time t is

$$\theta = \frac{\alpha}{2m\omega}\exp(-kt/2m)(k\sin\omega t + 2m\omega\cos\omega t)$$

where $\omega^2 = g/l - k^2/4m^2$ and l is the length of the pendulum.

• *Solution*

The equation of motion of the pendulum bob is

$$ml\,\ddot{\theta}= -mg\sin\theta - kl\,\dot{\theta}$$

and so, since θ remains small, we have

$$\ddot{\theta} + \frac{k}{m}\,\dot{\theta} + \frac{g}{l}\theta = 0$$

which has the solution

$$\theta = e^{-kt/2m}(A\sin\omega t + B\cos\omega t)$$

where

$$\omega^2 = \frac{g}{l} - \frac{1}{4}\frac{k^2}{m^2}.$$

Since $\theta = \alpha$ and $\dot{\theta}= 0$ when $t = 0$ we have $\alpha = B$ and $\omega A - kB/2m = 0$ and the result follows.

3.13 A particle of mass m hangs at rest at the end of an elastic string having natural length l and modulus of elasticity $\lambda = mln^2$. At time $t = 0$, when the particle is in equilibrium, the point of suspension begins to move so that the downward displacement is $d\sin pt$ where $p \neq n$. Find the length of the string at time t.

• *Solution*

Denoting the displacement of the particle from its position of equilibrium by x, we may write its equation of motion in the form $m\,\ddot{x} = mg - T$ where the tension T in the elastic string is given by $T = \frac{\lambda}{l}(a + x - d\sin pt)$ and $mg = \lambda a/l$.

Hence $m\,\ddot{x} = -\frac{\lambda}{l}(x - d\sin pt)$ and since the modulus $\lambda = mln^2$ we get

$$\ddot{x} + n^2 x = n^2 d \sin pt.$$

The general solution of this is

$$x = A \sin nt + B \cos nt + \frac{n^2 d}{n^2 - p^2} \sin pt.$$

Now $x = 0$ and $\dot{x} = 0$ when $t = 0$ and so $B = 0$ and $An + n^2 dp/(n^2 - p^2) = 0$. Hence

$$x = -\frac{ndp}{n^2 - p^2} \sin nt + \frac{n^2 d}{n^2 - p^2} \sin pt$$

and so the length of the string at time t, given by $l + a + x - d\sin pt$, is

$$l + \frac{g}{n^2} + \frac{pd}{n^2 - p^2}(p \sin pt - n \sin nt).$$

3.14 A particle of mass m moving along the x-axis is attracted towards the origin by a force $-m\omega^2 x\mathbf{i}$ and is acted upon by a disturbing force $mf \sin pt\mathbf{i}$ where $\mathbf{i}$ is a unit vector in the direction of the x-axis. If the particle is at rest at $x = a$ at time $t = 0$, show that the position of the particle at any subsequent time t is given by

$$x = \frac{f \sin pt}{\omega^2 - p^2} + a \cos \omega t - \frac{fp \sin \omega t}{\omega(\omega^2 - p^2)}.$$

- *Solution*

The equation of motion for the particle is

$$m\,\ddot{x}\,\mathbf{i} = -m\omega^2 x\mathbf{i} + mf \sin pt\mathbf{i}$$

giving

$$\ddot{x} + \omega^2 x = f \sin pt$$

whose general solution can be written in the form

$$x = \frac{f \sin pt}{\omega^2 - p^2} + A \sin \omega t + B \cos \omega t.$$

Now $x = a$ and $\dot{x} = 0$ when $t = 0$. Hence $B = a$ and $fp/(\omega^2 - p^2) + A\omega = 0$. The required result follows.

3.15 A particle of mass m is moving in a straight line, its displacement from a fixed point O of the line being x. It is acted upon by a restoring force $-m\omega^2 x\mathbf{i}$, a frictional force $-mk\,dx/dt\mathbf{i}$ such that $k^2 < 4\omega^2$, and a disturbing force $mf \cos \omega t\mathbf{i}$. If the particle is instantaneously at rest at O at time $t = 0$, find its displacement at any later time t.

- *Solution*

The equation of motion is

$$\ddot{x} + k\,\dot{x} + \omega^2 x = f\cos\omega t$$

whose general solution can be written in the form

$$x = \frac{f}{k\omega}\sin\omega t + e^{-kt/2}(A\cos\omega' t + B\sin\omega' t)$$

where

$$\omega'^2 = \omega^2 - k^2/4.$$

We have $x = 0$ and $\dot{x} = 0$ when $t = 0$ and so $A = 0$ and $f/k + B\omega' = 0$ giving for the displacement at time t:

$$x = \frac{f}{k}\left[\frac{\sin\omega t}{\omega} - \frac{e^{-kt/2}\sin\omega' t}{\omega'}\right].$$

3-4 Orbital motion

3.16 A planet is describing an ellipse about the sun at one focus. Show that its speed away from the sun is greatest when the radius vector to the planet is at right angles to the major axis of its orbit, and that it is then $2\pi ae/T(1 - e^2)^{\frac{1}{2}}$ where a is the length of the semi- major axis, e is the eccentricity and T is the periodic time.

- *Solution*

The equation of the ellipse in polar coordinates r, θ referred to the sun as focus is

$$\frac{l}{r} = 1 - e\cos\theta \tag{3.7}$$

where $l = a(1 - e^2)$ is the semi-latus rectum. Now $-lr^{-2}\,\dot{r} = e\sin\theta\,\dot{\theta}$ and so the speed away from the sun is

$$\dot{r} = -\frac{he\sin\theta}{l}$$

where $h = r^2\,\dot{\theta}$ is the areal constant. This is a maximum when $\sin\theta = -1$ so that $\theta = 3\pi/2$ or $-\pi/2$ and then $\dot{r} = he/l$.

Since the area of the ellipse is πab where $b = a\sqrt{1 - e^2}$ is the length of the semi-minor axis, and $h/2$ is the rate of description of area, the periodic time is given by $T = 2\pi ab/h$. It follows that the greatest speed of the planet away from the sun is

$$\dot{r} = \frac{2\pi eab}{Tl} = \frac{2\pi ea}{T\sqrt{1 - e^2}}.$$

3.17 The components of a double star revolve round each other once in time T, the greatest and least speeds in their relative orbit being v_1 and v_2 respectively. Show that the mean distance between the components is $T(v_1 v_2)^{\frac{1}{2}}/2\pi$ and that the total mass of the system is $T(v_1 v_2)^{\frac{3}{2}}/2\pi G$ where G is the constant of gravitation.

- *Solution*

If M_1 and M_2 are the masses of the two stars, their equations of motion are

$$M_1\ddot{\mathbf{r}}_1 = -\frac{GM_1M_2}{r^2}\hat{\mathbf{r}}, \quad M_2\ddot{\mathbf{r}}_2 = \frac{GM_1M_2}{r^2}\hat{\mathbf{r}}$$

where $\mathbf{r}_1$ and $\mathbf{r}_2$ are the position vectors of the stars referred to the origin of an inertial frame of reference and $\hat{\mathbf{r}}$ is a unit vector in the direction of $\mathbf{r} = \mathbf{r}_1 - \mathbf{r}_2$. Hence

$$\ddot{\mathbf{r}} = -\frac{k}{r^2}\hat{\mathbf{r}}$$

where $k = G(M_1 + M_2)$.

The speed of relative motion v is given by the standard formula

$$v^2 = k\left(\frac{2}{r} - \frac{1}{a}\right) \tag{3.8}$$

the length of the semi-major axis of the elliptical orbit being a. Since $r_{\max} = a(1 + e)$ and $r_{\min} = a(1 - e)$ it follows that

$$v_1^2 = k\left[\frac{2}{a(1-e)} - \frac{1}{a}\right] = \frac{k}{a}\left(\frac{1+e}{1-e}\right), \quad v_2^2 = k\left[\frac{2}{a(1+e)} - \frac{1}{a}\right] = \frac{k}{a}\left(\frac{1-e}{1+e}\right)$$

so that $v_1v_2 = k/a$.

Now *Kepler's third law of planetary motion* gives

$$\frac{T^2}{a^3} = \frac{4\pi^2}{k} \tag{3.9}$$

and so $T = 2\pi k/(v_1v_2)^{3/2}$. Hence

$$M_1 + M_2 = \frac{T(v_1v_2)^{3/2}}{2\pi G}.$$

Also the mean distance between the stars is given by $(r_{\max} + r_{\min})/2 = a$ where

$$a = \frac{k}{v_1v_2} = \frac{T(v_1v_2)^{1/2}}{2\pi}.$$

3.18 Two particles of masses m_1 and m_2 describing parabolic orbits about the sun collide and coalesce at a point distant b from the sun. Show that the orbit of the composite particle is an ellipse having major axis

$$\frac{(m_1 + m_2)^2}{4m_1m_2}b\operatorname{cosec}^2\frac{\alpha}{2}$$

where α is the angle between the directions of motion of the particles just before the collision.

Also show that if the parabolic orbits are coplanar and have latera recta l_1 and l_2 with angular momenta in the same sense, the latus rectum l of the orbit of the composite particle is given by $l^{\frac{1}{2}} = (m_1l_1^{\frac{1}{2}} + m_2l_2^{\frac{1}{2}})/(m_1 + m_2)$.

- *Solution*

Since the particles are describing parabolic paths, their speeds v just before they collide are equal and given by $v^2 = 2k/b$ where $k = GM$ and M is the mass of the sun. If the velocity of the composite particle, mass $m_1 + m_2$, just after the collision has magnitude V and makes an angle β with the direction of motion of the particle m_1, then by conservation of linear momentum

$$m_1 v + m_2 v \cos \alpha = (m_1 + m_2) V \cos \beta, \qquad m_2 v \sin \alpha = (m_1 + m_2) V \sin \beta$$

Hence

$$(m_1 + m_2)^2 V^2 = (m_1^2 + m_2^2 + 2m_1 m_2 \cos \alpha) v^2 < (m_1 + m_2)^2 v^2$$

if $\alpha \neq 0$ and so $V^2 < v^2 = 2GM/b$. It follows that the orbit of the composite particle is an ellipse.

Now from (3.8)

$$V^2 = GM \left(\frac{2}{b} - \frac{1}{a} \right)$$

where a is the length of the semi-major axis of the ellipse. Hence

$$\left(\frac{2}{b} - \frac{1}{a} \right)(m_1 + m_2)^2 = \frac{2}{b}(m_1^2 + m_2^2 + 2m_1 m_2 \cos \alpha)$$

which may be rewritten in the form

$$2a = \frac{(m_1 + m_2)^2}{4m_1 m_2} b \operatorname{cosec}^2 \frac{\alpha}{2}$$

giving the required result.

Let h_1, h_2 and h be the areal constants of the particles m_1, m_2 and the composite particle $m_1 + m_2$. Then

$$h_1 = GM l_1^{1/2}, \qquad h_2 = GM l_2^{1/2}, \qquad h = GM l^{1/2}$$

where l_1, l_2, l are the respective latera recta. Then by conservation of angular momentum we get

$$m_1 h_1 + m_2 h_2 = (m_1 + m_2) h$$

and so

$$m_1 l_1^{1/2} + m_2 l_2^{1/2} = (m_1 + m_2) l^{1/2}$$

which yields

$$l = \frac{l_1 m_1^2 + l_2 m_2^2 + 2(l_1 l_2)^{1/2} m_1 m_2}{(m_1 + m_2)^2}$$

producing the required result for $l^{1/2}$.

3.19 A particle P of mass m is attracted to a point O by a central force of magnitude mkr^{-3}. If P is projected with speed $(2k)^{\frac{1}{2}}/a$ from a point A distance a from O at an angle of $\pi/4$ to AO, show that the polar equation of the path is $ar^{-1} = 1 + \theta$.

Show also that the time taken for P to reach O is $a^2/k^{\frac{1}{2}}$

• *Solution*

For a central force F we have the differential equation

$$\frac{d^2u}{d\theta^2} + u = -\frac{F}{mh^2u^2} \tag{3.10}$$

where h is the areal constant, $u = r^{-1}$ and r, θ are polar coordinates referred to the centre of force O.

In the present example $F = -mku^3$ and so

$$\frac{d^2u}{d\theta^2} + \left(1 - \frac{k}{h^2}\right)u = 0.$$

Since the particle is projected with speed $\sqrt{2k}/a$ at a distance $r = a$ and an angle $\pi/4$ to AO, we have $h = \sqrt{2k}\sin(\pi/4) = \sqrt{k}$ so that $k/h^2 = 1$. Hence we get

$$\frac{d^2u}{d\theta^2} = 0$$

and so $u = A\theta + B$.

Taking $\theta = 0$ initially we obtain $B = 1/a$. Further, using

$$v^2 = h^2\left[\left(\frac{du}{d\theta}\right)^2 + u^2\right] \tag{3.11}$$

we have initially $2k/a^2 = h^2(A^2 + B^2)$ which gives $A = 1/a$. Hence $au = 1 + \theta$ or $ar^{-1} = 1 + \theta$.

Now $h = r^2 d\theta/dt$ and so $\sqrt{k} = -a\,dr/dt$. Therefore the time taken for the particle to reach O is

$$T = -\frac{a}{\sqrt{k}}\int_a^0 dr = \frac{a^2}{\sqrt{k}}.$$

3.20 A particle of mass m is moving under the action of a central force $F(u) = -mku^5$ where $u = r^{-1}$. Show that the speeds v_1, v_2 at the two points of the orbit of the particle where the directions of motion are perpendicular to the radius vector, that is the *apses*, satisfy $v_1^2 + v_2^2 = 2h^4/k$ where h is the areal constant.

• *Solution*

We have from (3.10)

$$\frac{d^2u}{d\theta^2} + u = -\frac{F(u)}{mh^2u^2} = \frac{ku^3}{h^2}.$$

Hence

$$\frac{d}{du}\left[\frac{1}{2}\left(\frac{du}{d\theta}\right)^2\right] + u = \frac{ku^3}{h^2}$$

and so

$$\frac{1}{2}\left(\frac{du}{d\theta}\right)^2 + \frac{1}{2}u^2 = \frac{ku^4}{4h^2} + \text{ constant.}$$

If $u = u_1$ and u_2 at the apses, where $du/d\theta = 0$, so that the directions of motion of the particle are perpendicular to the radius vector, we have

$$u_1^2 - \frac{ku_1^4}{2h^2} = u_2^2 - \frac{ku_2^4}{2h^2}$$

so that

$$u_1^2 + u_2^2 = \frac{2h^2}{k}.$$

Now using (3.11) we get, as required,

$$v_1^2 + v_2^2 = h^2(u_1^2 + u_2^2) = \frac{2h^4}{k}.$$

3.21 A particle of mass m is moving under the action of a central force $F(u) = -mku^5$ where $u = r^{-1}$. Show that

$$\frac{2h^2}{k}\left(\frac{du}{d\theta}\right)^2 = u^4 - \frac{2h^2u^2}{k} + \frac{4E}{mk}$$

where E is the total energy of the particle.

Show further that the orbit is a circle through the centre of force if $E = 0$.

- *Solution*

The kinetic energy of the particle is given by

$$T = \frac{1}{2}mh^2\left[\left(\frac{du}{d\theta}\right)^2 + u^2\right]$$

using (3.11), and its potential energy is

$$V(r) = mk\int_\infty^r \frac{dr}{r^5} = -\frac{mk}{4r^4}$$

chosen so that it vanishes in the limit $r \to \infty$. Hence the total energy $E = T + V$ is given by

$$E = \frac{1}{2}mh^2\left[\left(\frac{du}{d\theta}\right)^2 + u^2\right] - \frac{mku^4}{4}$$

so that

$$\frac{2h^2}{k}\left(\frac{du}{d\theta}\right)^2 = u^4 - \frac{2h^2u^2}{k} + \frac{4E}{mk}.$$

If $E = 0$ we see that

$$u^4 = \frac{2h^2}{k}\left[\left(\frac{du}{d\theta}\right)^2 + u^2\right].$$

Now $u = 1/r$ and so

$$\left(\frac{dr}{d\theta}\right)^2 = \frac{k}{2h^2} - r^2$$

from which it follows that

$$r = \sqrt{\frac{k}{2h^2}}\cos(\theta + \epsilon).$$

This is a circle through O of radius $\frac{1}{2}\sqrt{k/2h^2}$.

3-5 Motion of a charged particle in electric and magnetic fields

The planets move round the Sun under the force of gravity which is a *central force* directed towards the Sun. The electric force between charged particles is also a central force. On the other hand the force of a magnetic field $\mathbf{H}$ on a particle having electric charge q moving with velocity $\mathbf{v}$ is given by

$$\mathbf{F} = \frac{q}{c}\mathbf{v} \times \mathbf{H},$$

where c is the speed of light, which is *not* a central force.

Consider the following problems.

3.22 A particle of mass m and charge q is moving under the action of the force

$$\frac{q}{c}\dot{\mathbf{r}} \times \mathbf{H}$$

where $\mathbf{H}$ is a constant and uniform magnetic field vector, $\mathbf{r}$ is the position vector of the particle referred to the origin of an inertial frame of reference, and $\dot{\mathbf{r}} = dr/dt$ is the velocity of the particle.

If $\mathbf{r} = \mathbf{a}$ and $\dot{\mathbf{r}} = \mathbf{u}$ at time $t = 0$, where $\mathbf{u}.\mathbf{H} = 0$, find $\mathbf{r}$ as a function of the time t.

• *Solution*

An *inertial frame of reference* is one relative to which Newton's laws of motion hold and so Newton's second law of motion $md^2\mathbf{r}/dt^2 = \mathbf{F}$, where $\mathbf{F}$ is the force acting on the particle, gives

$$m\ddot{\mathbf{r}} = \frac{q}{c}\dot{\mathbf{r}} \times \mathbf{H}. \tag{3.12}$$

To solve this equation we need to obtain $\dot{\mathbf{r}}$. This can be done by integrating (3.12) with respect to t which gives

$$m\dot{\mathbf{r}} = \frac{q}{c}\mathbf{r} \times \mathbf{H} + \mathbf{b}$$

where $\mathbf{b}$ is a constant vector. Now we are given that $\mathbf{r} = \mathbf{a}$ and $\dot{\mathbf{r}} = \mathbf{u}$ at time $t = 0$ and so it follows that

$$\dot{\mathbf{r}} = \frac{q}{mc}(\mathbf{r} - \mathbf{a}) \times \mathbf{H} + \mathbf{u}. \tag{3.13}$$

We are also given that $\mathbf{u}.\mathbf{H} = 0$ and so $\dot{\mathbf{r}}.\mathbf{H} = 0$ from (3.13), using the property of the scalar triple product that it vanishes if any two vectors are equal so that $[(\mathbf{r} - \mathbf{a}) \times \mathbf{H}].\mathbf{H} = 0$. Integrating again with respect to time gives $\mathbf{r}.\mathbf{H} =$ constant and so

$$(\mathbf{r} - \mathbf{a}).\mathbf{H} = 0 \tag{3.14}$$

since the initial value of $\mathbf{r}.\mathbf{H}$ is $\mathbf{a}.\mathbf{H}$.

Let us set $\mathbf{R} = \mathbf{r} - \mathbf{a}$. Then we may rewrite (3.13) and (3.14) as

$$\dot{\mathbf{R}} = \mathbf{R} \times \omega + \mathbf{u}, \quad \mathbf{R}.\omega = 0 \tag{3.15}$$

where $\omega = q\mathbf{H}/mc$. It follows from (3.12) and (3.15) that

$$\ddot{\mathbf{R}} = \dot{\mathbf{R}} \times \omega = (\mathbf{R} \times \omega + \mathbf{u}) \times \omega$$

and using the formula for the vector triple product $(\mathbf{a} \times \mathbf{b}) \times \mathbf{c} = -\mathbf{c} \times (\mathbf{a} \times \mathbf{b}) = \mathbf{c}.\mathbf{a}\,\mathbf{b} - \mathbf{c}.\mathbf{b}\,\mathbf{a}$ we see that $(\mathbf{R} \times \omega) \times \omega = \mathbf{R}.\omega\,\omega - \omega^2\mathbf{R} = -\omega^2\mathbf{R}$ since $\mathbf{R}.\omega = 0$, and so we get

$$\ddot{\mathbf{R}} + \omega^2\mathbf{R} = \mathbf{u} \times \omega. \tag{3.16}$$

The general solution of this vector equation is composed of the sum of a particular integral $\omega^{-2}\mathbf{u} \times \omega$ and the general solution $\mathbf{A_1}\cos\omega t + \mathbf{A_2}\sin\omega t$ of the homogeneous equation, called the complementary function, given by

$$\mathbf{R} = \omega^{-2}\mathbf{u} \times \omega + \mathbf{A_1}\cos\omega t + \mathbf{A_2}\sin\omega t \tag{3.17}$$

where $\mathbf{A_1}$ and $\mathbf{A_2}$ are constant vectors. Since $\mathbf{R} = \mathbf{0}$ when $t = 0$ we have $\mathbf{A_1} = -\omega^{-2}\mathbf{u} \times \omega$. Also $\dot{\mathbf{R}} = \mathbf{u}$ when $t = 0$ and so $\mathbf{A_2} = \omega^{-1}\mathbf{u}$. It follows from (3.17) that

$$\mathbf{r} = \mathbf{a} + \omega^{-2}\mathbf{u} \times \omega(1 - \cos\omega t) + \omega^{-1}\mathbf{u}\sin\omega t. \tag{3.18}$$

3.23 Show that for the motion of a particle of charge q in a constant magnetic field $\mathbf{H}$, the quantity $\mathbf{H}.\mathbf{L} + q(\mathbf{r} \times \mathbf{H})^2/2c$ is constant, where $\mathbf{L}$ is the angular momentum of the particle about the origin.

- *Solution*

Since $\mathbf{L} = \mathbf{r} \times \mathbf{p}$ where $\mathbf{p} = m\dot{\mathbf{r}}$ and $\mathbf{r}$ is the position vector of the particle, we have

$$\mathbf{L} = \mathbf{r} \times m\dot{\mathbf{r}}$$

and so

$$\frac{d\mathbf{L}}{dt} = \mathbf{r} \times m\ddot{\mathbf{r}}.$$

Now the equation of motion of the charged particle is

$$m\ddot{\mathbf{r}} = \frac{q}{c}\dot{\mathbf{r}} \times \mathbf{H}.$$

Hence

$$\frac{d\mathbf{L}}{dt} = \mathbf{r} \times \left(\frac{q}{c}\dot{\mathbf{r}} \times \mathbf{H}\right)$$

and so

$$\frac{d(\mathbf{H}.\mathbf{L})}{dt} = \mathbf{H}.\left[\mathbf{r} \times \left(\frac{q}{c}\dot{\mathbf{r}} \times \mathbf{H}\right)\right] = \frac{q}{c}(\dot{\mathbf{r}} \times \mathbf{H}).(\mathbf{H} \times \mathbf{r}) = -\frac{d}{dt}\left[\frac{q}{2c}(\mathbf{r} \times \mathbf{H})^2\right]$$

that is

$$\frac{d}{dt}\left[\mathbf{H}.\mathbf{L} + \frac{q}{2c}(\mathbf{r} \times \mathbf{H})^2\right] = 0.$$

It follows that

$$\mathbf{H}.\mathbf{L} + \frac{q}{2c}(\mathbf{r} \times \mathbf{H})^2 = \text{constant}.$$

3.24 A particle of mass m and charge q moves in a constant magnetic field $\mathbf{H}$ under the action of a resistive force of magnitude $mk \times$speed. If the particle is projected from the origin with velocity $\mathbf{u}$ find the position vector of the particle when it eventually comes to rest.

- *Solution*

The equation of motion for the charged particle is

$$m\ddot{\mathbf{r}} = \frac{q}{c}\dot{\mathbf{r}} \times \mathbf{H} - mk\dot{\mathbf{r}}.$$

Hence

$$\dot{\mathbf{r}} = \frac{q}{mc}\mathbf{r} \times \mathbf{H} - k\mathbf{r} + \mathbf{u}.$$

and so the particle comes to rest when

$$\mathbf{r} \times \mathbf{H} = \frac{mc}{q}(k\mathbf{r} - \mathbf{u}).$$

Then we have

$$\mathbf{H} \times (\mathbf{r} \times \mathbf{H}) = \frac{mc}{q}(k\mathbf{H} \times \mathbf{r} - \mathbf{H} \times \mathbf{u}), \quad k\mathbf{H}.\mathbf{r} = \mathbf{H}.\mathbf{u}$$

Now $\mathbf{H} \times (\mathbf{r} \times \mathbf{H}) = \mathbf{H}^2\,\mathbf{r} - \mathbf{H}.\mathbf{r}\,\mathbf{H}$ and hence we get

$$H^2\,\mathbf{r} - \frac{1}{k}\mathbf{H}.\mathbf{u}\,\mathbf{H} = \frac{mc}{q}\left[\frac{mck}{q}(\mathbf{u} - k\mathbf{r}) - \mathbf{H} \times \mathbf{u}\right]$$

so that the position vector of the charged particle when it comes to rest is

$$\mathbf{r} = \frac{(mck/q)[(mck/q)\mathbf{u} + \mathbf{u} \times \mathbf{H}] + \mathbf{H}.\mathbf{u}\,\mathbf{H}}{k\left[H^2 + (mck/q)^2\right]}.$$

In the following problems we have an electric field as well as a magnetic field. Thus we use the *Lorentz force* on a particle having charge q moving with velocity $\mathbf{v}$ in an electric field $\mathbf{E}$ and magnetic field $\mathbf{H}$ given by

$$\mathbf{F} = q\mathbf{E} + \frac{q}{c}\mathbf{v} \times \mathbf{H} \qquad (3.19)$$

3.25 A particle of mass m and charge q enters an electromagnetic field with initial velocity $u(\alpha\mathbf{i} - \mathbf{j})$. The electromagnetic field is composed of a uniform electric field $E\mathbf{i}$ and a uniform magnetic field $H\mathbf{k}$ where $H = cE/u$. Find the position of the particle at time t and show that the particle will be unaffected by the field if $\alpha = 0$.

• *Solution*

The equation of motion of the charged particle is

$$m\ddot{\mathbf{r}} = q\mathbf{E} + \frac{q}{c}\dot{\mathbf{r}} \times \mathbf{H}$$

where $\mathbf{E} = E\mathbf{i}$ and $\mathbf{H} = H\mathbf{k}$. Hence

$$\ddot{x} = qE/m + \omega\,\dot{y},$$
$$\ddot{y} = -\omega\,\dot{x},$$
$$\ddot{z} = 0$$

where $\omega = qH/mc$.

Let us suppose that the particle is at the origin of the Cartesian coordinates x, y, z at time $t = 0$. Then we see at once that $z = 0$ for all time.

Also $\dot{y} = -\omega x - u$ since $x = 0$ and $\dot{y} = -u$ when $t = 0$, and so

$$\ddot{x} + \omega^2 x = \frac{q}{m}E - \omega u.$$

Hence, since $x = 0, \dot{x} = u\alpha$ when $t = 0$, we obtain

$$x = \frac{1}{\omega^2}\left(\frac{q}{m}E - \omega u\right)(1 - \cos\omega t) + \frac{u\alpha}{\omega}\sin\omega t$$

If $H = cE/u$ we have $\omega u = qE/m$ and thus

$$x = \frac{u\alpha}{\omega} \sin \omega t.$$

It follows that $\dot{y} = -u\alpha \sin \omega t - u$ and so

$$y = \frac{u\alpha}{\omega}(\cos \omega t - 1) - ut.$$

If $\alpha = 0$ this becomes $x = 0, y = -ut, z = 0$ so that the particle is unaffected by the field.

3.26 A particle of mass m and charge q is moving under the action of the Lorentz force $q\mathbf{E} + q\mathbf{v} \times \mathbf{H}/c$ where $\mathbf{E}$ and $\mathbf{H}$ are constant and uniform electric and magnetic field vectors respectively satisfying $\mathbf{E}.\mathbf{H} = 0$, $\mathbf{r}$ is the position vector of the particle referred to the origin of an inertial frame of reference, and $\mathbf{v}$ is the velocity of the particle.

If $\mathbf{r} = \mathbf{0}$ and $\mathbf{v} = \mathbf{0}$ at time $t = 0$ show that $\mathbf{v} = q\mathbf{E}t/m + \mathbf{r} \times \omega$ where $\omega = q\mathbf{H}/mc$ and

$$\mathbf{r} = \frac{q}{m\omega^2}\left[(1 - \cos \omega t)\mathbf{E} + \left(t - \frac{\sin \omega t}{\omega}\right)\mathbf{E} \times \omega\right].$$

Hence obtain $\dot{\mathbf{r}}$ and show that the maximum speed of the particle is $2cE/H$.

• *Solution*

The equation of motion of the charged particle is

$$m\ddot{\mathbf{r}} = q\mathbf{E} + \frac{q}{c}\dot{\mathbf{r}} \times \mathbf{H}.$$

Hence

$$\dot{\mathbf{r}} = \frac{q}{m}\mathbf{E}t + \mathbf{r} \times \omega$$

where $\omega = q\mathbf{H}/mc$, and so

$$\ddot{\mathbf{r}} = \frac{q}{m}\mathbf{E} + \left(\frac{q}{m}\mathbf{E}t + \mathbf{r} \times \omega\right) \times \omega,$$

that is

$$\ddot{\mathbf{r}} = \frac{q}{m}(\mathbf{E} + \mathbf{E} \times \omega\ t) + \mathbf{r}.\omega\ \omega - \omega^2\mathbf{r}$$

Now $\dot{\mathbf{r}}.\omega = 0$ since $\mathbf{E}.\omega = 0$ and so $\mathbf{r}.\omega = 0$. Hence

$$\ddot{\mathbf{r}} + \omega^2\mathbf{r} = \frac{q}{m}(\mathbf{E} + \mathbf{E} \times \omega\ t)$$

which has the general solution

$$\mathbf{r} = \mathbf{a} \cos \omega t + \mathbf{b} \sin \omega t + \frac{q}{m\omega^2}(\mathbf{E} + \mathbf{E} \times \omega\ t).$$

Since $\mathbf{r} = \mathbf{0}$ when $t = 0$ we have $\mathbf{a} = -q\mathbf{E}/m\omega^2$, and since $\dot{\mathbf{r}} = \mathbf{0}$ when $t = 0$ we also have $\mathbf{b} = -q\mathbf{E} \times \omega/m\omega^3$.

It follows that

$$\mathbf{r} = \frac{q}{m\omega^2}\mathbf{E}(1 - \cos \omega t) + \frac{q}{m\omega^2}\mathbf{E} \times \omega\ (t - \frac{\sin \omega t}{\omega}).$$

Hence we get

$$\dot{\mathbf{r}} = \frac{q}{m\omega}\mathbf{E} \sin \omega t + \frac{q}{m\omega^2}\mathbf{E} \times \omega\ (1 - \cos \omega t)$$

and so the speed of the particle is given by

$$v^2 = 2\left(\frac{cE}{H}\right)^2 (1 - \cos \omega t).$$

We see that the maximum speed is given by $v_{\text{max}} = 2cE/H$.

3-6 Rotating frames of reference

Consider two frames of reference S_1 and S_2 having a common origin O, with S_2 rotating relative to S_1 with angular velocity $\mathbf{\Omega}$.

If $\mathbf{r}$ is the position vector of a moving point P referred to O we have the velocity formula

$$\mathbf{v_1} = \mathbf{v_2} + \mathbf{\Omega} \times \mathbf{r} \qquad (3.20)$$

where $\mathbf{v_1} = (dr/dt)_1$ and $\mathbf{v_2} = (dr/dt)_2$ are the velocities of P relative to the frames S_1 and S_2 respectively.

Further we have the acceleration formula called the *Coriolis theorem*

$$\mathbf{a_1} = \mathbf{a_2} + \frac{d\mathbf{\Omega}}{dt} \times \mathbf{r} + 2\mathbf{\Omega} \times \mathbf{v_2} + \mathbf{\Omega} \times (\mathbf{\Omega} \times \mathbf{r}) \qquad (3.21)$$

where $\mathbf{a_1} = (dv_1/dt)_1$ and $\mathbf{a_2} = (dv_2/dt)_2$ are the accelerations of P relative to the frames S_1 and S_2 respectively.

The term $2\mathbf{\Omega} \times \mathbf{v_2}$ is known as the *Coriolis acceleration*.

3.27 Derive the (2.2) for the acceleration by considering a particle moving in the radial direction on a plane rotating with variable angular velocity about a fixed point of the plane, the axis of rotation being perpendicular to the plane.

• *Solution*

Let r, θ be the circular polar coordinates of the particle P relative to a two-dimensional frame of reference S_1 with origin O. Then if the particle is moving in the radial direction $\hat{\mathbf{r}}$ referred to a two-dimensional frame S_2 with origin O rotating with angular speed $\dot{\theta}$ relative to S_1, about an axis through O perpendicular to their common plane, we have

$$\mathbf{v_2} = \dot{r}\,\hat{\mathbf{r}}, \quad \mathbf{a_2} = \ddot{r}\,\hat{\mathbf{r}}.$$

Taking the angular velocity of S_2 relative to S_1 to be $\mathbf{\Omega} = \dot{\theta}\,\hat{\mathbf{k}}$ where $\hat{\mathbf{k}}$ is a unit vector perpendicular to the common plane, we get from the Coriolis theorem (3.21)

$$\mathbf{a_1} = \ddot{r}\,\hat{\mathbf{r}} + \ddot{\theta}\,r\hat{\mathbf{k}} \times \hat{\mathbf{r}} + 2\,\dot{\theta}\dot{r}\,\hat{\mathbf{k}} \times \hat{\mathbf{r}} + \dot{\theta}^2\,r\hat{\mathbf{k}} \times (\hat{\mathbf{k}} \times \hat{\mathbf{r}})$$

But $\hat{\mathbf{k}} \times \hat{\mathbf{r}} = \hat{\mathbf{s}}$ is a unit vector in the transverse direction and $\hat{\mathbf{k}} \times \hat{\mathbf{s}} = -\hat{\mathbf{r}}$ so that we get

$$\mathbf{a_1} = (\ddot{r} - r\dot{\theta}^2)\hat{\mathbf{r}} + (2\dot{r}\dot{\theta} + r\ddot{\theta})\hat{\mathbf{s}}$$

which is just the radial and transverse resolutes formula for the acceleration (2.2).

3.28 A particle is projected vertically upwards with speed V at a point on the Earth's surface having latitude λ. Show that due to the Earth's angular velocity Ω the particle will reach the ground at a distance $4\Omega V^3 \cos\lambda/3g^2$ due west of the point of projection.

• *Solution*

The Coriolis theorem (3.21) gives

$$\mathbf{a} = \ddot{\mathbf{r}} + 2\mathbf{\Omega} \times \dot{\mathbf{r}} + \mathbf{\Omega} \times (\mathbf{\Omega} \times \mathbf{r})$$

for the acceleration of the particle taking $\mathbf{r}$ to be its position vector referred to the point of projection and $\mathbf{\Omega}$ to be the constant angular velocity of the Earth.

Now the *apparent acceleration of gravity* $\mathbf{g}$ is given by

$$m\mathbf{g} = \mathbf{F} - m\mathbf{\Omega} \times (\mathbf{\Omega} \times \mathbf{r}) \tag{3.22}$$

where m is the mass of the particle and $\mathbf{F} = m\mathbf{a}$ is the force of gravity. Hence we have

$$\ddot{\mathbf{r}} = -2\mathbf{\Omega} \times \dot{\mathbf{r}} + \mathbf{g}$$

and so integrating with respect to time t gives

$$\dot{\mathbf{r}} = \mathbf{g}t - 2\mathbf{\Omega} \times \mathbf{r} + \mathbf{V} \tag{3.23}$$

where $\mathbf{V}$ is the initial velocity of the particle. Taking the unit vector in the upward direction at the point of projection to be $\mathbf{k}$ we have $\mathbf{V} = V\mathbf{k}$ and then integrating again with respect to time we get to zero order in Ω

$$\mathbf{r} = (Vt - \tfrac{1}{2}gt^2)\mathbf{k}$$

since $\mathbf{g} = -g\mathbf{k}$.

Hence to first order in Ω we have

$$\dot{\mathbf{r}} = (V - gt)\mathbf{k} - 2\mathbf{\Omega} \times \mathbf{k}\,(Vt - \tfrac{1}{2}gt^2)$$

and so integrating with respect to time we find that the position vector of the particle is given by

$$\mathbf{r} = (Vt - \tfrac{1}{2}gt^2)\mathbf{k} - 2\mathbf{\Omega} \times \mathbf{k}\,(\tfrac{1}{2}Vt^2 - \tfrac{1}{6}gt^3).$$

Now

$$\mathbf{\Omega} = \Omega\cos\lambda\,\mathbf{j} + \Omega\sin\lambda\,\mathbf{k}$$

where $\mathbf{j}$ is a unit vector directed towards the north at the point of projection. Hence $\mathbf{\Omega} \times \mathbf{k} = \Omega\cos\lambda\,\mathbf{i}$, where $\mathbf{i}$ is a unit vector pointing to the east, and so we get

$$\mathbf{r} = (Vt - \tfrac{1}{2}gt^2)\mathbf{k} - 2\Omega\cos\lambda(\tfrac{1}{2}Vt^2 - \tfrac{1}{6}gt^3)\mathbf{i}.$$

The particle reaches the ground when $t = 2V/g$ and so the particle is displaced a distance to the east given by $x = -\Omega\cos\lambda[V(2V/g)^2 - g(2V/g)^3/3]$, that is a distance $4\Omega V^3\cos\lambda/3g^2$ to the west, as required.

3.29 A missile is fired with velocity V towards the east from a point on the Earth's surface having latitude λ. If the angle of projection is α show that the increase in the range due to the rotation of the Earth is $4\Omega V^3\cos\lambda\sin 3\alpha/3g^2$.

• *Solution*

Using the formula (3.23) for the velocity obtained in the previous problem we have

$$\dot{\mathbf{r}} = \mathbf{g}t - 2\mathbf{\Omega} \times \mathbf{r} + \mathbf{V}$$

where $\mathbf{r}$ is the position vector of the missile referred to the point of projection.

Then to zero-order in Ω we get

$$\mathbf{r} = \tfrac{1}{2}\mathbf{g}t^2 + \mathbf{V}t$$

so that to first-order in Ω we obtain

$$\dot{\mathbf{r}} = \mathbf{g}t - 2\mathbf{\Omega}\times(\tfrac{1}{2}\mathbf{g}t^2 + \mathbf{V}t) + \mathbf{V}$$

which gives

$$\mathbf{r} = \tfrac{1}{2}\mathbf{g}t^2 + \mathbf{V}t - \mathbf{\Omega} \times (\tfrac{1}{3}\mathbf{g}t^3 + \mathbf{V}t^2).$$

Letting $\mathbf{i}, \mathbf{j}, \mathbf{k}$ be the unit vectors in the east, north and vertical directions respectively, we have $\mathbf{V} = V(\mathbf{i}\cos\alpha + \mathbf{k}\sin\alpha)$, $\mathbf{g} = -g\mathbf{k}$ and $\mathbf{\Omega} = \Omega(\mathbf{j}\cos\lambda + \mathbf{k}\sin\lambda)$ and so it follows that

$$\mathbf{r} = Vt(\mathbf{i}\cos\alpha + \mathbf{k}\sin\alpha) - \tfrac{1}{2}gt^2\mathbf{k} - \begin{vmatrix} \mathbf{i} & \mathbf{j} & \mathbf{k} \\ 0 & \Omega\cos\lambda & \Omega\sin\lambda \\ Vt^2\cos\alpha & 0 & Vt^2\sin\alpha - \tfrac{1}{3}gt^3 \end{vmatrix}. \quad (3.24)$$

Hence the coordinates of the missile are

$$\begin{aligned} x &= Vt\cos\alpha - \Omega\cos\lambda(Vt^2\sin\alpha - \tfrac{1}{3}gt^3), \\ y &= -\Omega Vt^2\sin\lambda\cos\alpha, \\ z &= Vt\sin\alpha - \tfrac{1}{2}gt^2 + \Omega Vt^2\cos\lambda\cos\alpha. \end{aligned}$$

Putting $z = 0$ we get for the *time of flight*

$$T = \frac{2V\sin\alpha}{g(1 - 2\Omega V\cos\lambda\cos\alpha/g)} \simeq \frac{2V\sin\alpha}{g}\left(1 + \frac{2\Omega V\cos\lambda\cos\alpha}{g}\right)$$

and so the *range* in the easterly direction is given by

$$R = \frac{2V^2\sin\alpha\cos\alpha}{g} + \frac{4\Omega V^3\sin\alpha\cos^2\alpha\cos\lambda}{g^2} - \frac{4\Omega V^3\sin^3\alpha\cos\lambda}{3g^2}.$$

Since $3\sin\alpha\cos^2\alpha - \sin^3\alpha = \sin 3\alpha$ we obtain to first-order in Ω

$$R = R_0 + \frac{4\Omega V^3\cos\lambda\sin 3\alpha}{3g^2}$$

where $R_0 = 2V^2\sin\alpha\cos\alpha/g$ is the range of the missile neglecting the rotation of the Earth.

The displacement in the northerly direction is of the first-order in Ω and so the increase in the range due to the rotation of the Earth, to first-order in Ω, is $4\Omega V^3\cos\lambda\sin 3\alpha/3g^2$ as required.

3.30 A stone is released from rest at a height h above the Earth's surface. Show that when the stone reaches the ground it will have been displaced towards the east by the distance

$$\frac{2}{3}\Omega\cos\lambda\sqrt{\frac{2h^3}{g}}.$$

• *Solution*

By putting $V = 0$ in (3.24) in the previous problem, and using $z = h$ when $t = 0$, we find that

$$\mathbf{r} = (h - \tfrac{1}{2}gt^2)\mathbf{k} + \tfrac{1}{3}t^3g\Omega\cos\lambda\,\mathbf{i}.$$

To zero-order in the angular velocity Ω of the Earth we have $h = \tfrac{1}{2}gt^2$ where t is the time taken to reach the ground. Hence $\mathbf{r} = \tfrac{1}{3}t^3g\Omega\cos\lambda\,\mathbf{i}$ is the position vector of the stone when it reaches the ground and so its displacement towards the east is $\tfrac{1}{3}(2h/g)^{3/2}g\Omega\cos\lambda$ which yields the required result.

4
VECTOR FIELD THEORY

We now turn our attention to some elementary problems in the theory of *vector fields*.

The subject of *vector field theory* is essential for the study of *potential theory* including *Newtonian gravitation* and *electrostatics*, for *electromagnetism* in general, for *fluid dynamics*, and for *quantum theory*: indeed for most branches of applied mathematics and mathematical physics.

We first need to remember that the gradient operator *nabla* is given by

$$\nabla = \mathbf{i}\frac{\partial}{\partial x} + \mathbf{j}\frac{\partial}{\partial y} + \mathbf{k}\frac{\partial}{\partial z} \tag{4.1}$$

using Cartesian coordinates, where $\mathbf{i}, \mathbf{j}, \mathbf{k}$ are the unit vectors in the directions of the x, y, z axes.

4.01 If $\mathbf{r}$ is the position vector of a point and $\mathbf{a}$ is a constant vector, show that

$$(i)\ \nabla \mathbf{a}.\mathbf{r} = \mathbf{a}, \quad (ii)\ \nabla.(\mathbf{a} \times \mathbf{r}) = 0, \quad (iii)\ \nabla \times (\mathbf{a} \times \mathbf{r}) = 2\mathbf{a}. \tag{4.2}$$

Further show that

$$\nabla(\mathbf{a_1}.\mathbf{r}\,\mathbf{a_2}.\mathbf{r}\ \mathbf{a_3}.\mathbf{r}) = \mathbf{a_2}.\mathbf{r}\,\mathbf{a_3}.\mathbf{r}\ \mathbf{a_1} + \mathbf{a_1}.\mathbf{r}\,\mathbf{a_3}.\mathbf{r}\,\mathbf{a_2} + \mathbf{a_1}.\mathbf{r}\ \mathbf{a_2}.\mathbf{r}\,\mathbf{a_3}$$

where $\mathbf{a_1}, \mathbf{a_2}, \mathbf{a_3}$ are constant vectors.

- *Solution*

(i) If $\mathbf{a} = a_1\mathbf{i} + a_2\mathbf{j} + a_3\mathbf{k}$ then

$$\nabla \mathbf{a}.\mathbf{r} = \nabla(a_1 x + a_2 y + a_3 z) = a_1\mathbf{i} + a_2\mathbf{j} + a_3\mathbf{k} = \mathbf{a}$$

This can be seen in another way. The gradient of a scalar function ϕ is given by the general formula

$$\nabla\phi = \frac{\partial\phi}{\partial n}\hat{\mathbf{n}}$$

where $\hat{\mathbf{n}}$ is a unit vector in the direction of the normal to a level surface of the function ϕ.

Now the level surface $\mathbf{a}.\mathbf{r} =$ constant is a plane for which $\mathbf{a}$ is in the direction of the normal. Since $\partial(\hat{\mathbf{n}}.\mathbf{r})/\partial n = 1$ it follows that $\nabla \mathbf{a}.\mathbf{r} = a\hat{\mathbf{a}} = \mathbf{a}$.

(ii) We have, making a cyclic interchange of the scalar product, that

$$\nabla.(\mathbf{a} \times \mathbf{r}) = -\mathbf{a}.(\nabla \times \mathbf{r}) = 0$$

since $\nabla \times \mathbf{r} = \mathbf{0}$.

(iii) Using the formula for the vector triple product we have

$$\nabla \times (\mathbf{a} \times \mathbf{r}) = \nabla.\mathbf{r}\,\mathbf{a} - \mathbf{a}.\nabla\mathbf{r} = 3\mathbf{a} - \mathbf{a} = 2\mathbf{a}$$

since

$$\nabla.\mathbf{r} = \frac{\partial}{\partial x}x + \frac{\partial}{\partial y}y + \frac{\partial}{\partial z}z = 3 \qquad (4.3)$$

and

$$\mathbf{a}.\nabla\mathbf{r} = \left(a_1\frac{\partial}{\partial x} + a_2\frac{\partial}{\partial y} + a_3\frac{\partial}{\partial z}\right)(x\mathbf{i} + y\mathbf{j} + z\mathbf{k})) = a_1\mathbf{i} + a_2\mathbf{j} + a_3\mathbf{k} = \mathbf{a}$$

Further we have

$$\nabla(\mathbf{a_1.r}\,\mathbf{a_2.r}\,\mathbf{a_3.r}) = \mathbf{a_2.r}\,\mathbf{a_3.r}\,\nabla\mathbf{a_1.r} + \mathbf{a_1.r}\,\mathbf{a_3.r}\,\nabla\mathbf{a_2.r} + \mathbf{a_1.r}\,\mathbf{a_2.r}\,\nabla\mathbf{a_3.r}$$

which gives the required result using part (i) of (4.2).

4.02 If $\mathbf{a}$ and $\mathbf{b}$ are constant vectors, derive an expression for

$$\mathbf{a}.\nabla\ \mathbf{b}.\nabla\frac{1}{r}$$

- *Solution*

In this problem we use the formula $\nabla f(r) = df/dr\ \hat{\mathbf{r}}$ which for $f(r) = r^n$ becomes $\nabla r^n = nr^{n-1}\hat{\mathbf{r}}$. Then we see that $\nabla r^{-1} = -r^{-2}\ \hat{\mathbf{r}} = -\ r^{-3}\mathbf{r}$ and so

$$\mathbf{a}.\nabla\mathbf{b}.\nabla\frac{1}{r} = -\mathbf{a}.\nabla\frac{\mathbf{b}.\mathbf{r}}{r^3} = -\mathbf{b}.\mathbf{r}\ \mathbf{a}.\nabla\frac{1}{r^3} - \frac{1}{r^3}\mathbf{a}.\nabla\mathbf{b}.\mathbf{r}$$

Now $\nabla r^{-3} = -3r^{-4}\hat{\mathbf{r}} = -3r^{-5}\mathbf{r}$ and $\nabla\mathbf{b}.\mathbf{r} = \mathbf{b}$ using part (i) of (4.2) of the previous problem, which lead to the result

$$\mathbf{a}.\nabla\mathbf{b}.\nabla\frac{1}{r} = \frac{3\mathbf{b}.\mathbf{r}\mathbf{a}.\mathbf{r}}{r^5} - \frac{\mathbf{a}.\mathbf{b}}{r^3}$$

4.03 If $\rho\mathbf{\Phi} = \mathrm{grad}\,\phi$ where ρ and ϕ are scalar functions of position and $\mathbf{\Phi}$ is a vector function, show that

$$\mathbf{\Phi}.\,\mathrm{curl}\,\mathbf{\Phi} = 0$$

- *Solution*

We have $\mathrm{curl}(\rho\mathbf{\Phi}) = \mathrm{curl\,grad}\,\phi = \nabla \times \nabla\phi = \mathbf{0}$. Hence $\rho\,\mathrm{curl}\,\mathbf{\Phi} - \mathbf{\Phi} \times \nabla\rho = \mathbf{0}$ and so $\rho\mathbf{\Phi}.\,\mathrm{curl}\,\mathbf{\Phi} = \mathbf{\Phi}.(\mathbf{\Phi} \times \nabla\rho) = 0$.

Hence $\mathbf{\Phi}.\,\mathrm{curl}\,\mathbf{\Phi} = 0$.

4.04 If $\mathbf{\Phi}$ is a vector function and ϕ is a scalar function of position, find the components of $\mathbf{\Phi}$ in the normal and tangential directions to a level surface of ϕ.

- *Solution*

We know that $\nabla\phi$ is a vector which is normal to the level surface $\phi = c$ where c is a constant. Hence the component of $\mathbf{\Phi}$ normal to $\phi = c$ is

$$\frac{(\mathbf{\Phi}.\nabla\phi)\nabla\phi}{(\nabla\phi)^2}$$

and so the component of $\mathbf{\Phi}$ tangential to $\phi = c$ is given by

$$\mathbf{\Phi} - \frac{(\mathbf{\Phi}.\nabla\phi)\nabla\phi}{(\nabla\phi)^2} = \frac{1}{(\nabla\phi)^2}[\nabla\phi.\nabla\phi\ \mathbf{\Phi} - \nabla\phi.\mathbf{\Phi}\ \nabla\phi] = \frac{\nabla\phi \times (\mathbf{\Phi} \times \nabla\phi)}{(\nabla\phi)^2}$$

4.05 If $\mathrm{div}\,\mathbf{\Phi} = 0$ show that $\mathrm{curl\,curl\,curl\,curl}\,\mathbf{\Phi} = \nabla^2\nabla^2\ \mathbf{\Phi}$.

• *Solution*

We have

$$\begin{aligned}
\text{curl curl curl curl } \boldsymbol{\Phi} &= \nabla \times \{\nabla \times [\nabla \times (\nabla \times \boldsymbol{\Phi})]\} \\
&= \nabla \times \{\nabla \times [\nabla\nabla.\boldsymbol{\Phi} - \nabla^2\boldsymbol{\Phi}]\} \\
&= -\nabla \times \{\nabla \times \nabla^2\boldsymbol{\Phi}\} \\
&= -\nabla\nabla.(\nabla^2\boldsymbol{\Phi}) + \nabla^2\nabla^2\boldsymbol{\Phi} \\
&= -\nabla\nabla^2\nabla.\boldsymbol{\Phi} + \nabla^2\nabla^2\boldsymbol{\Phi} \\
&= \nabla^2\nabla^2\boldsymbol{\Phi}
\end{aligned}$$

since $\nabla.\boldsymbol{\Phi} = 0$.

4.06 Show that

$$\text{grad}\left(\frac{\mathbf{c.r}}{r^n}\right) = \frac{\mathbf{c}}{r^n} - \frac{n\mathbf{c.r}\;\mathbf{r}}{r^{n+2}}$$

where $\mathbf{c}$ is a constant vector.

• *Solution*

We have

$$\text{grad}\left(\frac{\mathbf{c.r}}{r^n}\right) = \frac{1}{r^n}\nabla\mathbf{c.r} + \mathbf{c.r}\nabla\frac{1}{r^n}$$

Now $\nabla\mathbf{c.r} = \mathbf{c}$ and $\nabla r^{-n} = -nr^{-n-1}\hat{\mathbf{r}} = -nr^{-n-2}\mathbf{r}$ and so the required result follows at once.

4.07 If $\mathbf{A} = -\boldsymbol{\mu} \times \text{grad } r^{-1}$, where $\boldsymbol{\mu}$ is a constant vector, show that div $\mathbf{A} = 0$ and find $\mathbf{B} = \text{curl } \mathbf{A}$.

• *Solution*

We have div $\mathbf{A} = -\nabla.\left(\boldsymbol{\mu} \times \nabla\frac{1}{r}\right) = \boldsymbol{\mu}.\left(\nabla \times \nabla\frac{1}{r}\right) = 0$ since curl grad $\phi = \mathbf{0}$ generally. This proves the first result.

Further

$$\mathbf{B} = \text{curl } \mathbf{A} = -\nabla \times \left(\boldsymbol{\mu} \times \nabla\frac{1}{r}\right) = -\boldsymbol{\mu}\nabla^2\frac{1}{r} + \boldsymbol{\mu}.\nabla\nabla\frac{1}{r} = \boldsymbol{\mu}.\nabla\nabla\frac{1}{r}$$

since $\nabla^2 r^{-1} = 0$ for $r \neq 0$. But $\nabla r^{-1} = -r^{-3}\mathbf{r}$ and hence

$$\mathbf{B} = -\boldsymbol{\mu}.\nabla\frac{\mathbf{r}}{r^3} = -\mathbf{r}\;\boldsymbol{\mu}.\nabla\frac{1}{r^3} - \frac{1}{r^3}\boldsymbol{\mu}.\nabla\mathbf{r}$$

Now $\nabla r^{-3} = -3r^{-5}\mathbf{r}$ and $\boldsymbol{\mu}.\nabla\mathbf{r} = \boldsymbol{\mu}$. It follows that

$$\mathbf{B} = \frac{3\boldsymbol{\mu}.\mathbf{r}\;\mathbf{r}}{r^5} - \frac{\boldsymbol{\mu}}{r^3}$$

4.08 Show that

$$\int_R \nabla\phi\,d\tau = \oint_S \hat{\mathbf{n}}\phi\,dS, \qquad \int_R \nabla \times \boldsymbol{\Phi}\,d\tau = \oint_S \hat{\mathbf{n}} \times \boldsymbol{\Phi}\,dS \qquad (4.4)$$

where S is the boundary surface of a simply connected region R and $\hat{\mathbf{n}}$ is the unit vector in the direction of the outward normal to the surface.

• *Solution*

Here we use Gauss's divergence theorem

$$\int_R \text{div } \mathbf{A} \, d\tau = \oint_S \mathbf{A}.\hat{\mathbf{n}} \, dS \tag{4.5}$$

where $\mathbf{A}$ is a vector function of position and $\hat{\mathbf{n}}$ is the unit vector in the direction of the outward normal to the boundary surface S of the region R.

First we put $\mathbf{A} = \phi\mathbf{c}$ where $\mathbf{c}$ is a an arbitrary constant vector. Then we obtain

$$\mathbf{c}. \int_R \nabla\phi \, d\tau = \mathbf{c}. \oint_S \phi\hat{\mathbf{n}} \, dS$$

and since $\mathbf{c}$ is arbitrary, the first result of (4.4) follows.

Next we put $\mathbf{A} = \mathbf{c} \times \mathbf{\Phi}$ and then we see that div $\mathbf{A} = \nabla.(\mathbf{c} \times \mathbf{\Phi}) = -\mathbf{c}.(\nabla \times \mathbf{\Phi})$ and $\mathbf{A}.\hat{\mathbf{n}} = (\mathbf{c} \times \mathbf{\Phi}).\hat{\mathbf{n}} = -\mathbf{c}.(\hat{\mathbf{n}} \times \mathbf{\Phi})$.

Then we get

$$\mathbf{c}. \int_R \nabla \times \mathbf{\Phi} \, d\tau = \mathbf{c}. \oint_S \hat{\mathbf{n}} \times \mathbf{\Phi} \, dS$$

and since $\mathbf{c}$ is arbitrary, the second result of (4.4) follows.

4.09 Evaluate $\oint_S \hat{\mathbf{n}} \times (\mathbf{a} \times \mathbf{r}) \, dS$ where $\mathbf{a}$ is a constant vector and S is the boundary surface of a simply connected region R having volume V.

• *Solution*

Using the second result of (4.4) in the previous problem we have

$$\oint_S \hat{\mathbf{n}} \times \mathbf{A} \, dS = \int_R \nabla \times \mathbf{A} \, d\tau.$$

Setting $\mathbf{A} = \mathbf{a} \times \mathbf{r}$ gives

$$\oint_S \hat{\mathbf{n}} \times (\mathbf{a} \times \mathbf{r}) \, dS = \int_R \nabla \times (\mathbf{a} \times \mathbf{r}) \, d\tau.$$

But $\nabla \times (\mathbf{a} \times \mathbf{r}) = 2\mathbf{a}$ by part(iii) of (4.2) and so

$$\oint_S \hat{\mathbf{n}} \times (\mathbf{a} \times \mathbf{r}) \, dS = 2\mathbf{a} \int_R d\tau = 2\mathbf{a}V.$$

4.10 Evaluate $\oint_S \mathbf{a}.\mathbf{r} \, \hat{\mathbf{n}} \, dS$ where $\mathbf{a}$ is a constant vector and S is the boundary surface of a simply connected region R having volume V.

• *Solution*

Setting $\phi = \mathbf{c}.\mathbf{r}$ in the first result of (4.4) in problem 4.08 we get

$$\oint_S \mathbf{c}.\mathbf{r} \, \hat{\mathbf{n}} \, dS = \int_R \nabla\mathbf{c}.\mathbf{r} \, d\tau = \mathbf{c} \int_R d\tau = \mathbf{c}V.$$

4.11 Prove the generalization of Green's theorem due to Kelvin:

$$\int_R [\chi\nabla\phi.\nabla\psi + \phi\nabla.(\chi\nabla\psi)] \, d\tau = \oint_S \phi\chi\nabla\psi.\hat{\mathbf{n}} \, dS \tag{4.6}$$

or

$$\int_R [\chi\nabla\phi.\nabla\psi + \psi\nabla.(\chi\nabla\phi)] \, d\tau = \oint_S \psi\chi\nabla\phi.\hat{\mathbf{n}} \, dS \tag{4.7}$$

where S is the boundary surface of a simply connected region R.

• *Solution*

We set $\mathbf{A} = \phi\chi\nabla\psi$ in Gauss's divergence theorem

$$\int_R \text{div } \mathbf{A} \, d\tau = \oint_S \mathbf{A}.\hat{\mathbf{n}} \, dS$$

Now div $\mathbf{A} = \nabla.(\phi\chi\nabla\psi) = \chi\nabla\psi.\nabla\phi + \phi\nabla.(\chi\nabla\psi)$ and the first result (4.6) follows.
Similarly by setting $\mathbf{A} = \psi\chi\nabla\phi$ in Gauss's divergence theorem we obtain the second result (4.7).

4.12 If $\mathbf{H} = \frac{1}{2}\text{curl } \mathbf{F}$ and $\mathbf{F} = \text{curl } \mathbf{G}$ where $\mathbf{F}, \mathbf{G}, \mathbf{H}$ are vector functions of position, show that

$$\frac{1}{2}\int_R \mathbf{F}^2 \, d\tau = \frac{1}{2}\oint_S (\mathbf{G} \times \mathbf{F}).\hat{\mathbf{n}} \, dS + \int_R \mathbf{G}.\mathbf{H} \, d\tau$$

• *Solution*

This problem can be solved by setting $\mathbf{A} = \mathbf{G} \times \mathbf{F}$ in Gauss's divergence theorem (4.5). We get

$$\oint_S (\mathbf{G} \times \mathbf{F}).\hat{\mathbf{n}} \, dS = \int_R \text{div}(\mathbf{G} \times \mathbf{F}) \, d\tau$$

Now $\text{div}(\mathbf{G} \times \mathbf{F}) = \nabla.(\mathbf{G} \times \mathbf{F}) = \mathbf{F}.(\nabla \times \mathbf{G}) - \mathbf{G}.(\nabla \times \mathbf{F}) = \mathbf{F}.\text{curl } \mathbf{G} - \mathbf{G}.\text{curl } \mathbf{F}$.
Since $\text{curl } \mathbf{G} = \mathbf{F}$ and $\text{curl } \mathbf{F} = 2\mathbf{H}$ it follows that

$$\oint_S (\mathbf{G} \times \mathbf{F}).\hat{\mathbf{n}} \, dS = \int_R \mathbf{F}^2 \, d\tau - 2\int_R \mathbf{G}.\mathbf{H} \, d\tau$$

which gives the required result.

4.13 Show that

$$\int_R (\mathbf{F}.\text{curl curl } \mathbf{G} - \mathbf{G}.\text{curl curl } \mathbf{F}) \, d\tau = \oint_S (\mathbf{G}\times\text{curl } \mathbf{F} - \mathbf{F}\times\text{curl } \mathbf{G}).\hat{\mathbf{n}} \, dS$$

where $\mathbf{F}$ and $\mathbf{G}$ are vector functions of position.

• *Solution*

We apply Gauss's divergence theorem (4.5) to $\mathbf{A} = \mathbf{G}\times\text{curl } \mathbf{F} - \mathbf{F} \times \text{curl } \mathbf{G}$. Now we have

$$\text{div}(\mathbf{G} \times \text{curl } \mathbf{F}) - \text{div}(\mathbf{F} \times \text{curl } \mathbf{G})$$
$$= \nabla.[\mathbf{G} \times (\nabla \times \mathbf{F})] - \nabla.[\mathbf{F} \times (\nabla \times \mathbf{G})]$$
$$= -\mathbf{G}.[\nabla \times (\nabla \times \mathbf{F})] + (\nabla \times \mathbf{F}).(\nabla \times \mathbf{G}) + \mathbf{F}.[\nabla \times (\nabla \times \mathbf{G})] - (\nabla \times \mathbf{G}).(\nabla \times \mathbf{F})$$
$$= \mathbf{F}.\text{curl curl } \mathbf{G} - \mathbf{G}.\text{curl curl } \mathbf{F}$$

and the required result follows.

4.14 If γ is a closed curve bounding a surface S show that $\int_S (\nabla\phi\times\nabla\psi).d\mathbf{S} = \oint_\gamma \phi\nabla\psi.d\mathbf{s}$ where ϕ and ψ are scalar functions of position.

• *Solution*

In this problem we use *Stokes's theorem*

$$\oint_\gamma \mathbf{A}.d\mathbf{s} = \int_S \text{curl}\,\mathbf{A}.\hat{\mathbf{n}}\,dS \qquad (4.8)$$

where $\hat{\mathbf{n}}$ is a unidirectional normal unit vector to the surface S bounded by the closed curve γ.

Setting $\mathbf{A} = \phi\nabla\psi$ we get

$$\oint_\gamma \phi\nabla\psi.d\mathbf{s} = \int_S [\nabla \times (\phi\nabla\psi)].d\mathbf{S} = \int_S [\nabla\phi \times \nabla\psi + \phi(\nabla \times \nabla\psi)].d\mathbf{S}$$

But $\nabla \times \nabla\psi = 0$ and the required result follows.

4.15 Show that

$$\int_S \hat{\mathbf{n}} \times \nabla\phi\,dS = \oint_\gamma \phi\,d\mathbf{s}.$$

- *Solution*

Here we use Stokes's theorem (4.8) and put $\mathbf{A} = \phi\mathbf{c}$ where $\mathbf{c}$ is an arbitrary constant vector. Then we get

$$\int_S (\nabla \times \phi\mathbf{c}).\hat{\mathbf{n}}dS = \oint_\gamma \phi\mathbf{c}.d\mathbf{s}$$

which may be written, on making a cyclic interchange of the scalar product $(\nabla \times \phi\mathbf{c}).\hat{\mathbf{n}} = \mathbf{c}.(\hat{\mathbf{n}} \times \nabla\phi)$:

$$\mathbf{c}. \int_S \hat{\mathbf{n}} \times \nabla\phi\,dS = \mathbf{c}. \oint_\gamma \phi\,d\mathbf{s}$$

which leads to the result since $\mathbf{c}$ is arbitrary.

4.16 If $\text{div}\,\mathbf{E} = 4\pi\rho$ and $\mathbf{E} = -\,\text{grad}\,\phi$ show that $\frac{1}{2} \int \rho\phi\,d\tau = \frac{1}{8\pi} \int \mathbf{E}^2\,d\tau$ where $r\phi$ and $r^2\mathbf{E}$ are bounded as $r \to \infty$, the integrations being over all space.

- *Solution*

Since $\text{div}(\phi\mathbf{E}) = \phi\,\text{div}\,\mathbf{E} + \mathbf{E}.\nabla\phi$ we obtain, using Gauss's divergence theorem (4.5)

$$\int_R \text{div}\,\mathbf{A}\,d\tau = \int_S \mathbf{A}.\hat{\mathbf{n}}\,dS$$

with $\mathbf{A} = \phi\mathbf{E}$,

$$4\pi \int_R \rho\phi\,d\tau - \int_R \mathbf{E}^2\,d\tau = \int_S \phi\mathbf{E}.\hat{\mathbf{n}}\,dS.$$

As the boundary surface recedes to infinity the surface integral approaches zero and so the required result follows.

4.17 Show that the prolate spheroidal coordinates u, v, ϕ defined by
$x = a\sinh u \sin v \cos\phi,\quad y = a\sinh u \sin v \sin\phi,\quad z = a\cosh u \cos v$ where x, y, z are Cartesian coordinates and a is a constant, form an orthogonal system of *curvilinear coordinates*.

Obtain expressions for the volume element $d\tau$ and Laplace's operator ∇^2.

- *Solution*

We have $d\mathbf{r} = dx\mathbf{i} + dy\mathbf{j} + dz\mathbf{k} = ds_1\mathbf{e_1} + ds_2\mathbf{e_2} + ds_3\mathbf{e_3}$ where we write $ds_1 = h_1du, ds_2 = h_2dv, ds_3 = h_3d\phi$ and $\mathbf{e_1}, \mathbf{e_2}, \mathbf{e_3}$ are unit vectors. Since $\partial x/\partial u = a\cosh u \sin v \cos\phi$, $\partial x/\partial v = a\sinh u \cos v \cos\phi$, $\partial x/\partial\phi = -a\sinh u \sin v \sin\phi$ with similar expressions for the other partial differential coefficients, we find that

$$h_1\mathbf{e_1} = a(\cosh u \sin v \cos\phi\ \mathbf{i} + \cosh u \sin v \sin\phi\ \mathbf{j} + \sinh u \cos v\ \mathbf{k}),$$
$$h_2\mathbf{e_2} = a(\sinh u \cos v \cos\phi\ \mathbf{i} + \sinh u \cos v \sin\phi\ \mathbf{j} - \cosh u \sin v\ \mathbf{k}),$$
$$h_3\mathbf{e_3} = a(-\sinh u \sin v \sin\phi\ \mathbf{i} + \sinh u \sin v \cos\phi\ \mathbf{j})$$

from which it follows that $\mathbf{e_1}.\mathbf{e_2} = \mathbf{e_2}.\mathbf{e_3} = \mathbf{e_3}.\mathbf{e_1} = 0$ and so the system is orthogonal. Simple analysis shows that $h_1 = a\sqrt{\cosh^2u \sin^2v + \sinh^2u \cos^2v} = a\sqrt{\sin^2v + \sinh^2u}$, $h_2 = h_1$, $h_3 = a\sinh u \sin v$.

The volume element

$$d\tau = ds_1ds_2ds_3 = h_1h_2h_3dudvd\phi = a^3(\sin^2v + \sinh^2u)\sinh u \sin v\ dudvd\phi.$$

Further, since

$$\nabla^2 = \frac{1}{h_1h_2h_3}\left[\frac{\partial}{\partial u}\left(\frac{h_2h_3}{h_1}\frac{\partial}{\partial u}\right) + \frac{\partial}{\partial v}\left(\frac{h_3h_1}{h_2}\frac{\partial}{\partial v}\right) + \frac{\partial}{\partial\phi}\left(\frac{h_1h_2}{h_3}\frac{\partial}{\partial\phi}\right)\right] \quad (4.9)$$

we obtain

$$\nabla^2 = \frac{1}{a^2(\sin^2v + \sinh^2u)}\left[\frac{1}{\sinh u}\frac{\partial}{\partial u}\left(\sinh u \frac{\partial}{\partial u}\right) + \frac{1}{\sin v}\left(\sin v \frac{\partial}{\partial v}\right)\right] + \frac{1}{a^2\sinh^2u \sin^2v}\frac{\partial^2}{\partial\phi^2}$$

4.18 Show that the confocal parabolic coordinates ζ, η, ϕ defined by $x = (\zeta\eta)^{\frac{1}{2}}\cos\phi$, $y = (\zeta\eta)^{\frac{1}{2}}\sin\phi$, $z = \frac{1}{2}(\zeta - \eta)$ form an orthogonal system. Obtain expressions for the volume element $d\tau$ and Laplace's operator ∇^2.

- *Solution*

We have $d\mathbf{r} = dx\mathbf{i} + dy\mathbf{j} + dz\mathbf{k} = ds_1\mathbf{e_1} + ds_2\mathbf{e_2} + ds_3\mathbf{e_3}$ where we write $ds_1 = h_1d\zeta, ds_2 = h_2d\eta, ds_3 = h_3d\phi$ and $\mathbf{e_1}, \mathbf{e_2}, \mathbf{e_3}$ are unit vectors.
It can be seen that

$$h_1\mathbf{e_1} = \frac{1}{2}[(\eta/\zeta)^{1/2}(\cos\phi\ \mathbf{i} + \sin\phi\ \mathbf{j}) + \mathbf{k}],$$
$$h_2\mathbf{e_2} = \frac{1}{2}[(\zeta/\eta)^{1/2}(\cos\phi\ \mathbf{i} + \sin\phi\ \mathbf{j}) - \mathbf{k}],$$
$$h_3\mathbf{e_3} = (\zeta\eta)^{1/2}(-\sin\phi\ \mathbf{i} + \cos\phi\ \mathbf{j})$$

where $\mathbf{e_1}.\mathbf{e_2} = \mathbf{e_2}.\mathbf{e_3} = \mathbf{e_3}.\mathbf{e_1} = 0$ so that we have an orthogonal system of coordinates. Further $h_1 = \frac{1}{2}(\eta/\zeta + 1)^{1/2}, h_2 = \frac{1}{2}(\zeta/\eta + 1)^{1/2}, h_3 = (\zeta\eta)^{1/2}$.
It follows that the volume element

$$d\tau = ds_1ds_2ds_3 = h_1h_2h_3d\zeta d\eta d\phi = \frac{1}{4}(\zeta + \eta)d\zeta d\eta d\phi$$

and

$$\nabla^2 = \frac{4}{\zeta + \eta}\left[\frac{\partial}{\partial\zeta}\left(\zeta\frac{\partial}{\partial\zeta}\right) + \frac{\partial}{\partial\eta}\left(\eta\frac{\partial}{\partial\eta}\right) + \frac{\zeta + \eta}{4\zeta\eta}\frac{\partial^2}{\partial\phi^2}\right]$$

using the general formula (4.9) for ∇^2 given in problem 4.17.

4.19 Show that the parabolic cylindrical coordinates u, v, z defined by $x = \frac{1}{2}(u^2 - v^2)$, $y = uv$ form an orthogonal system. Obtain expressions for the volume element $d\tau$ and Laplace's operator ∇^2.

● *Solution*

We have $d\mathbf{r} = dx\mathbf{i} + dy\mathbf{j} + dz\mathbf{k} = ds_1\mathbf{e_1} + ds_2\mathbf{e_2} + ds_3\mathbf{e_3}$ where we write $ds_1 = h_1du, ds_2 = h_2dv, ds_3 = h_3dz$ and $\mathbf{e_1}, \mathbf{e_2}, \mathbf{e_3}$ are unit vectors.

It can be easily seen that $h_1\mathbf{e_1} = u\mathbf{i} + v\mathbf{j}, h_2\mathbf{e_2} = -v\mathbf{i} + u\mathbf{j}, h_3\mathbf{e_3} = \mathbf{k}$ where $\mathbf{e_1}.\mathbf{e_2} = \mathbf{e_2}.\mathbf{e_3} = \mathbf{e_3}.\mathbf{e_1} = 0$ so that the system of coordinates is orthogonal.

Further $h_1 = (u^2 + v^2)^{1/2}, h_2 = h_1, h_3 = 1$.

It follows that the volume element

$$d\tau = ds_1ds_2ds_3 = h_1h_2h_3dudvdz = (u^2 + v^2)dudvdz$$

and

$$\nabla^2 = \frac{1}{u^2 + v^2}\left[\frac{\partial^2}{\partial u^2} + \frac{\partial^2}{\partial v^2}\right] + \frac{\partial^2}{\partial z^2}$$

using the general formula (4.9) for ∇^2 given in problem 4.17.

4.20 Show that the elliptic cylindrical coordinates μ, ϕ, z defined by
$x = d\,\cosh\mu\,\cos\phi, \quad y = d\,\sinh\mu\,\sin\phi$ where d is a constant, form an orthogonal system.

Obtain expressions for the volume element $d\tau$ and Laplace's operator ∇^2.

● *Solution*

We have $d\mathbf{r} = dx\mathbf{i} + dy\mathbf{j} + dz\mathbf{k} = ds_1\mathbf{e_1} + ds_2\mathbf{e_2} + ds_3\mathbf{e_3}$ where we write $ds_1 = h_1d\mu, ds_2 = h_2d\phi, ds_3 = h_3dz$ and $\mathbf{e_1}, \mathbf{e_2}, \mathbf{e_3}$ are unit vectors.

It can be seen that

$$\begin{aligned}
h_1\mathbf{e_1} &= d\sinh\mu\cos\phi\,\mathbf{i} + d\cosh\mu\sin\phi\,\mathbf{j}, \\
h_2\mathbf{e_2} &= -d\cosh\mu\sin\phi\,\mathbf{i} + d\sinh\mu\cos\phi\,\mathbf{j}, \\
h_3\mathbf{e_3} &= \mathbf{k}
\end{aligned}$$

where $\mathbf{e_1}.\mathbf{e_2} = \mathbf{e_2}.\mathbf{e_3} = \mathbf{e_3}.\mathbf{e_1} = 0$ so that we have an orthogonal system of coordinates.

Further $h_1 = d(\sinh^2\mu + \sin^2\phi)^{1/2}, h_2 = h_1, h_3 = 1$.

It follows that the volume element

$$d\tau = ds_1ds_2ds_3 = h_1h_2h_3d\mu d\phi dz = d^2(\sinh^2\mu + \sin^2\phi)d\mu d\phi dz$$

and

$$\nabla^2 = \frac{1}{d^2(\sinh^2\mu + \sin^2\phi)}\left[\frac{\partial^2}{\partial\mu^2} + \frac{\partial^2}{\partial\phi^2}\right] + \frac{\partial^2}{\partial z^2}$$

using the general formula (4.9) for ∇^2 given in problem 4.17.

5
NEWTONIAN GRAVITATION

We now consider some problems on *potential theory* and in this chapter we will be concerned with Newtonian *gravitation*. The familiar Newton's law of force for the attraction between two particles having masses m_1 and m_2 is given by

$$F = \frac{Gm_1m_2}{r^2}$$

where r is the distance between the particles and G is the constant of gravitation.

Hence the gravitational field $\mathbf{F}$ due to a particle of mass m is

$$\mathbf{F} = -\frac{Gm}{r^2}\widehat{\mathbf{r}}, \tag{5.1}$$

where $\widehat{\mathbf{r}}$ is the unit vector in the radial direction, which may be written in the form $\mathbf{F} = -\nabla V$ where V is the *potential* given by

$$V = -\frac{Gm}{r} \tag{5.2}$$

This satisfies the equation named after Pierre Simon Laplace (1749-1827)

$$\nabla^2 V = 0 \tag{5.3}$$

provided $r \neq 0$.

In a region of space where the density of matter is ρ, the potential V given by $\mathbf{F} = -\nabla V$ satisfies *Poisson's equation*

$$\nabla^2 V = 4\pi G\rho \tag{5.4}$$

which becomes Laplace's equation in empty space.

For a finite distribution of matter having mass M we have $rV \to -GM$ as the radial distance $r \to \infty$.

The *theorem of total normal intensity* obtained by Karl Friedrich Gauss (1777-1855) states that the flux of the gravitational field $\mathbf{F}$ through a closed surface S is equal to $-4\pi GM$ where M is the mass enclosed by S, that is

$$\oint_S \mathbf{F}.\widehat{\mathbf{n}}\,dS = -4\pi GM \tag{5.5}$$

where $\mathbf{F}.\widehat{\mathbf{n}}$ is the component of the gravitational field $\mathbf{F}$ in the direction of the outward normal to the surface S given by the unit vector $\widehat{\mathbf{n}}$.

In the first problem we shall need the solution to Laplace's equation (5.3) for a spherically symmetrical distribution of matter.

5.01 A uniform spherical shell has density ρ and mass M, with outer radius a and inner radius b. Show that the potential is given by

$$-\frac{GM}{r} \qquad (r > a)$$

$$-2\pi G\rho(a^2 - b^2) \qquad (r < b)$$

where r is the distance from the centre.

- *Solution*

For $r > a$ the potential V satisfies Laplace's equation $\nabla^2 V = 0$. By spherical symmetry it follows that V is a function of r only and so Laplace's equation takes the form

$$\frac{d}{dr}\left(r^2 \frac{dV}{dr}\right) = 0 \qquad (5.6)$$

since the angular terms of Laplace's operator ∇^2 given by (0.9) in the Introduction produce zero contribution. Hence $r^2 dV/dr = -B$ and so

$$V(r) = A + \frac{B}{r} \qquad (5.7)$$

where A and B are constants. Now $rV \to -GM$ as $r \to \infty$ and so $A = 0$ and $B = -GM$. Hence

$$V(r) = -\frac{GM}{r} \qquad (r > a).$$

For $r < b$ we also have $\nabla^2 V = 0$ and so we may write

$$V(r) = C + \frac{D}{r}$$

where C and D are constants. But $V(r)$ is finite at the origin $r = 0$ and so $D = 0$ giving $V = C$. Now the potential at the $r = 0$ is

$$V(0) = -G \int_b^a \frac{\rho}{r}\, d\tau = -4\pi G\rho \int_b^a r\, dr = -4\pi G\rho \left[\frac{1}{2}r^2\right]_b^a = -2\pi G\rho(a^2 - b^2)$$

and hence

$$V(r) = -2\pi G\rho(a^2 - b^2) \qquad (r < b).$$

5.02 Find the potential inside and outside a solid uniform sphere of radius a and total mass M. Hence show that the gravitational energy of the sphere is

$$-\frac{3}{5}\frac{GM^2}{a}$$

- *Solution*

The potential outside the sphere is, as in the previous problem,

$$V(r) = -\frac{GM}{r} \qquad (r > a)$$

where r is the radial distance from the centre of the sphere.

Now suppose that $r \leq a$. Applying Gauss's theorem of total normal intensity (5.5) to a sphere S_r of radius r, we have

$$\oint_{S_r} F_r\, dS = -4\pi G\left(\frac{r^3}{a^3}M\right),$$

where $F_r = \mathbf{F}.\hat{\mathbf{r}}$ is the component of the gravitational field $\mathbf{F}$ in the direction of the radial unit vector $\hat{\mathbf{r}}$. By spherical symmetry it follows that

$$F_r = -\frac{GMr}{a^3}.$$

Now $F_r = -\partial V/\partial r$ where V is the potential, and so integrating over r we get, for $r \leq a$, that $V(r) = GMr^2/2a^3 + C$ where C is a constant. Now outside the sphere $V(r) = -GM/r$ and so by *continuity* at $r = a$ we have $GM/2a + C = -GM/a$ giving $C = -3GM/2a$. It follows that

$$V(r) = \frac{GM}{2a^3}(r^2 - 3a^2) \qquad (r \leq a).$$

The gravitational potential energy is given by

$$U = \frac{1}{2}\int \rho V \, d\tau \tag{5.8}$$

where ρ is the density of matter. Hence in the present problem we have

$$U = -\frac{1}{2}\frac{\rho GM}{2a}\int_0^a \left(3 - \frac{r^2}{a^2}\right) d\tau = -\frac{3GM^2}{4a^4}\int_0^a \left(3r^2 - \frac{r^4}{a^2}\right) dr$$

since $\rho = M/\left(\frac{4}{3}\pi a^3\right)$. Thus

$$U = -\frac{3GM^2}{4a^4}\left[r^3 - \frac{r^5}{5a^2}\right]_0^a = -\frac{3}{5}\frac{GM^2}{a} \tag{5.9}$$

which is analogous to the result (0.11) obtained in the introductory chapter in problem 0.05 on the electrostatic energy of a sphere of charge.

5.03 A solid sphere of mass M and radius a has a density which is proportional to the square of the distance r from the centre of the sphere. Find the potential as a function of r.

• *Solution*

Let the density of the sphere be given by $\rho = Kr^2$. Then the mass enclosed within a sphere S_r of radius r ($\leq a$) is

$$M(r) = \int_0^r Kr^2 d\tau = 4\pi \int_0^r Kr^4 dr = \frac{4\pi Kr^5}{5}$$

putting the volume element $d\tau = 4\pi r^2 dr$, and so $M = 4\pi Ka^5/5$ giving $K = 5M/4\pi a^5$. Hence $M(r) = Mr^5/a^5$.

Using Gauss's theorem of total normal intensity (5.5), that the flux of the gravitational field $\mathbf{F}$ through a closed surface S is equal to $-4\pi G\times$mass enclosed by S, we have

$$\oint_{S_r} F_r dS = -4\pi GM(r),$$

$F_r = \mathbf{F}.\hat{r}$ being the component of the gravitational field in the direction of the radial unit vector $\hat{r}$. By spherical symmetry it follows that

$$F_r = -\frac{GM(r)}{r^2} = -\frac{GMr^3}{a^5}.$$

Now $F_r = -\partial V/\partial r$ where V is the potential, and so integrating over r we get, for $r \leq a$, that $V(r) = GMr^4/4a^5 + C$ where C is a constant.

Outside the sphere

$$V(r) = -\frac{GM}{r} \qquad (r > a)$$

and so by continuity at $r = a$ we have $GM/4a + C = -GM/a$ giving $C = -5GM/4a$. It follows that

$$V(r) = \frac{GM}{4a^5}(r^4 - 5a^4) \qquad (r \le a).$$

5.04 A homogeneous sphere of density ρ_1 and radius b is surrounded by a spherical shell of density ρ_2 and thickness $a - b$. Find the potential as a function of the radial distance r.

- *Solution*

The mass enclosed by a spherical surface of radius r such that $b \le r \le a$ is

$$M(r) = \frac{4\pi}{3}[b^3\rho_1 + (r^3 - b^3)\rho_2].$$

By Gauss's theorem of total normal intensity (5.5), the gravitational field $F_r = \mathbf{F}.\hat{\mathbf{r}}$ in the radial direction is given by

$$F_r = -\frac{GM(r)}{r^2} = -\frac{4\pi G}{3}\left[\frac{b^3(\rho_1 - \rho_2)}{r^2} + \rho_2 r\right].$$

Since $F_r = -\partial V/\partial r$ we obtain by integration

$$V(r) = \frac{4\pi G}{3}\left[-\frac{b^3(\rho_1 - \rho_2)}{r} + \frac{1}{2}\rho_2 r^2\right] + C \qquad (b \le r \le a).$$

Now for $r > a$, $V(r) = -GM/r$ where the total mass is given by

$$M = \frac{4\pi}{3}[b^3(\rho_1 - \rho_2) + a^3\rho_2],$$

and so by continuity at $r = a$ we see that

$$\frac{4\pi G}{3}\left[-\frac{b^3(\rho_1 - \rho_2)}{a} + \frac{1}{2}a^2\rho_2\right] + C = -\frac{4\pi G}{3a}\left[b^3(\rho_1 - \rho_2) + a^3\rho_2\right].$$

Thus $C = -2\pi Ga^2\rho_2$ and so

$$V(r) = \frac{4\pi G}{3}\left[-\frac{b^3(\rho_1 - \rho_2)}{r} + \frac{1}{2}(r^2 - 3a^2)\rho_2\right] \qquad (b \le r \le a).$$

5.05 The density of a sphere of radius a and mean density $\bar{\rho}$ is a function of the distance r from the centre such that the attraction at any point inside the sphere is proportional to r^2. Express the density at a distance r from the centre in terms of $\bar{\rho}, a$ and r, and prove that the potential at an internal point is

$$-\frac{4\pi G\bar{\rho}}{9a}(4a^3 - r^3).$$

- *Solution*

Let us write the attractive force within the sphere as $F_r = -Kr^2$. Then the potential V satisfies

$$\frac{dV}{dr} = Kr^2$$

Using Poisson's equation (5.4), we see that the mass density ρ is given by

$$\rho = \frac{1}{4\pi G}\nabla^2 V = \frac{1}{4\pi Gr^2}\frac{d}{dr}\left(r^2\frac{dV}{dr}\right) = \frac{Kr}{\pi G}.$$

Hence the mass of the sphere is

$$M = \int \rho\, d\tau = 4\pi \int_0^a \rho r^2 dr = \frac{4K}{G}\int_0^a r^3 dr = \frac{Ka^4}{G}.$$

Now the mean density $\bar{\rho} = M/\frac{4}{3}\pi a^3$ and so $K = GM/a^4 = 4\pi G\bar{\rho}/3a$. Hence

$$\rho = \frac{4\bar{\rho}}{3a}r.$$

We now have $dV/dr = 4\pi G\bar{\rho}r^2/3a$ which gives

$$V = \frac{4\pi G\bar{\rho}}{9a}r^3 + C \qquad (r \leq a)$$

where C is a constant.

For $r > a$

$$V = -\frac{GM}{r} = -\frac{4\pi a^3 G\bar{\rho}}{3r}$$

and so, by continuity of the potential at $r = a$, we have $4\pi G\bar{\rho}a^2/9 + C = -4\pi a^2 G\bar{\rho}/3$ which yields $C = -16\pi G\bar{\rho}a^2/9$.

The required expression for the potential inside the sphere follows at once.

5.06 Find the potential in the three regions $0 < r < a$, $a < r < 2a$, $2a < r < \infty$ when the region $a < r < 2a$ contains matter of uniform density and total mass M.

If the spherical shell shrinks to a uniform solid sphere of the same mass and of radius a, find the energy liberated .

• *Solution*

The potential in the region $r > 2a$ is

$$V = -\frac{GM}{r} \qquad (r > 2a).$$

Now consider a spherical surface S_r of radius r where $a < r < 2a$. By Gauss's theorem of total normal intensity (5.5) we have

$$\oint_{S_r} F_r dS = -4\pi G\rho\, \frac{4}{3}\pi(r^3 - a^3).$$

Hence the potential satisfies

$$\frac{dV}{dr} = -F_r = \frac{4}{3}\pi G\rho\left(r - \frac{a^3}{r^2}\right)$$

giving

$$V = \frac{4}{3}\pi G \rho \left(\frac{r^2}{2} + \frac{a^3}{r} \right) + C$$

where C is a constant. Now the density of the matter in the region $a < r < 2a$ is $\rho = 3M/28\pi a^3$. Consequently by the continuity of the potential at $r = 2a$ we get $5GM/14a + C = -GM/2a$ and so $C = -6GM/7a$. It follows that

$$V = \frac{GM}{7a^3} \left(\frac{r^2}{2} - 6a^2 + \frac{a^3}{r} \right) \qquad (a < r < 2a).$$

The potential for $r < a$ must be constant. Hence by the continuity of the potential at $r = a$ we get

$$V = -\frac{9GM}{14a} \quad (r < a).$$

The gravitational energy of the original distribution of matter is

$$U = \frac{1}{2} \int \rho V \, d\tau = \frac{3GM^2}{98a^6} \int_a^{2a} \left(\frac{r^4}{2} - 6a^2 r^2 + a^3 r \right) dr$$

which gives

$$U = \frac{3GM^2}{98a^6} \left[\frac{r^5}{10} - 2a^2 r^3 + \frac{a^3 r^2}{2} \right]_a^{2a} = -\frac{141}{490} \frac{GM^2}{a}.$$

Now from (5.9) of problem 5.02 the potential energy of a uniform sphere of radius a and mass M is $-3GM^2/5a$ and so the energy set free by the collapse to this sphere is

$$\frac{GM^2}{a} \left(\frac{3}{5} - \frac{141}{490} \right) = \frac{153GM^2}{490a}.$$

Note that in the solutions to the problems in this chapter we have used:
(i) Gauss's theorem of total normal intensity,
(ii) Poisson's equation,
(iii) Laplace's equation,
(iv) the *continuity* of the potential at the boundaries separating different regions of space.

6
ELECTRICITY AND MAGNETISM

In the present chapter we shall be concerned with *electricity and magnetism,* including *current flow,* up to *Maxwell's electromagnetic equations.*

The pioneers in the field of electricity and magnetism were Charles Augustus Coulomb (1736-1806), Hans Christian Oersted (1777-1851), André Marie Ampère (1775-1836), Michael Faraday (1791-1867) and James Clerk Maxwell (1831-79).

The subject of electricity and magnetism provides a very large number of problems although we shall be able to look at only a relatively small representative selection.

In recent years the international system of units (SI units) based on the practical system of units has generally been used in electricity and magnetism. However, we shall use the *Gaussian system of units* since we want to be consistent with the units which are generally used in relativity and quantum theory.

6-1 Electrostatics

Our first area of discussion is *electrostatics* which is based on *Coulomb's law of force* between charged particles. This states that the force between two particles having charges q_1 and q_2 at a distance r apart is given by

$$\frac{kq_1q_2}{r^2}$$

where $k = 1$ when Gaussian units are used. As stated above we shall use Gaussian units in this book. It should be noted however that in SI units we have $k = (4\pi\epsilon_0)^{-1}$ where ϵ_0 is called the *permittivity* which results in 4π being replaced by ϵ_0^{-1} in subsequent formulae such as Gauss's theorem of total normal intensity (6.9).

The force is attractive for charges having different signs and repulsive if the charges have the same sign.

Hence an electric charge q produces the electric field

$$\mathbf{E} = \frac{q}{r^2}\hat{\mathbf{r}} \tag{6.1}$$

at a point with position vector $\mathbf{r} = r\hat{\mathbf{r}}$. Then $\mathbf{E} = -\nabla V$ where the potential V is given by

$$V = \frac{q}{r} \tag{6.2}$$

and satisfies Laplace's equation $\nabla^2 V = 0$ for $r \neq 0$.

An electric dipole composed of charges $-q$ and q has moment $\mu = q\delta\mathbf{s}$ where $\delta\mathbf{s}$ is the displacement vector of the charge q referred to the charge $-q$. It gives rise to the electrostatic potential

$$V = -q\delta\mathbf{s}.\nabla\frac{1}{r} = \frac{q\delta\mathbf{s}.\mathbf{r}}{r^3} = \frac{\mu.\mathbf{r}}{r^3}. \tag{6.3}$$

An electric dipole of moment μ in the direction of the polar axis of an r, θ coordinate system produces an electric field whose components in the radial and transverse directions are given by

$$E_r = \frac{2\mu \cos \theta}{r^3}, \quad E_\theta = \frac{\mu \sin \theta}{r^3} \tag{6.4}$$

respectively.

We start with a simple but instructive problem.

6.01 Find the electrostatic and gravitational forces in newtons between the electron and proton in a hydrogen atom in its ground state taking their distance apart to be the Bohr radius $a_0 = 5.29 \times 10^{-9} cm$ and the magnitude of their charges to be $q = 4.80 \times 10^{-10}$ electrostatic units (esu).

Hence determine the ratio of the gravitational force to the electrostatic force between the two particles given that the mass of an electron is $9.11 \times 10^{-28} gm$, the mass of the proton is a factor 1837 greater, and the gravitational constant is $G = 6.67 \times 10^{-8} cm^3 gm^{-1} s^{-2}$.

• *Solution*

The electrostatic force has magnitude in newtons N $(Kg \ m \ s^{-2})$

$$E = \frac{q^2}{a_0^2} = 8.23 \times 10^{-8} N.$$

The gravitational force has magnitude in newtons

$$F = \frac{GmM}{a_0^2} = 3.63 \times 10^{-47} N.$$

Hence $F/E = 4.4 \times 10^{-40}$ and so we see that the gravitational force is very weak compared with the electric force.

6.02 Two equal charges q are at opposite corners of a square of side l and an electric dipole of moment $\mu = 2\sqrt{2}\alpha q l$ is at a third corner directed towards one of the charges. Find the electric field strength at the fourth corner of the square.

• *Solution*

The diagonal of the square has length $\sqrt{2}l$. Hence for the system of dipole and two charges, the components of the electric field in the radial and transverse directions at the fourth corner of the square are given by

$$E_r = \frac{2\mu \cos(\pi/4)}{(\sqrt{2}l)^3} + \frac{2q \cos(\pi/4)}{l^2}, \quad E_\theta = \frac{\mu \sin(\pi/4)}{(\sqrt{2}l)^3}$$

using (6.1) and (6.4).

We then get that

$$E_r = \frac{\sqrt{2}(\alpha + 1)q}{l^2}, \quad E_\theta = \frac{\alpha q}{\sqrt{2}l^2}$$

and so

$$E = (E_r^2 + E_\theta^2)^{1/2} = \frac{q}{l^2}[2(\alpha + 1)^2 + \alpha^2/2]^{1/2}.$$

6.03 Find the electrostatic potential due to a quadrupole composed of a dipole having moment $\mu_1 = q\delta s_1$ where $\delta s_1 = i\delta x_1$, combined with a second dipole having the same magnitude but opposite sense and infinitesimally displaced relative to the first dipole by $\delta s_2 = i\delta x_2$.

• *Solution*

The potential due to the quadrupole is given by

$$V = -q\ \delta s_2 . \nabla\ \delta s_1 . \nabla \frac{1}{r}. \tag{6.5}$$

This gives

$$V = -q\delta x_2 \delta x_1\ i.\nabla\ i.\nabla \frac{1}{r} = q\delta x_2 \delta x_1\ i.\nabla \frac{i.r}{r^3} = q\delta x_2 \delta x_1 \left[\frac{1}{r^3} - \frac{3(i.r)^2}{r^5} \right]$$

using the result derived in problem 4.02, and so we get

$$V = -q\delta x_1 \delta x_2\ \frac{3\cos^2\theta - 1}{r^3}.$$

6.04 Two positive charges q_1 and q_2 are placed at points P_1 and P_2 respectively. Show that the tangent at infinity to the line of force which starts from q_1 making an angle α with $P_2 P_1$ produced, makes an angle $2\sin^{-1}[q_1^{\frac{1}{2}}(q_1 + q_2)^{-\frac{1}{2}}\sin(\alpha/2)]$ with $P_2 P_1$ and passes through the point P such that $P_1 P / P_2 P = q_2/q_1$.

• *Solution*

The equation of a line of force produced by charges q_i situated on a straight line is

$$\sum_i q_i \cos \theta_i = C \tag{6.6}$$

where θ_i is the angle between the line connecting the charge q_i to the field point and the straight line on which all the charges are placed, and C is a constant.

In the present problem the equation of the line of force is
$q_1 \cos\theta_1 + q_2 \cos\theta_2 = -q_1 \cos\alpha - q_2$.

The tangent to the line of force at ∞ makes an angle λ with $P_2 P_1$ where
$\theta_1 = \theta_2 = \pi - \lambda$ so that $q_1[-\cos\lambda + \cos\alpha] + q_2[-\cos\lambda + 1] = 0$ which yields
$\cos\lambda = (q_1 \cos\alpha + q_2)/(q_1 + q_2)$. Since $\cos\theta = 1 - 2\sin^2(\theta/2)$ the first result follows.

The tangent passes through the centre of charge P given by $q_1\ P_1 P = q_2\ P_2 P$ so that the second result follows.

6.05 Charges $3q, -q, -q$ are placed at A, B, C respectively, where B is the middle point of AC. Show that a line of force that starts from A making an angle α with AB such that $\alpha > \cos^{-1}(-\frac{1}{3})$ will not reach B or C and show that the asymptote of the line of force for which $\alpha = \cos^{-1}(-\frac{2}{3})$ is at right angles to AC.

• *Solution*

The equation of the line of force leaving A making an angle α with AB is, using (6.6),

$$3q(\cos\theta_1 - \cos\alpha) - q(\cos\theta_2 + 1) - q(\cos\theta_3 + 1) = 0.$$

Let us suppose that this line of force reaches C and makes an angle γ with AC. Then

$$3q(1 - \cos\alpha) - 2q - q(\cos\gamma + 1) = 0$$

giving $\cos\gamma = -3\cos\alpha$ and so $\alpha = \cos^{-1}(-\frac{1}{3}\cos\gamma)$. Now the maximum possible value of $\cos\gamma$ is 1 and thus $\alpha \leq \cos^{-1}(-\frac{1}{3})$. Hence if $\alpha > \cos^{-1}(-\frac{1}{3})$ the line of force will not reach C and will go to ∞.

Suppose $\cos\alpha = -\frac{2}{3}$. Then the line of force goes to ∞. At ∞ the line of force will make angle $\theta = \theta_1 = \theta_2 = \theta_3$ with AB where

$$3q\left(\cos\theta + \frac{2}{3}\right) - q(\cos\theta + 1) - q(\cos\theta + 1) = 0$$

giving $\cos\theta = 0$ and so $\theta = \pi/2$.

6.06 The two plates of a parallel plates condenser are each of area A, and the distance between them is d, this distance being small compared with A. Find the attraction between the plates when charged to a potential difference V, neglecting the effects produced by the edges.

- *Solution*

The force F on an element of area carrying charge of density σ is given by

$$F = 2\pi\sigma^2 \delta S \tag{6.7}$$

and the electric field strength in the outward normal direction to the surface of a conductor having surface charge density σ is given by

$$E_n = 4\pi\sigma. \tag{6.8}$$

In the case of a parallel plates condenser, Laplace's equation $\nabla^2 V = 0$ becomes $d^2V/dz^2 = 0$ so that the potential between the plates is given by $V = Az + B$ where z is a Cartesian coordinate chosen perpendicular to the plates and A, B are constants. We see that dV/dz is constant and hence must be V/d. From (6.8) the density of charge over the inner surface of the first plate is given by $\sigma = -(4\pi)^{-1}dV/dz$ and so $\sigma = -V/4\pi d$. The density of charge on the inner surface of the second plate is the same but opposite in sign: $V/4\pi d$. Hence the force on each plate is

$$2\pi\left(\frac{V}{4\pi d}\right)^2 A = \frac{AV^2}{8\pi d^2}.$$

6.07 A sheet of metal of thickness t is introduced between the two plates of a parallel plate condenser which are a distance d apart. Show that the capacity of the condenser is increased by

$$\frac{t}{4\pi d(d - t)}$$

per unit area.

- *Solution*

We introduce the metal sheet so that it is a distance h from plate 1. Suppose that plates 1 and 2 are at potentials V_1 and V_2 respectively. Let $-\sigma$ and σ be the charge densities on the inner surfaces of plates 1 and 2.

The potential between plate 1 and the metal sheet is $V = V_1 + 4\pi\sigma z$ and the potential between the metal sheet and plate 2 is $V = V_2 + 4\pi\sigma(z-d)$. Since the potential of the metal sheet is constant we have $V(z = h) = V(z = h+t)$ and so $V_1 + 4\pi\sigma h = V_2 + 4\pi\sigma(h+t-d)$. Hence $V_2 - V_1 = 4\pi\sigma(d - t)$ and so the capacity per unit area with the metal sheet is

$$C = \frac{\sigma}{V_2 - V_1} = \frac{1}{4\pi(d-t)}.$$

Since the capacity of the condenser in the absence of the metal sheet is $C_0 = 1/4\pi d$ per unit area we see that the increase in capacity per unit area is

$$C - C_0 = \frac{1}{4\pi}\left(\frac{1}{d-t} - \frac{1}{d}\right) = \frac{t}{4\pi d(d-t)}.$$

Alternatively, we may consider the condenser with the metal plate inserted as two condensers $C_1 = 1/4\pi h$ and $C_2 = 1/4\pi(d-t-h)$ in series, so that $1/C = 1/C_1 + 1/C_2 = 4\pi(h + d - t - h) = 4\pi(d - t)$ giving $C = 1/4\pi(d - t)$ again.

6.08 A circular disc of radius a is charged uniformly to surface density σ. Find the potential at distance z away from the disc along the axis of symmetry. Hence obtain the magnitude of the electric field and then, by letting $a \to \infty$, derive the electric field due to a charged infinite plane.

• *Solution*

The potential at a point on the axis of symmetry is given by

$$V = \sigma \int_0^a \frac{2\pi s\, ds}{r}$$

where s is the radius of a circular ring of width ds and $r^2 = z^2 + s^2$. Hence

$$V = 2\pi\sigma \int_0^a \frac{s\, ds}{\sqrt{z^2 + s^2}} = 2\pi\sigma\left[\sqrt{z^2 + s^2}\right]_0^a = 2\pi\sigma\left[\sqrt{a^2 + z^2} - z\right].$$

It follows that the component of the electric field along the axis of symmetry is given by

$$E_z = -\frac{\partial V}{\partial z} = 2\pi\sigma\left[1 - \frac{z}{\sqrt{a^2 + z^2}}\right].$$

We see that $E_z \to 2\pi\sigma$ as $a \to \infty$.

6.09 An infinite plane is charged to surface density σ, and P is a point distance h from the plane. Show that half of the total intensity $2\pi\sigma$ at P is due to the charge at points within a distance $2h$ from P.

• *Solution*

The contribution to the electric field intensity perpendicular to the plane from an annulus of radius s and width ds centred at a point on the plane opposite to P is given by

$$dE_z = \sigma\frac{2\pi s\, ds}{h^2 + s^2}\frac{h}{(h^2 + s^2)^{1/2}}.$$

Hence the contribution to the electric field intensity at P from the charge contained within a circle of radius a is

$$2\pi\sigma h \int_0^a \frac{s\,ds}{(h^2+s^2)^{3/2}} = 2\pi\sigma h \left[-\frac{1}{(h^2+s^2)^{1/2}}\right]_0^a = 2\pi\sigma h \left[\frac{1}{h} - \frac{1}{(h^2+a^2)^{1/2}}\right].$$

This is $\frac{1}{2}(2\pi\sigma)$ if $h/(h^2+a^2)^{1/2} = \frac{1}{2}$, that is $(h^2+a^2)^{1/2} = 2h$, and so half of the total intensity $2\pi\sigma$ at P is due to the charge within a distance $2h$ from P.

6.10 Three insulated concentric spherical conductors whose radii are a, b, c with $a < b < c$ have charges Q_a, Q_b, Q_c respectively. Find the potentials of the spheres and show that if the sphere of radius a is connected to earth, the potential of the sphere of radius c is reduced by

$$\frac{a}{c}\left(\frac{Q_a}{a} + \frac{Q_b}{b} + \frac{Q_c}{c}\right).$$

• *Solution*

In electrostatics, *Gauss's theorem of total normal intensity* takes the form

$$\oint_S \mathbf{E}.\hat{n}\,dS = 4\pi Q \tag{6.9}$$

where Q is the total charge contained by the surface S and $\hat{n}$ is the outward unit vector in the normal direction to S.

If the surface S is a sphere of radius r and we have spherical symmetry, this gives

$$\frac{\partial V}{\partial r} = -\frac{Q}{r^2}.$$

Now in charge free space the potential V satisfies Laplace's equation. In the case of spherical symmetry Laplace's equation becomes

$$\frac{1}{r^2}\frac{d}{dr}\left(r^2\frac{dV}{dr}\right) = 0$$

whose solution may be written as $V = A + B/r$ and thus we have

$$V = \frac{Q}{r} + A. \tag{6.10}$$

The potential inside the sphere of radius a is constant since it must be finite at the origin given by $r = 0$. Now the potential at the common centre of the spheres at $r = 0$ is $Q_a/a + Q_b/b + Q_c/c$ and hence the potential of the sphere a is given by

$$V_a = \frac{Q_a}{a} + \frac{Q_b}{b} + \frac{Q_c}{c}.$$

The potential outside the sphere c is $(Q_a + Q_b + Q_c)/r$ and so by *continuity* the potential of sphere c is

$$V_c = \frac{Q_a + Q_b + Q_c}{c}.$$

The potential between spheres a and b is given by $V = Q_a/r + A$. By putting $r = a$ we get $A = V_a - Q_a/a = Q_b/b + Q_c/c$ and so the potential of sphere b is

$$V_b = \frac{Q_a + Q_b}{b} + \frac{Q_c}{c}.$$

Suppose sphere a is earthed. Then the potential at the centre is zero and so, if Q'_a is now the charge on sphere a, we have $Q'_a/a + Q_b/b + Q_c/c = 0$. Since the potential of sphere c is now

$$V'_c = \frac{Q'_a + Q_b + Q_c}{c}$$

we get by simple algebra

$$V_c - V'_c = \frac{Q_a - Q'_a}{c} = \frac{a}{c}\left(\frac{Q_a}{a} + \frac{Q_b}{b} + \frac{Q_c}{c}\right).$$

6.11 An insulated spherical conductor which is formed of two hemispherical shells in contact, whose inner and outer radii are b_1 and b_2, has within it a concentric spherical conductor of radius a and outside it another concentric spherical conductor whose inner radius is c. If the middle conductor is charged and the inner and outer conductors are earthed, show that the two hemispherical shells will not separate if $2ac > b_1c + b_2a$.

- *Solution*

If V_0 is the potential of the middle shell then the potential between the inner shell and the middle shell is

$$V_1 = -\frac{b_1 V_0}{b_1 - a}\left(\frac{a}{r} - 1\right) \qquad (a \le r \le b_1)$$

and the potential between the middle shell and the outer shell is

$$V_2 = \frac{b_2 V_0}{c - b_2}\left(\frac{c}{r} - 1\right) \qquad (b_2 \le r \le c).$$

The electric field strength in the outward normal direction to the surface of a conductor having surface charge density σ is given by $E_n = 4\pi\sigma$.

Hence the charge density on surface $r = b_1$ is

$$\sigma_1 = \frac{1}{4\pi}\left(\frac{\partial V_1}{\partial r}\right)_{r=b_1-0} = \frac{1}{4\pi}\frac{b_1 V_0}{b_1 - a}\frac{a}{b_1^2}$$

and the charge density on surface $r = b_2$ is

$$\sigma_2 = -\frac{1}{4\pi}\left(\frac{\partial V_2}{\partial r}\right)_{r=b_2+0} = \frac{1}{4\pi}\frac{b_2 V_0}{c - b_2}\frac{c}{b_2^2}.$$

It follows from (6.7) that the force F on one of the hemispherical shells perpendicular to the plane of separation in the outward direction is

$$F = 2\pi[2\pi b_2^2\sigma_2^2 - 2\pi b_1^2\sigma_1^2]\int_0^{\pi/2}\cos\theta\sin\theta\,d\theta = 2\pi^2(b_2^2\sigma_2^2 - b_1^2\sigma_1^2).$$

This gives

$$F = \frac{1}{8}V_0^2\left[\frac{c^2}{(c - b_2)^2} - \frac{a^2}{(b_1 - a)^2}\right]$$

and if this force is negative the hemispheres will not separate. Thus we require $c^2(b_1 - a)^2 < a^2(c - b_2)^2$ which leads to the required result.

6.12 A condenser is formed of three coaxial cylinders of which the inner and outer are connected together. Show that the capacity per unit length is given by

$$C = \frac{1}{2}\frac{\ln(c/a)}{\ln(c/b)\ln(b/a)}$$

where a, b, c are the radii of the cylinders with $a < b < c$.

• *Solution*

The potential V outside an infinite cylindrically symmetrical distribution of charge q per unit length satisfies the Laplace's equation

$$\frac{1}{r}\frac{d}{dr}\left(r\frac{dV}{dr}\right) = 0$$

where r is the radial distance from the axis of symmetry .

Since Gauss's theorem of total normal intensity for this case gives

$$\frac{dV}{dr} = -\frac{2q}{r}$$

the potential takes the form

$$V = A - 2q\ln r \tag{6.11}$$

where A is a constant.

Suppose q_1 is the charge per unit length on the cylinder of radius a. Then the potential in the region between the cylinders a and b is

$$V_1 = A_1 - 2q_1\ln r \qquad (a \le r \le b).$$

Hence, if V_a and V_b are the potentials of cylinders a and b, we have

$$V_a = A_1 - 2q_1\ln a, \quad V_b = A_1 - 2q_1\ln b.$$

In the region between the cylinders b and c the potential is

$$V_2 = A_2 - 2(q_1 + q_2)\ln r \qquad (b \le r \le c)$$

where q_2 is the charge per unit length on cylinder b. Hence

$$V_b = A_2 - 2(q_1 + q_2)\ln b, \quad V_c = A_2 - 2(q_1 + q_2)\ln c$$

and so

$$V_b - V_c = 2(q_1 + q_2)\ln\frac{c}{b}.$$

Since $V_c = V_a$ it follows that

$$V_b - V_a = -2q_1\ln\frac{b}{a} = 2(q_1 + q_2)\ln\frac{c}{b}$$

giving

$$q_2 = -q_1\frac{\ln(c/a)}{\ln(c/b)}$$

and so the *capacity* = charge/(potential difference) of the system is given by

$$\frac{q_2}{V_b - V_a} = \frac{\ln(c/a)}{2\ln(b/a)\ln(c/b)}.$$

An alternative and quicker method for deriving the capacity of this system is to use the formula $C = C_1 + C_2$ for capacities C_1 and C_2 in parallel. Now the cylinders a and b have capacity $C_1 = 1/[2\ln(b/a)]$ per unit length while cylinders b and c have capacity $C_2 = 1/[2\ln(c/b)]$ per unit length. Hence the capacity of the system is

$$C = C_1 + C_2 = \frac{1}{2}\left(\frac{1}{\ln(b/a)} + \frac{1}{\ln(c/b)}\right) = \frac{\ln(c/a)}{2\ln(b/a)\ln(c/b)}$$

which is the same as the result obtained before.

6.13 Two equal charges q are placed at a distance d apart, and each of them is at a distance $d/2$ from an infinite conducting plane at zero potential. Find the force acting on each of the charges.

- *Solution*

Here we use the *method of images* based on a *uniqueness theorem* for point charges and conductors..

The effect of the presence of the earthed conducting plane can be represented by introducing two mirror image charges $-q$ at a distance d apart and at distances $d/2$ on the other side of the plane. Then the plane will be at zero potential and the uniqueness theorem will be satisfied.

The force acting on each of the charges q can be resolved into two components. The component towards the plane is

$$F_x = \frac{q^2}{d^2} + \frac{q^2}{2d^2}\cos\frac{\pi}{4}$$

and the component parallel to the plane is

$$F_y = \frac{q^2}{d^2} - \frac{q^2}{2d^2}\cos\frac{\pi}{4}.$$

Hence the force acting on each charge is

$$F = (F_x^2 + F_y^2)^{1/2} = \frac{q^2}{d^2}\left[\left(1+\frac{1}{2\sqrt{2}}\right)^2 + \left(1-\frac{1}{2\sqrt{2}}\right)^2\right]^{1/2} = \frac{3q^2}{2d^2}.$$

6.14 A sphere of radius a is earthed and positive charges q_1 and q_2 are placed on opposite sides of the sphere at distances $2a$ and $4a$ respectively from the centre and in a straight line passing through it. Show that the charge q_2 is repelled from the sphere if $q_2 < 25q_1/144$.

- *Solution*

A point charge q at a distance c from the centre of an earthed conducting sphere of radius a has an image charge $-aq/c$ at the inverse point distance a^2/c from the centre. The charge q and the image charge $-aq/c$ together produce zero potential over the surface of the sphere, and the uniqueness theorem is satisfied.

In the present problem, image charges $-q_1/2$ and $-q_2/4$ are situated at distances $a/2$ and $a/4$ from the centre respectively, on either side of the centre.

Then the force on charge q_2 away from the centre of the sphere is

$$F = \frac{q_1 q_2}{a^2}\left(\frac{1}{36} - \frac{1}{2}\frac{1}{(9/2)^2}\right) - \frac{q_2^2}{a^2}\frac{1}{4}\frac{1}{(15/4)^2}.$$

If this is positive, the charge q_2 will be repelled from the sphere. The condition for this is

$$q_1\left(\frac{1}{36} - \frac{2}{81}\right) - \frac{4q_2}{225} > 0$$

which yields the required result.

6.15 A point charge q is at a distance c from the centre of an earthed conducting sphere of radius a. Show that the surface density of charge on the spherical conductor is

$$-\frac{q}{4\pi}\frac{c^2-a^2}{a}\frac{1}{r_s^3} \tag{6.12}$$

where r_s is the distance of any point S of the sphere from the point charge q.

- *Solution*

The field produced by the presence of the spherical conductor is equivalent to that produced by an image charge $-aq/c$ at the image point distance a^2/c from the centre O of the sphere.

Let R be the radial distance from O to a field point P. Then, using (6.8) , the density of the surface charge is

$$\sigma = -\frac{1}{4\pi}\left(\frac{\partial V}{\partial R}\right)_{R=a+0}$$

where the potential is given by

$$V = q\left(\frac{1}{r} - \frac{a}{c}\frac{1}{r'}\right),$$

the distances from the charges q and $-aq/c$ to P being denoted by r and r' respectively.

By the cosine law in trigonometry we have $r^2 = R^2 + c^2 - 2Rc\cos\theta$ and $r'^2 = R^2 + (a^2/c)^2 - 2R(a^2/c)\cos\theta$ where θ is the angle between OP and the line joining O to the point charge. Hence

$$\frac{\partial V}{\partial R} = -q\left(\frac{1}{r^2}\frac{\partial r}{\partial R} - \frac{a}{c}\frac{1}{r'^2}\frac{\partial r'}{\partial R}\right) = -q\left(\frac{R - c\cos\theta}{r^3} - \frac{a}{c}\frac{R - (a^2/c)\cos\theta}{r'^3}\right).$$

Hence the surface density of charge is given by

$$\sigma = -\frac{q}{4\pi}\left[\frac{a}{c}\left(a - \frac{a^2}{c}\cos\theta\right)\frac{1}{r_s'^3} - (a - c\cos\theta)\frac{1}{r_s^3}\right]$$

which gives the required result since $r_s'/r_s = a/c$.

6.16 A conducting plane has a hemispherical boss of radius a, and at a distance c from the centre of the boss and along its axis there is a point charge q. If the plane and the boss are kept at zero potential, show that the charge induced on the boss is

$$-q\left[1 - \frac{c^2 - a^2}{c(c^2 + a^2)^{\frac{1}{2}}}\right].$$

- *Solution*

The effect of the conductor is to introduce image point charges $-aq/c$ and aq/c at distances a^2/c from the centre of the boss but on opposite sides of the centre, and a charge $-q$ at a distance c from the centre on the other side of the plane. Then the potential vanishes over the conductor.

If r and p are the distances from the charges q and $-q$ to a point P on the surface of the boss, the surface density of charge on the boss is, using (6.12),

$$\sigma = -\frac{q}{4\pi}\frac{c^2 - a^2}{a}\left(\frac{1}{r^3} - \frac{1}{p^3}\right).$$

Now by the cosine law in trigonometry $r^2 = a^2 + c^2 - 2ac\cos\theta$ and $p^2 = a^2 + c^2 + 2ac\cos\theta$ where θ is the angle made by the line joining the centre to P and the line on which the point charges lie. Hence

$$\sigma = -\frac{q}{4\pi}\frac{c^2 - a^2}{a}\left[(a^2 + c^2 - 2ac\cos\theta)^{-3/2} - (a^2 + c^2 + 2ac\cos\theta)^{-3/2}\right].$$

The total charge on the boss is given by $Q = \int\sigma\,dS = 2\pi a^2\int_0^{\pi/2}\sigma\sin\theta\,d\theta$, that is

$$Q = -\frac{q}{2}a(c^2 - a^2)\int_0^{\pi/2}\left[\frac{\sin\theta}{(a^2 + c^2 - 2ac\cos\theta)^{3/2}} - \frac{\sin\theta}{(a^2 + c^2 + 2ac\cos\theta)^{3/2}}\right]d\theta.$$

Carrying out the integration yields

$$Q = -\frac{q}{2}\frac{c^2 - a^2}{c}\left[-(a^2 + c^2 - 2ac\cos\theta)^{-1/2} - (a^2 + c^2 + 2ac\cos\theta)^{-1/2}\right]_0^{\pi/2}$$

and this gives the required result.

6.17 A conducting spherical shell of radius a is placed, insulated and without charge, in a uniform electric field of force E. If the sphere is divided into two hemispheres by a plane perpendicular to the field, find the forces required to prevent the hemispheres from separating.

• *Solution*

The potential at a point with spherical polar coordinates r, θ, ϕ referred to the direction of the electric field is

$$V = -Er\cos\theta + \frac{Ea^3\cos\theta}{r^2} \tag{6.13}$$

and so the surface density of charge is

$$\sigma = -\frac{1}{4\pi}\left(\frac{\partial V}{\partial r}\right)_{r=a+0} = \frac{3}{4\pi}E\cos\theta.$$

Hence the force acting on a hemispherical shell in the direction of the electric field intensity is

$$F = 2\pi a^2\int_0^{\pi/2}2\pi\sigma^2\cos\theta\sin\theta\,d\theta = \frac{9}{4}a^2E^2\int_0^{\pi/2}\cos^3\theta\sin\theta\,d\theta$$

and this gives $9a^2E^2/16$. The force due to the electric field on the other hemisphere is equal and opposite.

Counterbalancing forces have to be applied to prevent the hemispheres from separating.

6.18 Show that the complex potential w given by $z = c\coth(\frac{1}{2}w)$ corresponds to a pair of line charges of density $\pm\frac{1}{2}$ situated through the points $x = \pm c$, $y = 0$.

- *Solution*

The complex potential $w = V + iU$ due to a line charge λ per unit length passing through a point $z = z_0$ is

$$w = -2\lambda \ln(z - z_0). \tag{6.14}$$

In the present problem we have

$$z = c \coth(w/2) = c\frac{\exp(w/2) + \exp(-w/2)}{\exp(w/2) - \exp(-w/2)}$$

and so $\exp(w) = (z + c)/(z - c)$ giving

$$w = \ln\frac{z + c}{z - c}.$$

This complex potential corresponds to two line charges $\frac{1}{2}$ per unit length through the point $z = c$ and $-\frac{1}{2}$ per unit length through the point $z = -c$.

6.19 Show that the equipotentials given by constant V in the previous problem are circles centred at $x = c \coth V$, $y = 0$ having radii $c/|\sinh V|$. Hence find the capacity per unit length of a condenser formed by two conducting cylinders having radii a and b with parallel axes at a distance d ($> a + b$) apart.

- *Solution*

Using the formula

$$\coth(A + iB) = \frac{\sinh 2A - i \sin 2B}{\cosh 2A - \cos 2B}$$

we see that

$$x = \frac{c \sinh V}{\cosh V - \cos U}, \quad y = \frac{c \sin U}{\cos U - \cosh V}$$

from which it can be shown that

$$(x - c \coth V)^2 + y^2 = \frac{c^2}{\sinh^2 V}.$$

Hence the equipotentials given by $V =$ constant are circles centred at $x = c \coth V$, $y = 0$ having radii $c/|\sinh V|$.

Consider two cylindrical surfaces at potentials V_a and V_b having radii a and b respectively. Then

$$a = \frac{c}{|\sinh V_a|}, \quad b = \frac{c}{|\sinh V_b|}.$$

Suppose $V_a > 0$ and $V_b < 0$ so that the cylinders are outside each other. Then the distance between the axes of the cylinders is

$$d = c(\coth V_a + |\coth V_b|)$$

and by some elementary algebra we can show that

$$\frac{d^2 - a^2 - b^2}{2ab} = \cosh(V_a - V_b).$$

From the previous problem we know that the charge per unit length on cylinder a is $\frac{1}{2}$ and on cylinder b is $-\frac{1}{2}$

Hence the capacity per unit length of the condenser formed from the two cylinders is

$$C = \frac{\frac{1}{2}}{V_a - V_b} = \left[2\cosh^{-1}\frac{d^2 - a^2 - b^2}{2ab}\right]^{-1}.$$

6.20 Show that the capacity per unit length of a telegraph wire of radius a at height h above the surface of the earth is

$$C = \left[4\tanh^{-1}\left(\frac{h-a}{h+a}\right)^{\frac{1}{2}}\right]^{-1}.$$

• *Solution*

Put $d = b + h$ in the previous result and then let $b \to \infty$ to convert cylinder b into a plane at zero potential. Then

$$\frac{d^2 - a^2 - b^2}{2ab} \to \frac{h}{a}$$

and hence

$$C \to \frac{1}{2\cosh^{-1}h/a}.$$

Now let $h/a = \cosh x = 2\cosh^2\frac{1}{2}x - 1 = 2\sinh^2\frac{1}{2}x + 1$ which gives $\tanh^2\frac{1}{2}x = (h-a)/(h+a)$ and the required result follows.

6.21 Find the capacity per unit length of a condenser formed by two circular cylinders having radii a and b, one inside the other, with parallel axes at a distance d apart.

• *Solution*

This problem is very similar to problem 6.19 except that we now take $V_a > V_b > 0$ and $a < b$. Then the distance between the axes of the cylinders is $d = c(\coth V_b - \coth V_a)$ and $(a^2 + b^2 - d^2)/2ab = \cosh(V_a - V_b)$.

Now the charge per unit length on cylinder a is $\frac{1}{2}$ and so the capacity of the condenser formed by the two cylinders is

$$C = \frac{\frac{1}{2}}{V_a - V_b} = \left[2\cosh^{-1}\frac{a^2 + b^2 - d^2}{2ab}\right]^{-1}.$$

6.22 Find the capacity per unit length of a condenser formed by two confocal elliptic cylinders having semi-major and semi-minor axes a_1, b_1 and a_2, b_2 respectively.

• *Solution*

To solve this problem we use the complex potential $w = V + iU$ given by $z = c\cosh(w/A)$ where A and c are constants. Then we have

$$x = c\cosh\frac{V}{A}\cos\frac{U}{A}, \quad y = c\sinh\frac{V}{A}\sin\frac{U}{A}.$$

Hence

$$\frac{x^2}{c^2\cosh^2(V/A)} + \frac{y^2}{c^2\sinh^2(V/A)} = 1$$

so that the equipotential surfaces given by $V =$ constant form a system of confocal ellipses with semi-axes of lengths $a = c\cosh(V/A)$ and $b = c\sinh(V/A)$. Then we have

$x = a\cos(U/A)$ and $y = b\sin(U/A)$ where U/A is the auxiliary angle. As we describe the ellipse given by constant V, we have that U/A increases by 2π and so the change in U is given by $[U] = 2\pi A$.

If the potentials of the cylinders are V_1 and V_2 it follows that

$$V_2 - V_1 = A\ln\left(\frac{a_2 + b_2}{a_1 + b_1}\right).$$

The charge on the outer surface of the cylinder at potential V_1 is $Q = -[U]/4\pi = -A/2$ per unit length and hence the capacity per unit length of the cylinders is

$$C = \frac{Q}{V_1 - V_2} = \left[2\ln\left(\frac{a_2 + b_2}{a_1 + b_1}\right)\right]^{-1}.$$

In the case of a pair of coaxial circular cylinders for which $a_1 = b_1, a_2 = b_2$ we get $C = [2\ln(a_2/a_1)]^{-1}$.

6.23 Find the capacity per unit length between a flat strip of width $2c$ and an elliptic cylinder whose foci coincide with the edges of the strip and whose semi-major axis is a.

• *Solution*

We use the result obtained in the previous problem. When an ellipse degenerates into a straight line connecting the foci a distance apart $2c$ we get $a_1 = c$ and $b_1 = 0$. Hence we get for the capacity

$$C = \left[2\ln\frac{a + b}{c}\right]^{-1}.$$

Now $b^2 = a^2 - c^2$ and so

$$C = \left[2\ln\left(\frac{a + \sqrt{a^2 - c^2}}{c}\right)\right]^{-1} = \left[2\cosh^{-1}\frac{a}{c}\right]^{-1}.$$

6.24 Two planes intersecting at an angle α are raised to potentials $V_0/2$ and $-V_0/2$. Find the equipotential surfaces.

• *Solution*

This problem can be solved by using the complex potential $w = A - iB\ln z$ where $w = V + iU$ and A, B are constants. Putting $z = re^{i\theta}$ we get $V + iU = A + B\theta - iB\ln r$ and so $V = A + B\theta$ and $U = -B\ln r$. Hence the equipotentials are given by $\theta =$ constant.

We take $\theta = 0$ to be the first plane and $\theta = \alpha$ to be the second plane. Then we have $V_0/2 = A$ and $-V_0/2 = A + B\alpha$ giving $B = -V_0/\alpha$. Hence the equipotential surfaces are given by

$$V = \frac{V_0}{2}\left(1 - \frac{2\theta}{\alpha}\right)$$

where θ is a constant angle referred to the plane at potential $V_0/2$.

6.25 The inner surface of a thin spherical conductor has a constant charge Q, and an outer concentric thin spherical conductor is at zero potential. If the outer conductor contracts from radius b to radius c under the internal forces show that the work done by the electrical forces is

$$\frac{1}{2}Q^2\left(\frac{1}{c} - \frac{1}{b}\right).$$

• *Solution*

The potential between the spherical conductors takes the form given by (6.10):

$$V = \frac{Q}{r} + A$$

where r is the radial distance from the common centre and A is a constant. It follows that the potential of the inner sphere of radius a is given by

$$V_a = Q\left(\frac{1}{a} - \frac{1}{b}\right).$$

The electrical energy of a conductor at potential V having charge Q is given by $\frac{1}{2}QV$. Hence the initial electrical energy is

$$E_i = \frac{1}{2}QV_a = \frac{1}{2}Q^2\left(\frac{1}{a} - \frac{1}{b}\right).$$

After the contraction of the outer sphere to radius c the electrical energy is

$$E_f = \frac{1}{2}Q^2\left(\frac{1}{a} - \frac{1}{c}\right)$$

and so the loss in the electrical energy is

$$E_i - E_f = \frac{1}{2}Q^2\left(\frac{1}{c} - \frac{1}{b}\right)$$

which is the work done by the electrical forces.

6.26 V_P is the potential at any point P between two closed equipotentials V_1, V_2 where V_1 contains V_2. If a charge q is put at P and both equipotentials are replaced by earthed conducting shells, show that the charges Q_1 and Q_2 induced on the two surfaces are given by

$$\frac{Q_1}{V_2 - V_P} = \frac{Q_2}{V_P - V_1} = \frac{q}{V_1 - V_2}.$$

• *Solution*

Here we use the *reciprocal theorem* which states that if charges $Q_1, Q_2, ..., Q_n$ on a system of conductors produce potentials $V_1, V_2, ..., V_n$ and if charges $Q'_1, Q'_2, ..., Q'_n$ produce potentials $V'_1, V'_2, ..., V'_n$, then

$$\sum_{i=1}^{n} Q_i V'_i = \sum_{i=1}^{n} Q'_i V_i. \tag{6.15}$$

Now by this theorem we see that

$$Q_1 V_1 + Q_2 V_2 + q V_P = 0.$$

Since all the tubes of force from q must end on the two conductors we have $Q_1 + Q_2 = -q$. Hence

$$Q_1(V_1 - V_P) + Q_2(V_2 - V_P) = 0.$$

Also

$$Q_1(V_1 - V_2) - q(V_2 - V_P) = 0$$

and so the required result follows.

6.27 A conducting sphere of radius a is embedded in the centre of a sphere of material of radius b and dielectric constant K. Find the capacity of the spherical conductor.

• *Solution*

We apply Gauss's theorem of total normal intensity in the presence of a dielectric material to a sphere S_r of radius r where $a < r < b$. Then we get

$$\oint_{S_r} \mathbf{D}.d\mathbf{S} = 4\pi Q \tag{6.16}$$

where Q is the free charge on the conductor and $\mathbf{D} = K\mathbf{E}$ is the displacement vector in the material having dielectric constant K. Hence $E = Q/Kr^2$ and since $E = -\partial V/\partial r$ the potential V in the dielectric is given by

$$V = \frac{1}{K}\frac{Q}{r} + A \qquad (a < r < b)$$

where A is a constant.

Outside the dielectric the potential is given by

$$V = \frac{Q}{r} \qquad (r > b)$$

and so by continuity at $r = b$ we find that $A = (Q/b)(1 - 1/K)$ giving

$$V = \frac{1}{K}\frac{Q}{r} + \frac{Q}{b}\left(1 - \frac{1}{K}\right) \qquad (a < r < b).$$

Hence the capacity of the spherical conductor is

$$C = \frac{Q}{V_a} = \frac{abK}{b + a(K - 1)}.$$

6.28 Prove that when a material having dielectric constant K completely fills the region between the plates of a spherical condenser the capacity is K times the capacity in the absence of dielectric.

• *Solution*

Suppose the radii of the spherical plates are a and b with $a < b$. From the previous problem we have

$$V = \frac{1}{K}\frac{Q}{r} + A \qquad (a < r < b)$$

and so

$$V_a - V_b = \frac{Q}{K}\left(\frac{1}{a} - \frac{1}{b}\right).$$

Hence the capacity of the spherical condenser is

$$C = \frac{Q}{V_a - V_b} = \frac{Kab}{b - a}$$

which is K times the capacity without the dielectric.

6.29 The plane surface separating two dielectrics with constants K_1 and K_2 has free charge density σ per unit area. The electric field intensities on the two sides of the boundary are E_1 and E_2 at angles θ_1 and θ_2 to the normal. Prove that $K_2 \cot \theta_2 = K_1 \cot \theta_1 \, [1 + 4\pi\sigma(K_1 E_1 \cos \theta_1)^{-1}]$.

• *Solution*

The boundary conditions normal and tangential to the surface of separation are respectively

$$(D_n)_2 - (D_n)_1 = 4\pi\sigma, \quad (E_s)_2 = (E_s)_1$$

where the first boundary condition follows from Gauss's theorem of total normal intensity (6.16) in the presence of a dielectric.

Thus we have $K_2 E_2 \cos\theta_2 - K_1 E_1 \cos\theta_1 = 4\pi\sigma$ and $E_2 \sin\theta_2 = E_1 \sin\theta_1$.

Hence $K_2 E_1 \sin\theta_1 \cot\theta_2 - K_1 E_1 \cos\theta_1 = 4\pi\sigma$ and the result follows.

6.30 A spherical condenser consists of two conducting spheres of radii a and c with the outer sphere earthed. The region between the conductors is filled with two dielectrics separated by a spherical surface of radius b, the inner material having dielectric constant K_1 and the outer material having dielectric constant K_2. Find the capacity of the condenser.

• *Solution*

By applying Gauss's theorem of total normal intensity (6.16) as in problem 6.27 we have

$$V_r = \frac{Q}{K_1}\frac{1}{r} + A_1 \quad (a < r < b)$$

and

$$V_r = \frac{Q}{K_2}\left(\frac{1}{r} - \frac{1}{c}\right) \quad (b < r < c)$$

since $V_c = 0$ because the outer sphere is earthed. By continuity at $r = b$ we also have

$$A_1 = \frac{Q}{K_2}\left(\frac{1}{b} - \frac{1}{c}\right) - \frac{Q}{K_1}\frac{1}{b}$$

and so

$$V_a = \frac{Q}{K_1}\left(\frac{1}{a} - \frac{1}{b}\right) + \frac{Q}{K_2}\left(\frac{1}{b} - \frac{1}{c}\right).$$

Hence the capacity of the spherical condenser is

$$C = \frac{Q}{V_a - V_c} = \frac{K_1 K_2 abc}{K_1 a(c - b) + K_2 c(b - a)}.$$

6.31 An infinitely long cylindrical shell of internal radius a and external radius b is composed of material of dielectric constant K. It is placed uncharged in a uniform electric field E perpendicular to its axis. Prove that the field within the region $0 < r < a$ is reduced by the factor

$$\left\{1 + \left[\frac{(K-1)^2}{4K}\left(1 - \frac{a^2}{b^2}\right)\right]\right\}^{-1}.$$

• *Solution*

The potential may be written in the form $V = V_0 + V'$ where $V_0 = -Er\cos\theta$ is the potential in the absence of the cylindrical shell of dielectric, and r, θ are polar angles referred to the electric field vector as polar axis.

We have $\nabla^2 V' = 0$ at all points. For a cylindrically symmetrical configuration the solution of Laplace's equation may be written in the form

$$V(r,\theta) = \sum_{n=1}^{\infty}(A_n r^n + B_n r^{-n})(a_n \cos n\theta + b_n \sin n\theta) + a_0 \ln r. \qquad (6.17)$$

Since V is finite at $r = 0$ we see that $V' = Ar \cos\theta$ for $r \leq a$, where A is a constant, because the continuity conditions can only be satisfied if the angular dependence is $\cos\theta$. Also $V' \to 0$ as $r \to \infty$ and so $V' = Dr^{-1} \cos\theta$ for $r \geq b$ where D is a constant.

Within the material we must have $V' = Br \cos\theta + Cr^{-1} \cos\theta$ where B and C are constants.

By the continuity of V and $K\partial V/\partial r$ at $r = a$ we get $(A - E)a = (B - E)a + Ca^{-1}$ and $A - E = K(B - E - Ca^{-2})$, while by continuity at $r = b$ we get $(B - E)b + Cb^{-1} = -Eb + Db^{-1}$ and $K(B - E - Cb^{-2}) = -E - Db^{-2}$.

Since the field in the region $0 < r < a$ is reduced by the factor $(E - A)/E$ by the presence of the dielectric material, some simple algebra yields the result required.

6.32 Obtain the potential due to a uniformly charged ring of radius a having total charge Q.

● *Solution*

The potential of an axially symmetric distribution of charge may be expanded in the form

$$V(r,\theta) = \sum_{n=0}^{\infty}(A_n r^n + B_n r^{-n-1})P_n(\cos\theta) \qquad (6.18)$$

where r, θ are polar coordinates referred to the axis of symmetry as polar axis, $P_n(\mu)$ is a Legendre polynomial and A_n, B_n are constants.

Taking the origin O at the centre of the ring, the potential at a point on the polar axis at a distance z from O is given by

$$V(z) = \frac{Q}{(z^2 + a^2)^{1/2}}.$$

Since

$$(z^2 + a^2)^{-1/2} = \sum_{n=0}^{\infty} \gamma_n(z, a)P_n(0) \qquad (6.19)$$

where

$$\begin{aligned}\gamma_n(r, s) &= r^n/s^{n+1} \quad (r < s) \\ &= s^n/r^{n+1} \quad (r > s)\end{aligned} \qquad (6.20)$$

we see that the potential due to the ring is given by

$$V(r,\theta) = Q\sum_{n=0}^{\infty} \gamma_n(r, a)P_n(0)P_n(\cos\theta) \qquad (6.21)$$

where we have used $P_n(1) = 1$ for all n.

We note that $P_n(0) = 0$ for odd n, $P_0 = 1$ and that

$$P_n(0) = (-1)^{n/2}\frac{1.3.5...(n-1)}{2.4.6...n}. \qquad (6.22)$$

for even n.

6.33 Show that the electric field strength at the surface of an earthed conducting sphere of radius b enclosing a concentric ring of radius a and charge Q is

$$\frac{Q}{b^2} + Q \sum_{n=1}^{\infty} (-1)^n \frac{1.3...(2n-1)}{2.4...2n} \frac{4n+1}{b^2} \left(\frac{a}{b}\right)^{2n} P_{2n}(\cos\theta).$$

• *Solution*

From the result (6.21) obtained in the previous problem, the potential due to the ring for $r > a$ is

$$V_{ring}(r,\theta) = \frac{Q}{a} \sum_{n=0}^{\infty} P_{2n}(0) \left(\frac{a}{r}\right)^{2n+1} P_{2n}(\cos\theta).$$

The potential due to the induced charge on the surface of the sphere of radius b must be finite at the origin so that we may put

$$V_{sphere} = \frac{Q}{a} \sum_{n=0}^{\infty} A_{2n} r^{2n} P_{2n}(\cos\theta)$$

and because the sphere is earthed we must have $V_{ring} + V_{sphere} = 0$ at $r = b$ which gives

$$A_{2n} = -P_{2n}(0) \frac{1}{b^{2n}} \left(\frac{a}{b}\right)^{2n+1}.$$

Hence

$$V = V_{ring} + V_{sphere} = \frac{Q}{a} \sum_{n=0}^{\infty} P_{2n}(0) \left[\left(\frac{a}{r}\right)^{2n+1} - \left(\frac{a}{b}\right)^{2n+1} \left(\frac{r}{b}\right)^{2n} \right] P_{2n}(\cos\theta)$$

and so the electric field strength at the surface of the sphere is

$$E_r = -\left(\frac{\partial V}{\partial r}\right)_{r=b} = Q \sum_{n=0}^{\infty} P_{2n}(0) \frac{4n+1}{b^2} \left(\frac{a}{b}\right)^{2n} P_{2n}(\cos\theta)$$

which yields the result using (6.22).

6.34 A ring of radius a having a uniform line density of charge λ per unit length surrounds an earthed concentric spherical conductor of radius b. Find the density of charge at any point of the surface of the conductor

• *Solution*

Using the result (6.21) given in problem 6.32 we know that in the absence of the sphere the potential of the ring for $r \leq a$ is given by

$$V_{ring} = 2\pi\lambda \sum_{n=0}^{\infty} P_{2n}(0) \left(\frac{r}{a}\right)^{2n} P_{2n}(\cos\theta)$$

taking the axis of symmetry to be the polar axis for the polar coordinates r, θ and the origin at the centre of the ring.

The potential due to the induced charge on the sphere $r = b$ may be written in the form

$$V_{sphere} = 2\pi\lambda \sum_{n=0}^{\infty} A_{2n} \frac{1}{r^{2n+1}} P_{2n}(\cos\theta) \quad (r \geq b).$$

Since the sphere is earthed we must have $V_{ring} + V_{sphere} = 0$ for $r = b$ and so

$$A_{2n} = -P_{2n}(0) \left(\frac{b}{a}\right)^{2n} b^{2n+1}.$$

Hence the potential $V = V_{ring} + V_{sphere}$ due to the ring and the induced charge on the sphere is

$$V = 2\pi\lambda \sum_{n=0}^{\infty} P_{2n}(0) \left[\left(\frac{r}{a}\right)^{2n} - \left(\frac{b}{a}\right)^{2n} \left(\frac{b}{r}\right)^{2n+1}\right] P_{2n}(\cos\theta) \quad (b \le r \le a)$$

giving for the surface density of charge σ on the sphere $r = b$

$$\sigma = -\frac{1}{4\pi} \left(\frac{\partial V}{\partial r}\right)_{r=b} = -\frac{\lambda}{2b} \sum_{n=0}^{\infty} P_{2n}(0)(4n+1) \left(\frac{b}{a}\right)^{2n} P_{2n}(\cos\theta).$$

6.35 A dielectric sphere is surrounded by a thin circular wire of radius a carrying a charge Q. Show that the potential within the sphere is given by

$$\frac{Q}{a} \left[1 + \sum_{n=1}^{\infty} (-1)^n \frac{1+4n}{1+2n(1+K)} \frac{1.3.5...(2n-1)}{2.4.6...2n} \left(\frac{r}{a}\right)^{2n} P_{2n}(\cos\theta)\right]$$

where K is the dielectric constant of the sphere.

- *Solution*

Let b be the radius of the dielectric sphere. Then for $r < b$ we may write the potential V in the form

$$V = \frac{Q}{a} \sum_{n=0}^{\infty} A_{2n} \left(\frac{r}{a}\right)^{2n} P_{2n}(\cos\theta)$$

where the polar axis is the axis of symmetry and the origin is at the centre.

For $b < r < a$ we have

$$V = \frac{Q}{a} \sum_{n=0}^{\infty} P_{2n}(0) \left(\frac{r}{a}\right)^{2n} P_{2n}(\cos\theta) + \frac{Q}{a} \sum_{n=0}^{\infty} B_{2n} \left(\frac{a}{r}\right)^{2n+1} P_{2n}(\cos\theta)$$

where the first term arises from the ring and the second term is the perturbation term due to the dielectric sphere.

Now V and $K\partial V/\partial r$ are continuous at $r = b$ and so we get

$$A_{2n} \left(\frac{b}{a}\right)^{2n} = P_{2n}(0) \left(\frac{b}{a}\right)^{2n} + B_{2n} \left(\frac{a}{b}\right)^{2n+1}$$

and

$$2nK A_{2n} \frac{b^{2n-1}}{a^{2n}} = 2n \frac{b^{2n-1}}{a^{2n}} P_{2n}(0) - (2n+1)B_{2n} \frac{a^{2n+1}}{b^{2n+2}}.$$

Hence

$$(2n+1+2nK)A_{2n} = (4n+1)P_{2n}(0)$$

and the result follows using (6.22).

6.36 A point charge q is placed at a point A distance a from the infinite plane face of a semi-infinite body of material having dielectric constant K. Find the force acting on the point charge.

● *Solution*

We use the method of images and put an image charge q' at the image point A' in the dielectric distance a from the plane face. Let P be a field point. Putting $r = AP$ and $r' = A'P$, the potential outside the dielectric is

$$V_P = \frac{q}{r} + \frac{q'}{r'}.$$

We represent the field within the dielectric by putting a point charge q'' at the point A. Then the potential at a field point P inside the dielectric is given by

$$V_P = \frac{q''}{r}.$$

Continuity of the potential at the plane face of the dielectric then requires that $q + q' = q''$.

Also $K\partial V/\partial n$, where the derivative is in the normal direction to the plane face, is continuous there. Now at a point on the interface, $\partial r/\partial n = a/r$ and so we get

$$\frac{q}{r^2}\frac{a}{r} - \frac{q'}{r^2}\frac{a}{r} = K\frac{q''}{r^2}\frac{a}{r}$$

from which it follows that $q - q' = Kq''$. Hence

$$q' = -q\frac{K-1}{K+1}, \quad q'' = \frac{2q}{K+1}. \tag{6.23}$$

Then all the conditions of the appropriate uniqueness theorem are satisfied.

The force acting on the charge q due to the presence of the dielectric is

$$\frac{qq'}{(2a)^2} = -\frac{q^2}{(2a)^2}\frac{K-1}{K+1}.$$

which is attractive.

6.37 A spherical conductor of internal radius b which is uncharged and insulated, surrounds a spherical conductor of radius a, the distance between their centres being δ which is small. If the charge on the inner conductor is Q, find the potential at points between the conductors and show that the surface density of charge at a point P on the inner conductor is

$$\frac{Q}{4\pi}\left(\frac{1}{a^2} - \frac{3\delta\cos\theta}{b^3 - a^3}\right)$$

where θ is the angle that the radius through P makes with the line of centres and terms in δ^2 are neglected.

● *Solution*

Taking the origin at the centre of the sphere of radius a, we can write the equation of the sphere of radius b in the approximate form $r \simeq b + \delta P_1(\cos\theta)$ where δ is the distance between the centres.

The potential between the spheres may be written in the form

$$V = A + \frac{B}{r} + \left(Cr + \frac{D}{r^2}\right)P_1(\cos\theta)$$

where A, B are constants and C, D are constants of order δ.

Suppose V_a and V_b are the potentials of the spheres a and b respectively. Then

$$V_a = A + \frac{B}{a} + \left(Ca + \frac{D}{a^2}\right)P_1$$

and, since $r^{-1} \simeq b^{-1}(1 - b^{-1}\delta P_1)$ on sphere b, we have to order δ

$$V_b = A + \frac{B}{b}\left(1 - \frac{\delta}{b}P_1\right) + \left(Cb + \frac{D}{b^2}\right)P_1.$$

Hence $V_a = A + a^{-1}B, Ca + a^{-2}D = 0, V_b = A + b^{-1}B, Cb + b^{-2}D - \delta b^{-2}B = 0$. Since the charge on sphere a is Q we have $B = Q$ and so

$$C = \frac{Q\delta}{b^3 - a^3}, \quad D = -\frac{Q\delta a^3}{b^3 - a^3}.$$

Hence the surface density of charge on the sphere a is given by

$$\sigma = -\frac{1}{4\pi}\left(\frac{\partial V}{\partial r}\right)_{r=a} = \frac{Q}{4\pi}\left(\frac{1}{a^2} - \frac{3\delta\cos\theta}{b^3 - a^3}\right).$$

since $P_1 = \cos\theta$.

6.38 The equation of the surface of a conductor is $r = a(1 + \delta P_n)$ where δ is small and the conductor is placed in a uniform field of force parallel to the polar axis. Show that the surface density of the induced charge at any point is greater than it would be if the surface were spherical by

$$\frac{3\delta En}{4\pi(2n+1)}[(n+1)P_{n+1} + (n-2)P_{n-1}].$$

• *Solution*

The general axially symmetric solution to Laplace's equation $\nabla^2 V = 0$ in polar coordinates r, θ is given by (6.18) to be

$$V = \sum_{n=0}^{\infty}\left(A_n r^n + \frac{B_n}{r^{n+1}}\right)P_n(\cos\theta).$$

Hence for a conducting surface $r = a(1 + \delta P_n)$ at zero potential situated in a uniform field E acting in the direction of the polar axis the potential is given by, to first order in δ,

$$V = \left(\frac{a^3}{r^2} - r\right)E\cos\theta + \frac{3\delta a^{n+1}En}{2n+1}\frac{1}{r^n}P_{n-1} + \frac{3\delta a^{n+3}E(n+1)}{2n+1}\frac{1}{r^{n+2}}P_{n+1}$$

since $(2n+1)\cos\theta\, P_n = (n+1)P_{n+1} + nP_{n-1}$.

Now $\sigma = -(4\pi)^{-1}(\partial V/\partial r)_{surface}$ and so it follows that the surface density of induced charge σ over the earthed conductor is, to the first order in δ,

$$\sigma = \frac{3E}{4\pi}\cos\theta + \frac{3\delta E}{4\pi(2n+1)}\left[(n^2 - 2n)P_{n-1} + \{(n+1)(n+2) - 2(n+1)\}P_{n+1}\right]$$

which yields the required result since the surface density of charge for a sphere is $(3E/4\pi)\cos\theta$.

6-2 Magnetism

The magnetic potential due to a magnetic dipole of moment $\mathbf{M}$ is

$$\phi = -\mathbf{M}.\nabla\frac{1}{r} = \frac{\mathbf{M}.\mathbf{r}}{r^3} \tag{6.24}$$

where $\mathbf{r}$ is the position vector of the field point referred to the dipole.

Hence the magnetic field due to the dipole is, using the result derived in problem 4.06,

$$\mathbf{H} = -\nabla\phi = -\nabla\frac{\mathbf{M}.\mathbf{r}}{r^3} = -\frac{\mathbf{M}}{r^3} + \frac{3(\mathbf{M}.\mathbf{r})\ \mathbf{r}}{r^5}. \tag{6.25}$$

The potential energy of a dipole of moment $\mathbf{M}$ in a magnetic field $\mathbf{H}$ is given by $-\mathbf{M}.\mathbf{H}$ and the force acting on the dipole is $\mathbf{F} = \nabla(\mathbf{M}.\mathbf{H}) = \mathbf{M}.\nabla\mathbf{H}$. Also the couple acting on the dipole is given by $\mathbf{N} = \mathbf{M} \times \mathbf{H}$.

The energy of two dipoles having moments $\mathbf{M_1}$ and $\mathbf{M_2}$ is given by $W = -\mathbf{M_2}.\mathbf{H_1}$ or $W = -\mathbf{M_1}.\mathbf{H_2}$ where $\mathbf{H_1}$ and $\mathbf{H_2}$ are the magnetic fields due to the respective dipoles. Hence their mutual potential energy is

$$W = \frac{\mathbf{M_1}.\mathbf{M_2}}{r^3} - \frac{3(\mathbf{M_1}.\mathbf{r})(\mathbf{M_2}.\mathbf{r})}{r^5} \tag{6.26}$$

where $\mathbf{r}$ is the position vector of $\mathbf{M_2}$ referred to $\mathbf{M_1}$.

6.39 Show that the potential energy of two coplanar magnetic dipoles each of moment M at a distance r apart making angles θ_1 and θ_2 with the line joining them is given by

$$\frac{M^2}{r^3}(\sin\theta_1 \sin\theta_2 - 2\cos\theta_1\cos\theta_2).$$

Further show that there are four configurations for which the couples acting on the dipoles vanish and prove that just one of these four equilibrium configurations is stable.

• *Solution*

The potential energy of the two coplanar dipoles is, using (6.26),

$$W = \frac{M^2}{r^3}[\cos(\theta_1 - \theta_2) - 3\cos\theta_1\cos\theta_2]$$

which leads directly to the result given above.

The dipoles are in equilibrium when

$$\frac{\partial W}{\partial\theta_1} = \frac{M^2}{r^3}(2\sin\theta_1\cos\theta_2 + \cos\theta_1\sin\theta_2) = 0,$$

$$\frac{\partial W}{\partial\theta_2} = \frac{M^2}{r^3}(2\cos\theta_1\sin\theta_2 + \sin\theta_1\cos\theta_2) = 0.$$

Thus for equilibrium we have $\cos\theta_1\sin\theta_2 = 0$ and $\sin\theta_1\cos\theta_2 = 0$ whose solutions are

(i) $\theta_1 = 0, \theta_2 = 0$; (ii) $\theta_1 = 0, \theta_2 = \pi$; (iii) $\theta_1 = \pi/2, \theta_2 = \pi/2$; (iv) $\theta_1 = \pi/2, \theta_2 = 3\pi/2$.

W has an absolute maximum or minimum when $\partial W/\partial\theta_1 = \partial W/\partial\theta_2 = 0$ if

$$\frac{\partial^2 W}{\partial\theta_1^2}\frac{\partial^2 W}{\partial\theta_2^2} > \left(\frac{\partial^2 W}{\partial\theta_1\partial\theta_2}\right)^2.$$

Further W is a minimum or a maximum according as $\partial^2 W/\partial\theta_1^2 > 0$ or $\partial^2 W/\partial\theta_1^2 < 0$ respectively.

Now

$$\frac{\partial^2 W}{\partial\theta_1^2} = \frac{\partial^2 W}{\partial\theta_2^2} = \frac{M^2}{r^3}(2\cos\theta_1\cos\theta_2 - \sin\theta_1\sin\theta_2),$$

$$\frac{\partial^2 W}{\partial\theta_1\partial\theta_2} = \frac{M^2}{r^3}(\cos\theta_1\cos\theta_2 - 2\sin\theta_1\sin\theta_2).$$

Hence for case (i) there is a stable equilibrium for all rotations; for case (ii) there is an unstable equilibrium for all rotations; and for cases (iii) and (iv) there are unstable equilibriums for certain rotations.

6.40 Two small magnets lie in a plane with their axes making angles θ_1 and θ_2 with the line connecting them. Show that the couple acting on the second magnet vanishes if $\tan\theta_1 + 2\tan\theta_2 = 0$.

• *Solution*

The couple acting on a dipole $\mathbf{M_2}$ situated in a magnetic field $\mathbf{H}$ is $\mathbf{N} = \mathbf{M_2} \times \mathbf{H}$ and so, using the formula (6.25) for the magnetic field of a dipole we get

$$\mathbf{N} = \frac{\mathbf{M_1} \times \mathbf{M_2}}{r^3} + \frac{3(\mathbf{M_1}.\mathbf{r})\mathbf{M_2} \times \mathbf{r}}{r^5}.$$

Hence the magnitude of the couple is given by

$$N = \frac{M_1 M_2}{r^3}[\sin(\theta_1 - \theta_2) + 3\cos\theta_1\sin\theta_2].$$

It follows that the couple vanishes if $\sin\theta_1\cos\theta_2 + 2\cos\theta_1\sin\theta_2 = 0$ which yields the result required.

6.41 Two small magnets of equal moments $\mathbf{M}$ have their centres fixed and can turn about them in a uniform magnetic field $\mathbf{H}$ whose direction is perpendicular to the line joining their centres. Show that the position in which the magnets both point in the direction of the lines of force of the magnetic field is in stable equilibrium if $H > 3M/r^3$ where r is the distance between their centres.

• *Solution*

If the magnets make angles θ_1 and θ_2 with the line joining them, their potential energy is given by

$$W = \frac{M^2}{r^3}(\sin\theta_1\sin\theta_2 - 2\cos\theta_1\cos\theta_2) - MH(\sin\theta_1 + \sin\theta_2)$$

so that

$$\frac{\partial W}{\partial\theta_1} = \frac{M^2}{r^3}(\cos\theta_1\sin\theta_2 + 2\sin\theta_1\cos\theta_2) - MH\cos\theta_1,$$

$$\frac{\partial W}{\partial \theta_2} = \frac{M^2}{r^3}(\sin\theta_1\cos\theta_2 + 2\cos\theta_1\sin\theta_2) - MH\cos\theta_2.$$

Hence $\partial W/\partial\theta_1 = \partial W/\partial\theta_2 = 0$ when $\theta_1 = \theta_2 = \pi/2$ and so we have equilibrium when the dipoles point in the direction of the magnetic field lines.

Differentiating a second time and setting $\theta_1 = \theta_2 = \pi/2$ gives $\partial^2 W/\partial\theta_1^2 = \partial^2 W/\partial\theta_2^2 = -M^2/r^3 + MH$ and $\partial^2 W/\partial\theta_1\partial\theta_2 = -2M^2/r^3$ and so we get

$$\frac{\partial^2 W}{\partial\theta_1^2}\frac{\partial^2 W}{\partial\theta_2^2} - \left(\frac{\partial^2 W}{\partial\theta_1\partial\theta_2}\right)^2 = \left(MH - \frac{M^2}{r^3}\right)^2 - \left(\frac{2M^2}{r^3}\right)^2 = \left(MH + \frac{M^2}{r^3}\right)\left(MH - \frac{3M^2}{r^3}\right)$$

which is positive if $H > 3M/r^3$. Since $\partial^2 W/\partial\theta_1^2$ is also positive then, this is the condition for stable equilibrium.

6.42 A very small magnet of moment **M** is placed with its axis perpendicular to the infinite plane face of a semi-infinite block of material. If the magnet is a distance d from the plane face, show that the magnet is attracted to the block with a force

$$\frac{3}{8}\left(\frac{\mu - 1}{\mu + 1}\right)\frac{M^2}{d^4}$$

where μ is the permeability of the material.

- *Solution*

The effect of the material can be represented by an image magnetic dipole

$$\mathbf{M}' = \frac{\mu - 1}{\mu + 1}\mathbf{M}$$

situated at the mirror image point distant d from the plane face. This is analogous to the case of a point charge placed near a block of dielectric, discussed in problem 6.36.

The force between dipoles **M** and **M'** is given by $\mathbf{F} = \mathbf{M}.\nabla\mathbf{H}'$ where

$$\mathbf{H}' = -\frac{\mathbf{M}'}{r^3} + \frac{3\mathbf{M}'.\mathbf{r}\ \mathbf{r}}{r^5}$$

is the magnetic field of the dipole **M'** using (6.25). Hence

$$F = 2MM'\frac{\partial}{\partial z}\left(\frac{1}{z^3}\right) = -\frac{6MM'}{z^4}$$

where $z = 2d$ is the distance between the dipoles and so the result follows.

6.43 A spherical shell of permeability μ having inner radius a and outer radius b is situated in a uniform magnetizing field **H**. Find the induced magnetic field in the inner region $r < a$.

- *Solution*

The magnetic potential in the absence of the shell may be written $\phi_0 = -Hr\cos\theta$ in spherical polar coordinates.

If ϕ' is the contribution to the total potential ϕ due to the presence of the shell, we have $\phi = \phi_0 + \phi'$. Then ϕ and the normal component of the magnetic field $B_n = -\mu\partial\phi/\partial r$ are continuous at $r = a$ and $r = b$.

Since ϕ' is finite at $r = 0$ and vanishes for large r we must have

$$\begin{aligned}
\phi' &= Ar\cos\theta \quad (r < a) \\
&= Br\cos\theta + Cr^{-2}\cos\theta \quad (a \leq r \leq b) \\
&= Dr^{-2}\cos\theta \quad (r > b).
\end{aligned}$$

Since ϕ is continuous at $r = a$ and $r = b$ we have $Aa = Ba + Ca^{-2}$ and $Bb + Cb^{-2} = Db^{-2}$ respectively. Also $B_r = -\mu\partial\phi/\partial r$ is continuous at $r = a$ and $r = b$ and so we have $H - A = \mu(H - B + 2Ca^{-3})$ and $\mu(H - B + 2Cb^{-3}) = H + 2Db^{-3}$ respectively. These equation yield

$$A = \frac{2H(\mu - 1)^2(1 - a^3/b^3)}{(1 + 2\mu)(2 + \mu) - 2(\mu - 1)^2 a^3/b^3}$$

and so the potential in the inner region $r < a$ is

$$\phi = (A - H)r\cos\theta = -\frac{9\mu Hz}{(1 + 2\mu)(2 + \mu) - 2(\mu - 1)^2 a^3/b^3}$$

where $z = r\cos\theta$.

Hence the magnetic field in the inner region $r < a$ is given by

$$-\frac{\partial\phi}{\partial z} = \frac{9\mu H}{9\mu + 2(\mu - 1)^2(1 - a^3/b^3)}.$$

6.44 Find the potential of the induced magnetic field produced by the presence of a solid circular cylinder of radius a and permeability μ situated in a magnetizing field whose potential is $A(x^2 - y^2)$.

- *Solution*

In the absence of the cylinder the potential is $\phi_0 = A(x^2 - y^2) = Ar^2\cos 2\theta$ in circular polar coordinates.

Let ϕ' be the contribution to the total potential ϕ from the presence of the cylinder. Then we must have

$$\phi' = Br^2\cos 2\theta \quad (r < a), \qquad \phi' = \frac{C\cos 2\theta}{r^2} \quad (r > a).$$

By the continuity of $\phi = \phi_0 + \phi'$ at $r = a$ we get $C = Ba^4$ and by the continuity of $\mu\partial\phi/\partial r$ at $r = a$ we obtain $2Aa - 2C/a^3 = 2\mu(A + B)a$ which give

$$C = -a^4\left(\frac{\mu - 1}{\mu + 1}\right)A.$$

Hence the magnetic potential outside the cylinder is

$$\phi = A\left[r^2 - a^4\left(\frac{\mu - 1}{\mu + 1}\right)\frac{1}{r^2}\right]\cos 2\theta \quad (r > a)$$

and inside the cylinder it is

$$\phi = \frac{2Ar^2\cos 2\theta}{\mu + 1} \quad (r < a).$$

6-3 Electric current flow

In the following few problems we shall be using the *Biot-Savart law* which gives the magnetic field **H** due to a current I flowing in a circuit γ as the integral

$$\mathbf{H} = \frac{I}{c} \oint_\gamma \frac{d\mathbf{s} \times \mathbf{r}}{r^3} \qquad (6.27)$$

where c is the speed of light and $\mathbf{r}$ is the position vector of the field point referred to the element of arc $d\mathbf{s}$.

6.45 A current I is flowing through a circuit of wire having the form of a regular plane polygon of $2n$ sides, the distance between parallel sides being $2a$. Find the magnetic field at the centre of the wire.

• *Solution*

We consider one side of the polygon. Its perpendicular distance from the centre C of the polygon is a and it subtends an angle π/n at C.

The magnetic field at the centre C due to this side can be written as

$$\frac{I}{c} \int \frac{\cos\theta}{a^2 + s^2} \, ds$$

where ds is an element of length of the side distant $s = a\tan\theta$ from the centre of the side.

Now $ds/d\theta = a\sec^2\theta = (a^2 + s^2)/a$ and so the magnetic field at C due to one side of the polygon is

$$\frac{2I}{ac} \int_0^{\pi/2n} \cos\theta \, d\theta = \frac{2I}{ac} \sin\frac{\pi}{2n}.$$

Since the polygon has $2n$ sides, the magnetic field at the centre of the polygon is

$$\frac{4nI}{ac} \sin\frac{\pi}{2n}.$$

6.46 A current I is flowing in an elliptical circuit with semi-axes of lengths a and b. Find the magnetic field at the centre of the ellipse in the form of an integral.

• *Solution*

We can take parametric equations of the ellipse in the form $x = a\cos\theta$, $y = b\sin\theta$ referred to its centre C. Then the position vector of a point of the ellipse referred to C as origin of a rectangular Cartesian frame is given by $\mathbf{r} = a\cos\theta \, \mathbf{i} + b\sin\theta \, \mathbf{j}$ where $\mathbf{i}$ and $\mathbf{j}$ are unit vectors in the directions of the x and y axes. Then we have $d\mathbf{r} = (-a\sin\theta \, \mathbf{i} + b\cos\theta \, \mathbf{j})d\theta$ and so

$$\frac{\mathbf{r} \times d\mathbf{r}}{r^3} = \frac{ab(\cos^2\theta \, \mathbf{i} \times \mathbf{j} - \sin^2\theta \, \mathbf{j} \times \mathbf{i})d\theta}{r^3} = \frac{abk \, d\theta}{r^3}$$

where **k** is a unit vector perpendicular to the ellipse.

The Biot-Savart law gives

$$\mathbf{H} = \frac{I}{c} \oint \frac{d\mathbf{r} \times (-\mathbf{r})}{r^3} = \frac{abI}{c} \int_0^{2\pi} \frac{d\theta}{r^3} \mathbf{k}$$

since $-\mathbf{r}$ is directed from the circuit to C. Now $r^2 = a^2 \cos^2\theta + b^2 \sin^2\theta$ and so the magnetic field at the centre of the ellipse in the z direction parallel to $\mathbf{k}$ is

$$H_z = \frac{4abI}{c} \int_0^{\pi/2} \frac{d\theta}{(a^2 \cos^2\theta + b^2 \sin^2\theta)^{3/2}}.$$

6.47 A wire is wound in a flat open spiral whose polar equation is $r = a\theta/2\pi n$ where n is the total number of turns and a is the radial distance from the centre to the outer end of the spiral. If the wire carries a current I show that the axial component of the magnetic field at a distance z along the axis of the spiral is

$$\frac{2\pi nI}{ac} \left\{ -\frac{a}{(a^2 + z^2)^{1/2}} + \ln \left[\frac{a + (a^2 + z^2)^{1/2}}{z} \right] \right\}.$$

- *Solution*

The axial component of the magnetic field at a distance z along the axis, due to an element of arc $d\theta$ of the spiral distance r from the origin, is given by

$$dH_z = \frac{I}{c} \frac{r\,d\theta}{(z^2 + r^2)} \frac{r}{(z^2 + r^2)^{1/2}}.$$

Now $dr/d\theta = a/2\pi n$ and so the axial component of the magnetic field due to the entire spiral is

$$B_z = \frac{2\pi nI}{ac} \int_0^a \frac{r^2\,dr}{(z^2 + r^2)^{3/2}} = \frac{2\pi nI}{ac} \left\{ \left[-\frac{r}{(z^2 + r^2)^{1/2}} \right]_0^a + \int_0^a \frac{dr}{(z^2 + r^2)^{1/2}} \right\}$$

and since $\int (z^2 + r^2)^{-1/2} dr = \ln[r + (z^2 + r^2)^{1/2}]$ the result follows.

6.48 A current I flows around a circular wire of radius a and a current I' flows in a long straight wire in the same plane. Show that the mutual attraction is $4\pi II'(\sec\alpha - 1)/c^2$ where 2α is the angle subtended by the circle at the nearest point of the straight wire.

- *Solution*

The force acting on an element ds of a circuit carrying a current I due to a magnetic field $\mathbf{H}$ is $(I/c)ds \times \mathbf{H}$. Now the magnetic field due to the current I' flowing in the straight wire has magnitude $H = 2I'/cr$ at a perpendicular distance r from the wire. Hence the attractive force on the circular wire due to the magnetic field H is

$$F = \int \frac{I}{c} H \cos\theta\,ds = \frac{2II'}{c^2} \int \frac{\cos\theta\,ds}{r}$$

where θ is the angle the radius vector from the centre of the circle C makes with the perpendicular from C to the straight wire. Thus

$$F = \frac{4II'a}{c^2} \int_0^\pi \frac{\cos\theta\,d\theta}{d - a\cos\theta}$$

where d is the perpendicular distance from C to the straight wire.

Now

$$\int_0^\pi \frac{d\theta}{d - a\cos\theta} = \frac{\pi}{(d^2 - a^2)^{1/2}}$$

and so

$$F = \frac{4\pi II'}{c^2}\left[\frac{d}{(d^2 - a^2)^{1/2}} - 1\right] = \frac{4\pi II'}{c^2}\left(\frac{1}{\cos\alpha} - 1\right)$$

since $\cos\alpha = (d^2 - a^2)^{1/2}/d$, and the result follows.

6.49 If the Earth is modeled as a uniformly magnetized sphere of radius a, show that a circular current I flowing from west to east and surrounding the Earth along the parallel of north latitude λ is acted on by a force towards the north pole given by $6\pi a(IH/c)\sin\lambda\cos^2\lambda$ where H is the magnitude of the intensity of the magnetic field of the Earth at the equator.

• *Solution*

Let us suppose that the Earth has a dipole of moment M directed towards the south geographical pole and take the polar axis of the coordinate system in the north-south direction. Then for $r > a$ the magnetic potential is $\phi = M\cos\theta/r^2$ giving for the $\hat{r}$ and $\hat{\theta}$ components of the magnetic field

$$H_r = -\frac{\partial\phi}{\partial r} = \frac{2M\cos\theta}{r^3}, \quad H_\theta = -\frac{1}{r}\frac{\partial\phi}{\partial\theta} = \frac{M\sin\theta}{r^3}.$$

At the equator, $\theta = \pi/2$ and $r = a$, we have $H_r = 0$ and $H_\theta = M/a^3$ and so $M = a^3 H$. Hence on the surface of the Earth $H_r = 2H\cos\theta$ and $H_\theta = H\sin\theta$.

Now consider an element ds of the current. The force acting on it is $\mathbf{F} = (I/c)ds \times \mathbf{H}$ and so we have

$$\mathbf{F} = -\frac{I}{c}ds\hat{\phi} \times (2H\cos\theta\,\hat{r} + H\sin\theta\,\hat{\theta}) = \frac{IH}{c}ds(-2\cos\theta\,\hat{\theta} + \sin\theta\,\hat{r})$$

where $\theta = \pi/2 + \lambda$ since the current is in the parallel of north latitude λ.

By symmetry there cannot be a force acting on the current perpendicular to its axis. The force acting on the wire in the direction of the north pole is

$$F_z = -\frac{IH}{c}\oint(2\cos\theta\sin\theta + \sin\theta\cos\theta)\,ds = \frac{IH}{c}(2\pi a\cos\lambda)(3\sin\lambda\cos\lambda)$$

which yields the required result.

6.50 A circular wire of radius a and resistance R is made to spin in a magnetic field H with angular velocity ω about an axis in the plane of the wire and perpendicular to H. Show that the mean rate of dissipation of energy by the induced electric current in the wire is $\pi^2 a^4\omega^2 H^2/2Rc^2$ approximately.

• *Solution*

Suppose that the plane of the wire is perpendicular to the magnetic field at time $t = 0$. Then the normal to the plane of the wire makes an angle ωt with the field lines at any subsequent time t. Hence at time t the magnetic flux through the circular wire is $N = \pi a^2 H\cos\omega t$.

Hence from *Faraday's law of induction*

$$IR = -\frac{1}{c}\frac{dN}{dt} \tag{6.28}$$

we have that the induced current I in the wire is given by

$$IR = \frac{1}{c}\pi a^2 H\omega \sin \omega t$$

and so the rate of dissipation of energy at time t is

$$W = I^2 R = \frac{\pi^2 a^4 \omega^2 H^2 \sin^2 \omega t}{Rc^2}$$

giving the required result since the mean value of $\sin^2 \omega t$ is $\frac{1}{2}$.

6.51 A circular coil of wire of radius a and resistance R is fixed in a plane and a thin bar magnet of moment M and length $2b$ is perpendicular to this plane and moves along the axis of the coil. Show that when the centre of the magnet is a small distance z from the centre of the coil, the flux of magnetic induction passing through the coil is approximately

$$2\pi M \left(\frac{1}{d} - \frac{3a^2 z^2}{2d^5} \right)$$

where $d^2 = a^2 + b^2$. If $z = z_0 \sin \omega t$ find the current induced in the coil.

• *Solution*

We shall regard the magnet as being composed of two magnetic poles $-m, m$ at a distance $2b$ apart.

If $\mathbf{r}$ is the position vector, referred to the magnetic pole m, of a point P radial distance ρ from the centre of the coil, the field at P is $\mathbf{H} = m\mathbf{r}/r^3$.

Let $\hat{n}dS$ be the vector area of an annulus of radius ρ and width $d\rho$ of the circular cross section of the coil. Then

$$\mathbf{H}.\hat{n}dS = \frac{m\mathbf{r}.\hat{n}}{r^3}dS = \frac{m(z-b)}{[(z-b)^2 + \rho^2]^{3/2}}2\pi\rho d\rho.$$

Hence the magnetic flux through the coil due to the magnetic pole m is given by

$$N_+ = \int \mathbf{H}.\hat{n}dS = 2\pi m \int_0^a \frac{z-b}{[(z-b)^2 + \rho^2]^{3/2}}\rho d\rho = 2\pi m \left\{ 1 - \frac{z-b}{[(z-b)^2 + a^2]^{1/2}} \right\}.$$

Similarly the magnetic flux due to the magnetic pole $-m$ is

$$N_- = -2\pi m \left\{ 1 - \frac{z+b}{[(z+b)^2 + a^2]^{1/2}} \right\}$$

and so the total flux is

$$N = N_+ + N_- = 2\pi m \left\{ \frac{z+b}{[(z+b)^2 + a^2]^{1/2}} - \frac{z-b}{[(z-b)^2 + a^2]^{1/2}} \right\}.$$

Now putting $d^2 = a^2 + b^2$ and expanding using the binomial theorem assuming that z is small, we find approximately

$$N = 2\pi M \left(\frac{1}{d} - \frac{3a^2 z^2}{2d^5} \right)$$

since $M = 2bm$.

If $z = z_0 \sin \omega t$ we get

$$N = 2\pi M \left(\frac{1}{d} - \frac{3a^2 z_0^2 \sin^2 \omega t}{2d^5} \right).$$

and so, using Faraday's law of induction (6.28), the induced current in the coil is given by

$$I = -\frac{1}{cR}\frac{dN}{dt} = \frac{3\pi M a^2 z_0^2 \omega \sin 2\omega t}{cRd^5}.$$

6.52 A thin bar magnet of moment M and length $2b$ lies along the axis of circular wire of radius a and has a small longitudinal oscillation $z = z_0 \sin \omega t$ about the centre of the wire. If R and L are the resistance and inductance of the wire, find the induced current in the wire.

• *Solution*

From the previous problem the electromotive force in the wire is

$$\mathcal{E} = \mathsf{Im}\left[\frac{3\pi M a^2 z_0^2 \omega}{cd^5} \exp(2i\omega t) \right]$$

where $d^2 = a^2 + b^2$ and Im denotes the imaginary part.

Now $\mathcal{E} = IR + LdI/dt$ and so, putting the current $I = \mathsf{Im}[A\exp(2i\omega t)]$, we obtain $3\pi M a^2 z_0^2 \omega / cd^5 = A(R + 2i\omega L)$ giving

$$A = \frac{3\pi M a^2 z_0^2 \omega}{cd^5}\frac{R - 2i\omega L}{R^2 + 4\omega^2 L^2} = \frac{3\pi M a^2 z_0^2 \omega}{cd^5 \sqrt{R^2 + 4\omega^2 L^2}} \exp(-i\alpha)$$

where $\tan\alpha = 2\omega L/R$ and so

$$I = \frac{3\pi M a^2 z_0^2 \omega}{c(a^2 + b^2)^{5/2}}\frac{\sin(2\omega t - \alpha)}{\sqrt{R^2 + 4\omega^2 L^2}}.$$

6.53 A wire forms a regular hexagon and the vertices are joined to the centre by wires each of which has a resistance r/n where r is the resistance of each side of the hexagon. Show that the resistance to a current entering at a vertex and leaving by the opposite vertex is $2(n+3)r/[(n+1)(n+4)]$.

• *Solution*

To solve this problem we shall use *Kirchoff's laws*:

(1) The algebraic sum of currents flowing out of a junction in a network is zero.

(2) For any closed circuit in a network, the algebraic sum of the products of the current and the resistance is zero.

Let A, B, C, D, E, F, A be the vertices of the hexagon in order round the hexagon, and let O be its centre.

Suppose a current I enters at the vertex A and leaves at the vertex D. Let x be the currents in the wires AB, AF, CD, ED, let y be the currents in the wires AO, OD, and let z be the currents in the wires BO, FO, OC, OE.

Applying Kirchoff's first law at the junction A gives $I = 2x + y$, applying Kirchoff's second law to the loop AOB gives $y/n = x + z/n$, and applying Kirchoff's second law to the loop BOC gives $x - z = 2z/n$.

These equations yield

$$y = \frac{n(n+3)}{(n+4)(n+1)}I.$$

Suppose R is the equivalent resistance of the hexagon. Then we have $IR = 2yr/n$ which gives the required result.

6.54 A rectangular network having sides a and na is divided into n square meshes of side a by $n-1$ parallel wires. A current I enters and leaves by adjacent vertices of the rectangle which are also the vertices of the first mesh. If the network is composed of uniform wires of resistance r/a per unit length, find the equivalent resistance of the network.

• *Solution*

Let I_s be the current flowing in the sth square mesh. Then applying Kirchoff's second law to the sth square mesh we get $-I_{s-1}r + 4I_s r - I_{s+1}r = 0$ where r is the resistance of a wire of length a. Thus we have

$$-I_{s-1} + 4I_s - I_{s+1} = 0.$$

Putting $I_s = A\alpha^s$ we obtain $-\alpha^{s-1} + 4\alpha^s - \alpha^{s+1} = 0$ and so we have $\alpha^2 - 4\alpha + 1 = 0$ which gives $\alpha = 2 \pm \sqrt{3}$.

Hence $I_s = A\alpha^s + B\alpha^{-s}$ where $\alpha = 2 + \sqrt{3}$ since $2 - \sqrt{3} = (2 + \sqrt{3})^{-1}$.

Now $I = I_0 = A + B$ and $0 = I_{n+1} = A\alpha^{n+1} + B\alpha^{-(n+1)}$ giving $I = A + B$ and $A\alpha^{2n+2} + B = 0$. This yields

$$A = \frac{I}{1 - \alpha^{2n+2}}, \quad B = -\frac{\alpha^{2n+2}I}{1 - \alpha^{2n+2}}$$

and so

$$I_s = \frac{(\alpha^s - \alpha^{2n+2-s})I}{1 - \alpha^{2n+2}}.$$

Let R be the equivalent resistance of the network. Then $RI = r(I_0 - I_1) = r(I - I_1)$ and hence

$$R = r\left(1 - \frac{I_1}{I}\right) = r\left[1 - \frac{\alpha - \alpha^{2n+1}}{1 - \alpha^{2n+2}}\right] = \frac{r(1-\alpha)(1 + \alpha^{2n+1})}{1 - \alpha^{2n+2}}.$$

6.55 A voltage pulse

$$U(t) = \begin{cases} 0 & (t < 0) \\ U_0 & (0 \leq t \leq t_0) \\ 0 & (t > t_0) \end{cases}$$

is applied across a circuit composed of a resistance R and an inductance L connected in series. Find the voltage across the inductance.

• *Solution*

If I is the current in the circuit we have

$$L\frac{dI}{dt} + RI = U(t).$$

Let us suppose that $I = 0$ for $t < 0$. For $t > t_0$ we see that $I = A\exp(-Rt/L)$ where A is a constant. Now for $0 \le t \le t_0$ the solution to the above equation can be written as

$$I = \frac{U_0}{R} + B\exp\left(-\frac{Rt}{L}\right)$$

where, by continuity at $t = 0$ we have $U_0/R + B = 0$, and by continuity at $t = t_0$ we have $U_0/R + B\exp(-Rt_0/L) = A\exp(-Rt_0/L)$. Hence

$$I = \begin{cases} 0 & (t < 0) \\ (U_0/R)[1 - \exp(-Rt/L)] & (0 \le t \le t_0) \\ (U_0/R)[\exp(Rt_0/L) - 1]\exp(-Rt/L) & (t > t_0) \end{cases}$$

and so the potential difference across the inductance is given by

$$L\frac{dI}{dt} = \begin{cases} 0 & (t < 0) \\ U_0\exp(-Rt/L) & (0 \le t \le t_0) \\ U_0[1 - \exp(Rt_0/L)]\exp(-Rt/L) & (t > t_0) \end{cases} .$$

6.56 A circuit is composed of two impedances connected in parallel. Each impedance is composed of a self-inductance, a resistance and a capacity connected in series. If their values are L_1, R_1, C_1 and L_2, R_2, C_2 respectively, find the current produced by an electromotive force $a\sin\omega t$ across the circuit.

- *Solution*

In this problem we introduce *complex impedances* given by $Z_1 = R_1 + iX_1$ and $Z_2 = R_2 + iX_2$ where $X_1 = L_1\omega - 1/(C_1\omega)$ and $X_2 = L_2\omega - 1/(C_2\omega)$.

The complex electromotive force is given by $\mathcal{E} = a\exp(i\omega t)$ and the complex currents through the impedances Z_1 and Z_2 are given by $I_1 = \mathcal{E}/Z_1$ and $I_2 = \mathcal{E}/Z_2$. Then the total complex current is given by $I = I_1 + I_2$.

We let the total complex current be given by $I = \mathcal{E}/Z$ where $1/Z = 1/Z_1 + 1/Z_2$ since the impedances are in parallel and so

$$I = a\exp(i\omega t)\left(\frac{1}{Z_1} + \frac{1}{Z_2}\right).$$

Now we can write $Z_1 = |Z_1|\exp(i\alpha_1 t)$ and $Z_2 = |Z_2|\exp(i\alpha_2 t)$ where $|Z_1| = \sqrt{R_1^2 + X_1^2}$, $\tan\alpha_1 = X_1/R_1$ and $|Z_2| = \sqrt{R_2^2 + X_2^2}$, $\tan\alpha_2 = X_2/R_2$.

Hence

$$I = a\exp(i\omega t)\left[\frac{\exp(-i\alpha_1)}{|Z_1|} + \frac{\exp(-i\alpha_2 t)}{|Z_2|}\right].$$

The imaginary part of I is the actual current passing through the circuit and this is given by

$$a\left[\frac{\sin(\omega t - \alpha_1)}{\sqrt{R_1^2 + +(L_1\omega - 1/C_1\omega)^2}} + \frac{\sin(\omega t - \alpha_2)}{\sqrt{R_2^2 + +(L_2\omega - 1/C_2\omega)^2}}\right].$$

6.57 A condenser of capacity C is connected by wires of resistance r so that it is in parallel with a coil of self-inductance L, the resistance of the coil and its connecting wires being R. If this configuration forms part of a circuit in which there is an electromotive force of frequency p, show that it can be replaced by an equivalent wire having only pure resistance provided $R^2 - L/C = p^2LC(r^2 - L/C)$.

• *Solution*

The two complex impedances are $Z_1 = r - i/(Cp)$ and $Z_2 = R + iLp$ so that the total complex impedance is Z where $Z^{-1} = Z_1^{-1} + Z_2^{-1}$. Hence we have

$$Z = \frac{Z_1 Z_2}{Z_1 + Z_2} = \frac{Rr + L/C + i[Lrp - R/(Cp)]}{R + r + i[Lp - 1/(Cp)]}.$$

The imaginary part of Z vanishes if
$(R+r)[Lrp - R/(Cp)] - [Lp - 1/(Cp)](Rr + L/C) = 0$ and then simple algebra shows that the configuration can be replaced by an equivalent wire having just pure resistance if $R^2 - L/C = LCp^2(r^2 - L/C)$.

6.58 A circuit is composed of two condensers having capacities C_1 and C_2 connected in series with an inductance L. Initially there is no current flow and the charge on C_1 is Q and the charge on C_2 is zero. Find the current through the inductance at any subsequent time t.

• *Solution*

The equation representing the flow of current I in the circuit is

$$L\frac{dI}{dt} + \frac{Q_1}{C_1} + \frac{Q_2}{C_2} = 0$$

where Q_1 and Q_2 are the charges on the two condensers. Now $I = dQ_1/dt = dQ_2/dt$ and so

$$L\frac{d^2I}{dt^2} + I\left(\frac{1}{C_1} + \frac{1}{C_2}\right) = 0$$

whose general solution may be written as

$$I = A\sin\left[\left(\frac{C_1 + C_2}{LC_1C_2}\right)^{1/2} t + \epsilon\right]$$

where A and ϵ are constants. Initially $I = 0$ and so $\epsilon = 0$. Hence we have

$$\frac{Q_1}{C_1} + \frac{Q_2}{C_2} = -AL\left(\frac{C_1 + C_2}{LC_1C_2}\right)^{1/2} \cos\left[\left(\frac{C_1 + C_2}{LC_1C_2}\right)^{1/2} t\right].$$

But $Q_1 = Q$ and $Q_2 = 0$ when $t = 0$ and so $A = -Q\{C_2/[LC_1(C_1 + C_2)]\}^{1/2}$. Hence

$$I = -Q\left[\frac{C_2}{LC_1(C_1 + C_2)}\right]^{1/2} \sin\left[\left(\frac{C_1 + C_2}{LC_1C_2}\right)^{1/2} t\right].$$

6.59 A circuit is composed of a condenser of capacity C, a wire of resistance R and a coil of self-inductance $L = 2R^2C$ and negligible resistance. The coil is in parallel with the wire and the condenser is in series with them. At zero time the charge on the condenser is Q_0 and there is no current passing through the coil. Show that the charge on the condenser at a later time t is $Q_0 \exp(-kt)(\cos kt - \sin kt)$ where $k = 1/2RC$.

• *Solution*

If I_1 is the current through the inductance L and I_2 is the current through the resistance R we have

$$L\frac{dI_1}{dt} + \frac{Q}{C} = 0, \quad RI_2 + \frac{Q}{C} = 0$$

and $I = I_1 + I_2$ where I is the current through the condenser C. Hence $R(I-I_1)+Q/C = 0$ and so $RdI/dt + I/C + RQ/LC = 0$ since $dQ/dt = I$. It follows that

$$\frac{d^2Q}{dt^2} + \frac{1}{RC}\frac{dQ}{dt} + \frac{Q}{LC} = 0.$$

To solve this equation we put $Q = a\exp(nt)$ and then we get

$$n^2 + \frac{n}{RC} + \frac{1}{LC} = 0$$

giving

$$n = \frac{1}{2}\left[-\frac{1}{RC} \pm \sqrt{\frac{1}{(RC)^2} - \frac{4}{LC}}\right].$$

But $L = 2R^2C$ and so $n = -k \pm ik$ where $k = 1/2RC$.

Hence the general solution takes the form $Q = \exp(-kt)(A\cos kt + B\sin kt)$ where A and B are constants. Now $Q = Q_0$ and $I_1 = 0$ when $t = 0$. Thus $I = dQ/dt = -Q_0/RC$ when $t = 0$. This gives $A = Q_0$ and $-kA + kB = -Q_0/RC$ so that $B = Q_0[1 - 1/(kRC)] = -Q_0$. The required result follows at once.

6.60 Two closed circuits of negligible resistance each contain a condenser chosen so that they are both tuned to a frequency n when the circuits are far apart. Show that when they are close together and are coupled by a mutual inductance M, the frequencies of the two principal oscillations are $n(1 \pm m)^{-1/2}$ where $m^2 = M^2/L_1L_2$ and L_1, L_2 are the self inductances.

• *Solution*

When the coils are far apart we have for each coil

$$L\frac{d^2I}{dt^2} + \frac{I}{C} = 0$$

where I is the current. Putting $I = A\exp(int)$ we get $-n^2L + 1/C = 0$ and hence $n^2 = 1/(L_1C_1) = 1/(L_2C_2)$.

Now suppose that the coils are close together and that there is a mutual inductance M between them. Then we have

$$L_1\frac{d^2I_1}{dt^2} + M\frac{d^2I_2}{dt^2} + \frac{I_1}{C_1} = 0, \quad L_2\frac{d^2I_2}{dt^2} + M\frac{d^2I_1}{dt^2} + \frac{I_2}{C_2} = 0$$

where I_1 and I_2 are the currents in the respective closed circuits.

Putting $I_1 = A_1\exp(ipt)$ and $I_2 = A_2\exp(ipt)$ we get

$$(-p^2L_1 + 1/C_1)A_1 - p^2MA_2 = 0,$$
$$-p^2MA_1 + (-p^2L_2 + 1/C_2)A_2 = 0.$$

For consistency we must have

$$(-p^2L_1 + 1/C_1)(-p^2L_2 + 1/C_2) = p^4M^2$$

giving $(p^2 - n^2)^2 = p^4 m^2$ where $m^2 = M^2/L_1 L_2$. Hence $p^2 - n^2 = \mp p^2 m$ or $p^2 = n^2/(1 \pm m)$ and this gives the result required.

6.61 An electron of mass m in a dielectric is bound by a force directed towards the origin given by $-m\omega_0^2 \mathbf{r}$, the motion being damped by a velocity dependent resistance producing a retardation $\Gamma \dot{\mathbf{r}}$ where $\mathbf{r}$ is the position vector of the electron. If the dielectric is placed in an electromagnetic field of frequency ω show that the radiation emitted by the electron is out of phase with the stimulating radiation by an angle δ such that at the resonance frequency $\omega = \omega_0$

$$\frac{d\delta}{d\omega} = \frac{2}{\Gamma} \qquad (\omega = \omega_0)$$

• *Solution*

The equation of motion of the electron can be written in the complex form

$$m(\ddot{\mathbf{r}} + \Gamma \dot{\mathbf{r}} + \omega_0^2 \mathbf{r}) = \mathbf{E_0} \exp(i\omega t).$$

Putting $\mathbf{r} = \mathbf{a} \exp(i\omega t)$ we get $m(-\omega^2 + i\omega\Gamma + \omega_0^2)\mathbf{a} = \mathbf{E_0}$ and so a particular solution of the equation of motion is

$$\mathbf{r} = \frac{\mathbf{E_0}}{m} \frac{\exp i(\omega t - \delta)}{\sqrt{\omega^2 \Gamma^2 + (\omega_0^2 - \omega^2)^2}}$$

where $\tan \delta = \Gamma \omega/(\omega_0^2 - \omega^2)$. It follows that

$$\frac{d\delta}{d\omega} = \frac{\Gamma(\omega_0^2 + \omega^2)}{\Gamma^2 \omega^2 + (\omega_0^2 - \omega^2)^2}$$

which is $2/\Gamma$ at the resonance frequency $\omega = \omega_0$.

6-4 Electromagnetism

The equations of the electromagnetic field are

(i) $\operatorname{div} \mathbf{D} = 4\pi\rho$, (ii) $\operatorname{div} \mathbf{B} = 0$, (iii) $\operatorname{curl} \mathbf{H} = \dfrac{4\pi \mathbf{j}}{c} + \dfrac{1}{c}\dfrac{\partial \mathbf{D}}{\partial t}$, (iv) $\operatorname{curl} \mathbf{E} = -\dfrac{1}{c}\dfrac{\partial \mathbf{B}}{\partial t}$ (6.29)

where $\mathbf{E}$ and $\mathbf{H}$ are the electric and magnetic fields respectively, $\mathbf{D} = K\mathbf{E}$ is the electric displacement, $\mathbf{B} = \mu\mathbf{H}$ is the magnetic induction, $\mathbf{j}$ is the current density and ρ is the charge density.

These equations were given in 1865 by Maxwell in his treatise entitled *A Dynamical Theory of Electromagnetic Field*.

6.62 Maxwell's equations *in vacuo* take the form $\operatorname{div} \mathbf{E} = 0$, $\operatorname{div} \mathbf{H} = 0$, $\operatorname{curl} \mathbf{H} = (1/c)\partial \mathbf{E}/\partial t$, $\operatorname{curl} \mathbf{E} = -(1/c)\partial \mathbf{H}/\partial t$

In this case show that $\mathbf{E}$ and $\mathbf{H}$ satisfy the wave equations

$$\nabla^2 \mathbf{E} = \frac{1}{c^2}\frac{\partial^2 \mathbf{E}}{\partial t^2}, \quad \nabla^2 \mathbf{H} = \frac{1}{c^2}\frac{\partial^2 \mathbf{H}}{\partial t^2}$$

• *Solution*

We have that

$$\operatorname{curl}\operatorname{curl}\mathbf{E} = -\frac{1}{c}\frac{\partial \operatorname{curl}\mathbf{H}}{\partial t} = -\frac{1}{c^2}\frac{\partial^2 \mathbf{E}}{\partial t^2}.$$

Now $\operatorname{curl}\operatorname{curl}\mathbf{E} = \nabla \times (\nabla \times \mathbf{E}) = \nabla\nabla.\mathbf{E} - \nabla^2\mathbf{E} = -\nabla^2\mathbf{E}$ since $\nabla.\mathbf{E} = \operatorname{div}\mathbf{E} = 0$, and so the first result follows.

Also

$$\operatorname{curl}\operatorname{curl}\mathbf{H} = \frac{1}{c}\frac{\partial \operatorname{curl}\mathbf{E}}{\partial t} = -\frac{1}{c^2}\frac{\partial^2 \mathbf{H}}{\partial t^2}.$$

But $\operatorname{curl}\operatorname{curl}\mathbf{H} = -\nabla^2\mathbf{H}$ since $\operatorname{div}\mathbf{H} = 0$ and so the second result follows.

6.63 Show that the *Poynting vector* representing the energy flux

$$\mathbf{S} = \frac{c}{4\pi}\mathbf{E} \times \mathbf{H} \tag{6.30}$$

and the energy density of the electromagnetic field

$$U = \frac{KE^2}{8\pi} + \frac{\mu H^2}{8\pi} = \frac{\mathbf{D}.\mathbf{E}}{8\pi} + \frac{\mathbf{B}.\mathbf{H}}{8\pi} \tag{6.31}$$

satisfy

$$\oint_S \mathbf{S}.\hat{\mathbf{n}}\, dS + \int_R \mathbf{E}.\mathbf{j}\, d\tau = -\frac{\partial}{\partial t}\int_R U\, d\tau \tag{6.32}$$

where the region R is bounded by the closed surface S.

- *Solution*

From Maxwell's equations (6.29) we have

$$\mathbf{H}.\operatorname{curl}\mathbf{E} - \mathbf{E}.\operatorname{curl}\mathbf{H} + \frac{4\pi}{c}\mathbf{E}.\mathbf{j} = -\frac{1}{c}\left(\mathbf{E}.\frac{\partial \mathbf{D}}{\partial t} + \mathbf{H}.\frac{\partial \mathbf{B}}{\partial t}\right).$$

Now $\nabla.(\mathbf{E} \times \mathbf{H}) = \mathbf{H}.\operatorname{curl}\mathbf{E} - \mathbf{E}.\operatorname{curl}\mathbf{H}$ and so

$$\frac{c}{4\pi}\operatorname{div}(\mathbf{E} \times \mathbf{H}) + \mathbf{E}.\mathbf{j} = -\frac{1}{4\pi}\left(\mathbf{E}.\frac{\partial \mathbf{D}}{\partial t} + \mathbf{H}.\frac{\partial \mathbf{B}}{\partial t}\right) = -\frac{\partial}{\partial t}\left(\frac{KE^2}{8\pi} + \frac{\mu H^2}{8\pi}\right)$$

giving $\operatorname{div}\mathbf{S} + \mathbf{E}.\mathbf{j} = -\partial U/\partial t$.

Integrating over the region of volume space R we then get

$$\int_R \operatorname{div}\mathbf{S}\, d\tau + \int_R \mathbf{E}.\mathbf{j}\, d\tau = -\frac{\partial}{\partial t}\int_R U\, d\tau$$

from which the result follows by using Gauss's divergence theorem.

6.64 If V is a scalar potential and $\mathbf{A}$ is a vector potential such that

$$\mathbf{B} = \operatorname{curl}\mathbf{A}, \quad \mathbf{E} = -\nabla V - \frac{1}{c}\frac{\partial \mathbf{A}}{\partial t} \tag{6.33}$$

and we choose the *gauge* so that

$$\operatorname{div}\mathbf{A} + \frac{K\mu}{c}\frac{\partial V}{\partial t} = 0 \tag{6.34}$$

show that V and $\mathbf{A}$ satisfy the wave equations

$$\nabla^2 V - \frac{K\mu}{c^2}\frac{\partial^2 V}{\partial t^2} = -\frac{4\pi\rho}{K}, \quad \nabla^2\mathbf{A} - \frac{K\mu}{c^2}\frac{\partial^2 \mathbf{A}}{\partial t^2} = -\frac{4\pi\mu\mathbf{j}}{c} \tag{6.35}$$

• *Solution*

Since $\mathbf{D} = K\mathbf{E}$ and $\mathbf{B} = \mu\mathbf{H}$ we have

$$\mathbf{H} = \frac{1}{\mu}\operatorname{curl}\mathbf{A}, \quad \mathbf{D} = -K\left(\nabla V + \frac{1}{c}\frac{\partial\mathbf{A}}{\partial t}\right)$$

It follows from the Maxwell equation (iii) of (6.29) that

$$\frac{1}{\mu}\operatorname{curl}\operatorname{curl}\mathbf{A} = \frac{4\pi\mathbf{j}}{c} - \frac{K}{c}\nabla\frac{\partial V}{\partial t} - \frac{K}{c^2}\frac{\partial^2\mathbf{A}}{\partial t^2}$$

and from the Maxwell equation (i) that

$$\nabla^2 V + \frac{1}{c}\operatorname{div}\frac{\partial\mathbf{A}}{\partial t} = -\frac{4\pi\rho}{K}$$

Now using $\operatorname{curl}\operatorname{curl}\mathbf{A} = \nabla\operatorname{div}\mathbf{A} - \nabla^2\mathbf{A}$ and the gauge condition (6.34) we get the required wave equations (6.35) for the scalar and vector potentials.

6.65 If $\mathbf{P_0}$ and $\mathbf{I_0}$ are externally maintained polarizations such that $\mathbf{D} = K\mathbf{E} + 4\pi\mathbf{P_0}$ and $\mathbf{B} = \mu\mathbf{H} + 4\pi\mathbf{I_0}$ where the induced polarizations of the medium are included in the dielectric constant K and the permeability μ, verify that we may write

$$\mathbf{E} = \operatorname{grad}\operatorname{div}\mathbf{\Pi} - \frac{K\mu}{c^2}\frac{\partial^2\mathbf{\Pi}}{\partial t^2} - \frac{\mu}{c}\operatorname{curl}\frac{\partial\mathbf{\Pi}^*}{\partial t}$$

and

$$\mathbf{H} = \operatorname{grad}\operatorname{div}\mathbf{\Pi}^* - \frac{K\mu}{c^2}\frac{\partial^2\mathbf{\Pi}^*}{\partial t^2} + \frac{K}{c}\operatorname{curl}\frac{\partial\mathbf{\Pi}}{\partial t}$$

where the *Hertz vectors* $\mathbf{\Pi}$ and $\mathbf{\Pi}^*$ are solutions of the wave equations

$$\nabla^2\mathbf{\Pi} - \frac{K\mu}{c^2}\frac{\partial^2\mathbf{\Pi}}{\partial t^2} = -\frac{4\pi\mathbf{P_0}}{K}, \quad \nabla^2\mathbf{\Pi}^* - \frac{K\mu}{c^2}\frac{\partial^2\mathbf{\Pi}^*}{\partial t^2} = -\frac{4\pi\mathbf{I_0}}{\mu}$$

• *Solution*

We have

$$\operatorname{div}\mathbf{D} = K\operatorname{div}\mathbf{E} + 4\pi\operatorname{div}\mathbf{P_0} = K\nabla^2\operatorname{div}\mathbf{\Pi} - K\operatorname{div}\frac{K\mu}{c^2}\frac{\partial^2\mathbf{\Pi}}{\partial t^2} + 4\pi\operatorname{div}\mathbf{P_0} = 0$$

and so the Maxwell equation (i) of (6.29) with $\rho = 0$ is verified.
  Also

$$\operatorname{div}\mathbf{B} = \mu\operatorname{div}\mathbf{H} + 4\pi\operatorname{div}\mathbf{I_0} = \mu\nabla^2\operatorname{div}\mathbf{\Pi}^* - \mu\operatorname{div}\frac{K\mu}{c^2}\frac{\partial^2\mathbf{\Pi}^*}{\partial t^2} + 4\pi\operatorname{div}\mathbf{I_0} = 0$$

and so the Maxwell equation (ii) is verified.
  Further

$$\operatorname{curl}\mathbf{H} = -\frac{K\mu}{c^2}\operatorname{curl}\frac{\partial^2\mathbf{\Pi}^*}{\partial t^2} + \frac{K}{c}\frac{\partial}{\partial t}\nabla\times(\nabla\times\mathbf{\Pi}) = -\frac{K\mu}{c^2}\operatorname{curl}\frac{\partial^2\mathbf{\Pi}^*}{\partial t^2} + \frac{K}{c}\frac{\partial}{\partial t}\nabla\nabla.\mathbf{\Pi} - \frac{K}{c}\frac{\partial}{\partial t}\nabla^2\mathbf{\Pi}$$

and so we get

$$\operatorname{curl}\mathbf{H} = \frac{K}{c}\frac{\partial}{\partial t}\left(\operatorname{grad}\operatorname{div}\mathbf{\Pi} - \frac{\mu}{c}\operatorname{curl}\frac{\partial\mathbf{\Pi}^*}{\partial t} - \frac{K\mu}{c^2}\frac{\partial^2\mathbf{\Pi}}{\partial t^2}\right) = \frac{1}{c}\frac{\partial\mathbf{D}}{\partial t}$$

using the equations for $\mathbf{\Pi}$ and $\mathbf{D}$. Thus the Maxwell equation (iii) with $\mathbf{j} = 0$ is verified.

Lastly we have

$$\operatorname{curl} \mathbf{E} = -\frac{K\mu}{c^2} \operatorname{curl} \frac{\partial^2 \mathbf{\Pi}}{\partial t^2} - \frac{\mu}{c} \frac{\partial}{\partial t} \nabla \times (\nabla \times \mathbf{\Pi}^*) = -\frac{1}{c} \frac{\partial \mathbf{B}}{\partial t}$$

using a similar analysis to the above, and thus the Maxwell equation (iv) is verified.

6.66 Find the electric and magnetic fields *in vacuo* due to an oscillating dipole at the origin given by $\exp(-i\omega t)\mathbf{p}$ where $\mathbf{p}$ is a constant vector, and show that as $r \to \infty$

$$\mathbf{E} \to -\frac{\omega^2 p}{c^2 r} \exp[i(kr - \omega t)] \sin\theta \,\widehat{\theta}, \quad \mathbf{H} \to -\frac{\omega^2 p}{c^2 r} \exp[i(kr - \omega t)] \sin\theta \,\widehat{\phi}$$

• *Solution*

For this case the Hertz vector $\mathbf{\Pi}$ satisfies the equation

$$\nabla^2 \mathbf{\Pi} - \frac{1}{c^2} \frac{\partial^2 \mathbf{\Pi}}{\partial t^2} = -4\pi \exp(-i\omega t)\mathbf{p}\,\delta(\mathbf{r}).$$

where $\delta(\mathbf{r})$ is a generalized function known as the Dirac delta function which satisfies the conditions $\delta(\mathbf{r}) = 0$ for $\mathbf{r} \neq \mathbf{0}$ and $\int \psi(\mathbf{r}')\delta(\mathbf{r}')d\tau' = \psi(\mathbf{0})$, the integration being over all space.

Now the outgoing wave solution of the equation $(\nabla^2 + k^2)\mathbf{f}(\mathbf{r}) = \mathbf{F}(\mathbf{r})$ has the form $\mathbf{f}(\mathbf{r}) = \int G(\mathbf{r}, \mathbf{r}')\mathbf{F}(\mathbf{r}') \, d\tau'$ where the Green function is given by $G(\mathbf{r}, \mathbf{r}') = -(4\pi)^{-1} \exp(ik\,|\mathbf{r} - \mathbf{r}'|)/\,|\mathbf{r} - \mathbf{r}'|$. Hence for $\mathbf{F}(\mathbf{r}') = \mathbf{C}\delta(\mathbf{r}')$ we obtain $\mathbf{f}(\mathbf{r}) = -\mathbf{C}(4\pi r)^{-1} \exp(ikr)$ and thus

$$\mathbf{\Pi} = \frac{1}{r} \exp[i(kr - \omega t)]\,\mathbf{p}$$

where $k = \omega/c$.

Then we have

$$\mathbf{E} = \operatorname{grad}\operatorname{div}\mathbf{\Pi} + k^2\mathbf{\Pi}, \quad \mathbf{H} = -\frac{i\omega}{c} \operatorname{curl}\mathbf{\Pi}.$$

Now

$$\operatorname{grad}\operatorname{div}\mathbf{\Pi} = \exp(-i\omega t)\,\mathbf{p}.\nabla\left[\nabla\frac{\exp(ikr)}{r}\right]$$

and

$$\operatorname{curl}\mathbf{\Pi} = -\exp(-i\omega t)\,\mathbf{p} \times \nabla\left[\frac{\exp(ikr)}{r}\right].$$

Hence

$$\mathbf{E} = \exp(-i\omega t)\left\{\mathbf{p}.\nabla\left[\nabla\frac{\exp(ikr)}{r}\right] + k^2\mathbf{p}\frac{\exp(ikr)}{r}\right\}$$

and

$$\mathbf{H} = \frac{i\omega}{c} \exp(-i\omega t)\mathbf{p} \times \nabla\left[\frac{\exp(ikr)}{r}\right].$$

We now use spherical polar coordinates r, θ, ϕ taking $\mathbf{p}$ in the direction of the polar axis so that $\widehat{\mathbf{p}} = \widehat{\mathbf{r}}\cos\theta - \widehat{\theta}\sin\theta$ and

$$\nabla = \widehat{\mathbf{r}}\frac{\partial}{\partial r} + \widehat{\theta}\frac{1}{r}\frac{\partial}{\partial \theta} + \widehat{\phi}\frac{1}{r\sin\theta}\frac{\partial}{\partial \phi}.$$

We have $\mathbf{p}.\nabla\nabla f(r) = \mathbf{p}.\nabla(\mathbf{r}r^{-1}df/dr) = (r^{-1}df/dr)\mathbf{p}.\nabla\mathbf{r} + \mathbf{r}\,\mathbf{p}.\nabla(r^{-1}df/dr)$
$= (r^{-1}df/dr)\mathbf{p} + \mathbf{p}r\cos\theta\,d/dr(r^{-1}df/dr)$.
Then putting $f(r) = r^{-1}\exp(ikr)$ we get

$$\mathbf{E} = p\exp(-i\omega t)$$
$$\times \left\{\cos\theta\left[\frac{d^2}{dr^2}\left(\frac{\exp(ikr)}{r}\right) + k^2\frac{\exp(ikr)}{r}\right]\widehat{\mathbf{r}} - \frac{\sin\theta}{r}\left[\frac{d}{dr}\left(\frac{\exp(ikr)}{r}\right) + k^2\exp(ikr)\right]\widehat{\theta}\right\}$$

Also we have

$$\mathbf{H} = \frac{i\omega p}{c}\exp(-i\omega t)\sin\theta\frac{d}{dr}\left(\frac{\exp(ikr)}{r}\right)\widehat{\phi}$$

since $\widehat{\mathbf{r}} \times \widehat{\theta} = \widehat{\phi}$.

Now carrying out the differentiations with respect to r we obtain

$$\mathbf{E} = p\exp[i(kr - \omega t)]\left[2\cos\theta\left(\frac{1}{r^3} - \frac{ik}{r^2}\right)\widehat{\mathbf{r}} - \sin\theta\left(\frac{ik}{r^2} - \frac{1}{r^3} + \frac{k^2}{r}\right)\widehat{\theta}\right]$$

and

$$\mathbf{H} = \frac{i\omega p}{c}\exp[i(kr - \omega t)]\sin\theta\left(\frac{ik}{r} - \frac{1}{r^2}\right)\widehat{\phi}$$

which give the required results in the limit $r \to \infty$ since $k = \omega/c$.

6.67 A linear quadrupole having moment $a^2q\cos\omega t$ is composed of point charges $-q, 2q, -q$ at the points $z = -a$, $z = 0$, $z = a$. Find the electric and magnetic fields *in vacuo* at large distances and hence obtain the average rate at which energy is radiated from the quadrupole.

• *Solution*

This quadrupole is composed of two dipoles $p\cos\omega t$ and $-p\cos\omega t$ centred at $z = -a/2$ and $z = a/2$ respectively, where $p = qa$. Using the results obtained in the previous problem we see that for large r

$$E_\theta = H_\phi = -a\frac{\partial}{\partial z}\left[-\frac{\omega^2 p}{c^2 r}\cos(kr - \omega t)\sin\theta\right]$$

which, to leading order in $1/r$, yields

$$E_\theta = H_\phi = \frac{qa^2\omega^3}{c^3 r}\sin(kr - \omega t)\sin\theta\cos\theta.$$

Using the Poynting vector $(c/4\pi)\mathbf{E} \times \mathbf{H}$ we see that the rate at which energy is radiated is given by the integral over a spherical surface of large radius r:

$$\left(\frac{c}{4\pi}\right)2\pi r^2\int_0^\pi E_\theta H_\phi\sin\theta\,d\theta.$$

Hence the average rate at which energy is radiated is

$$\frac{q^2a^4\omega^6}{c^6}\frac{c}{4}\int_0^\pi\sin^3\theta\cos^2\theta\,d\theta = \frac{q^2a^4\omega^6}{c^6}\frac{c}{15}$$

which may be rewritten in terms of the wavelength $\lambda = 2\pi/k$ of the radiation as

$$\frac{64\pi^6 a^4 q^2 c}{15\lambda^6}.$$

7
FLUID DYNAMICS

The equation obtained by Leonhard Euler (1707-1783) which describes the motion of an *inviscid fluid* is

$$\frac{d\mathbf{v}}{dt} = \mathbf{F} - \frac{1}{\rho}\nabla p \qquad (7.1)$$

where $\mathbf{v}$ is the velocity of the fluid referred to an inertial frame of reference, ρ is the density of the fluid, p is the pressure in the fluid, and $\mathbf{F}$ is the force per unit mass acting on the fluid.

Also *conservation of mass* gives the *continuity equation*

$$\frac{\partial \rho}{\partial t} + \operatorname{div}(\rho\mathbf{v}) = 0 \qquad (7.2)$$

which may be rewritten in the alternative form

$$\frac{d\rho}{dt} + \rho\operatorname{div}\mathbf{v} = 0 \qquad (7.3)$$

using $d\rho/dt = \partial\rho/\partial t + \mathbf{v}.\nabla\rho$.

For an incompressible fluid $d\rho/dt = 0$ and so the continuity equation then becomes $\operatorname{div}\mathbf{v} = 0$.

7.01 A liquid is in equilibrium under the action of an external force

$$\mathbf{F} = a[(y+z)\mathbf{i} + (z+x)\mathbf{j} + (x+y)\mathbf{k}]$$

where a is a constant. Find the equations of the surfaces of equal pressure.

• *Solution*

Since the liquid is in equilibrium we have $d\mathbf{v}/dt = \mathbf{0}$ and since a liquid is incompressible we have that ρ is constant. Hence we may write Euler's equation (7.1) for this problem as

$$\mathbf{F} = \nabla\left(\frac{p}{\rho}\right).$$

In rectangular Cartesian coordinates x, y, z we have

$$a(y+z) = \frac{\partial}{\partial x}\left(\frac{p}{\rho}\right), \quad a(z+x) = \frac{\partial}{\partial y}\left(\frac{p}{\rho}\right), \quad a(x+y) = \frac{\partial}{\partial z}\left(\frac{p}{\rho}\right)$$

and so on integrating we obtain

$$\begin{aligned} p/\rho &= ax(y+z) + f(y,z) \\ &= ay(z+x) + g(z,x) \\ &= az(x+y) + h(x,y) \end{aligned}$$

where $f(y,z), g(z,x), h(x,y)$ are functions to be determined.

For these equations to be consistent we must have

$$\frac{p}{\rho} = a(xy + yz + zx) + b$$

where b is a constant

Thus the surfaces of equal pressure are given by

$$xy + yz + zx = c$$

where c is a constant, and these are hyperboloids.

7.02 For a conservative system of external forces obtain Helmholtz's formula

$$\frac{d}{dt}\left(\frac{\omega}{\rho}\right) = \frac{1}{\rho}\omega.\nabla\mathbf{v} \tag{7.4}$$

where ρ is the density of the fluid and $\omega = \nabla \times \mathbf{v}$ is the *vorticity*.

• *Solution*

First we note that $\mathbf{v} \times (\nabla \times \mathbf{v}) = \frac{1}{2}\nabla(\mathbf{v}.\mathbf{v}) - \mathbf{v}.\nabla\mathbf{v}$ and so we have for the total derivative following the fluid

$$\frac{d\mathbf{v}}{dt} = \frac{\partial\mathbf{v}}{\partial t} + \mathbf{v}.\nabla\mathbf{v} = \frac{\partial\mathbf{v}}{\partial t} + \frac{1}{2}\nabla\mathbf{v}^2 + \omega \times \mathbf{v}$$

where $\omega = \nabla \times \mathbf{v}$ is the *vorticity*.

Since the external forces are conservative we may put $\mathbf{F} = -\nabla\Omega$ and hence we may write Euler's equation (7.1) as

$$\frac{\partial\mathbf{v}}{\partial t} + \omega \times \mathbf{v} = -\nabla\left(\frac{1}{2}v^2 + \Omega\right) - \frac{1}{\rho}\nabla p. \tag{7.5}$$

Now $\nabla P = \rho^{-1}\nabla p$ where $P = \int^p \rho^{-1}dp$ and so Euler's equation can be rewritten in the form

$$\frac{\partial\mathbf{v}}{\partial t} + \omega \times \mathbf{v} = -\nabla U \tag{7.6}$$

where $U = \Omega + \frac{1}{2}v^2 + P$. Hence taking the vector product with ∇ we get

$$\frac{\partial\omega}{\partial t} + \nabla \times (\omega \times \mathbf{v}) = \mathbf{0}.$$

But

$$\nabla \times (\omega \times \mathbf{v}) = \omega\nabla.\mathbf{v} + \mathbf{v}.\nabla\omega - \mathbf{v}\nabla.\omega - \omega.\nabla\mathbf{v} \tag{7.7}$$

where $\nabla.\omega = \nabla.(\nabla \times \mathbf{v}) = 0$ and so we have

$$\frac{\partial\omega}{\partial t} + \mathbf{v}.\nabla\omega + \omega\nabla.\mathbf{v} = \omega.\nabla\mathbf{v}$$

Now

$$\frac{d}{dt}\left(\frac{\omega}{\rho}\right) = \omega\frac{d}{dt}\left(\frac{1}{\rho}\right) + \frac{1}{\rho}\frac{d\omega}{dt} = -\frac{\omega}{\rho^2}\frac{d\rho}{dt} + \frac{1}{\rho}\frac{d\omega}{dt}$$

and the continuity equation (7.3) gives $d\rho/dt = -\rho\,\mathrm{div}\,\mathbf{v}$ so that

$$\rho\frac{d}{dt}\left(\frac{\omega}{\rho}\right) = \omega\nabla.\mathbf{v} + \frac{\partial\omega}{\partial t} + \mathbf{v}.\nabla\omega$$

since

$$\frac{d\omega}{dt} = \frac{\partial\omega}{\partial t} + \mathbf{v}.\nabla\omega,$$

which leads to the Helmholtz formula (7.4).

7.03 Show that in the steady motion of an incompressible fluid with velocity **v** under the action of conservative forces

$$\mathbf{v}.\nabla\omega - \omega.\nabla\mathbf{v} = 0$$

where $\omega = \nabla \times \mathbf{v}$.

- *Solution*

Since the external forces are conservative we may put $\mathbf{F} = -\nabla\Omega$ and hence we may rewrite Euler's equation in the form (7.5).

For the steady motion of an incompressible fluid $\partial\mathbf{v}/\partial t = 0$ and ρ is a constant. Thus in the present problem we have

$$\omega \times \mathbf{v} = -\nabla\left(\frac{1}{2}v^2 + \Omega + \frac{p}{\rho}\right) \tag{7.8}$$

and this gives $\nabla \times (\omega \times \mathbf{v}) = 0$ since $\nabla \times \nabla U = 0$. Now

$$\nabla \times (\omega \times \mathbf{v}) = \omega\nabla.\mathbf{v} + \mathbf{v}.\nabla\omega - \omega.\nabla\mathbf{v} - \mathbf{v}\nabla.\omega$$

But for an incompressible fluid the continuity equation is $\nabla.\mathbf{v} = 0$. Further $\nabla.\omega = 0$ and so the required result follows.

7.04 If the velocity potential in a fluid is given by $\phi = a(x+y)t + bxy$ where t is the time, determine the equation of the stream lines and show that the stream lines are not the same as the path lines.

- *Solution*

The *stream lines* are given by

$$\frac{dx}{u} = \frac{dy}{v} = ds$$

where ds is an element of arc and the components of the velocity vector $\mathbf{v} = -\nabla\phi$ are given by

$$u = -\frac{\partial\phi}{\partial x} = -(at + by), \quad v = -\frac{\partial\phi}{\partial y} = -(at + bx).$$

It follows that at a fixed time t the stream lines are given by

$$\frac{dx}{ds} = u = -(at + by), \quad \frac{dy}{ds} = v = -(at + bx).$$

Hence

$$\frac{d^2x}{ds^2} - b^2x = abt$$

whose solution may be written

$$x = -\frac{at}{b} + \alpha(t)\exp(-bs) + \beta(t)\exp(bs)$$

where α and β are arbitrary functions of time and this gives

$$y = -\frac{a}{b}t + \alpha(t)\exp(-bs) - \beta(t)\exp(bs).$$

Suppose that $x = x_0$, $y = y_0$ at the point $s = 0$ at time t. Then

$$x = -\frac{at}{b} + \left[\frac{at}{b} + \frac{1}{2}(x_0 + y_0)\right]\exp(-bs) + \frac{1}{2}(x_0 - y_0)\exp(bs),$$

$$y = -\frac{at}{b} + \left[\frac{at}{b} + \frac{1}{2}(x_0 + y_0)\right]\exp(-bs) - \frac{1}{2}(x_0 - y_0)\exp(bs)$$

for the stream lines at time t.

Hence the equation of a stream line at time t is

$$\left(x + \frac{at}{b}\right)^2 - \left(y + \frac{at}{b}\right)^2 = 2(x_0 - y_0)\left[\frac{at}{b} + \frac{1}{2}(x_0 + y_0)\right]$$

which is a rectangular hyperbola with moving centre $(-at/b, -at/b)$ and moving foci.

Path lines are given by

$$\frac{dx}{dt} = u = -(at + by), \quad \frac{dy}{dt} = v = -(at + bx)$$

and so we have

$$\frac{d^2x}{dt^2} - b^2x = -a(1 - bt)$$

whose solution may be written in the form

$$x = \frac{a}{b^2}(1 - bt) + c\exp(-bt) + d\exp(bt)$$

where c and d are constants, which gives also

$$y = -\frac{1}{b}\left(\frac{dx}{dt} + at\right) = \frac{a}{b^2}(1 - bt) + c\exp(-bt) - d\exp(bt).$$

Hence the equation of the path lines is

$$\left(x - \frac{a}{b^2} + \frac{at}{b}\right)^2 - \left(y - \frac{a}{b^2} + \frac{at}{b}\right)^2 = 4cd$$

which is a rectangular hyperbola with moving centre at $(a/b^2 - at/b, a/b^2 - at/b)$.

We see that the path lines and the stream lines are different.

If $a = 0$ so that there is no dependence on the time t, the stream lines and the path lines coincide and become the rectangular hyperbolas

$$x^2 - y^2 = x_0^2 - y_0^2.$$

7.05 A liquid rotates with constant angular velocity Ω about a vertical axis under the action of gravity. Show that the vorticity at any point of the fluid is 2Ω and that the free surface is a paraboloid of revolution with its axis vertical.

• *Solution*

For a liquid rotating with angular velocity $\boldsymbol{\Omega}$ the velocity is given by $\mathbf{v} = \boldsymbol{\Omega} \times \mathbf{r}$ and so the vorticity $\boldsymbol{\omega} = \nabla \times \mathbf{v}$ becomes

$$\boldsymbol{\omega} = \nabla \times (\boldsymbol{\Omega} \times \mathbf{r}) = 2\boldsymbol{\Omega}$$

using (iii) of (4.2).

For the steady motion of a liquid under gravity (7.8) gives

$$\boldsymbol{\omega} \times \mathbf{v} = -\nabla \left(\Omega + \frac{1}{2}v^2 + \frac{p}{\rho} \right)$$

where $\Omega = gz$, taking the z-axis in the direction of the upward vertical, and ρ is the density.

Now

$$\boldsymbol{\omega} \times \mathbf{v} = 2\boldsymbol{\Omega} \times (\boldsymbol{\Omega} \times \mathbf{r}) = 2\boldsymbol{\Omega}\boldsymbol{\Omega}.\mathbf{r} - 2\Omega^2 \mathbf{r}$$

and

$$v^2 = (\boldsymbol{\Omega} \times \mathbf{r}).(\boldsymbol{\Omega} \times \mathbf{r}) = \boldsymbol{\Omega}.[\mathbf{r} \times (\boldsymbol{\Omega} \times \mathbf{r})] = \Omega^2 r^2 - (\boldsymbol{\Omega}.\mathbf{r})^2.$$

Setting $\boldsymbol{\Omega} = \Omega \mathbf{k}$ and $\mathbf{r} = x\mathbf{i} + y\mathbf{j} + z\mathbf{k}$ we obtain

$$\boldsymbol{\omega} \times \mathbf{v} = -2\Omega^2(x\mathbf{i} + y\mathbf{j}), \quad v^2 = \Omega^2(x^2 + y^2).$$

Hence

$$\frac{\partial}{\partial x}\left(\Omega + \frac{1}{2}v^2 + \frac{p}{\rho} \right) = 2\Omega^2 x, \quad \frac{\partial}{\partial y}\left(\Omega + \frac{1}{2}v^2 + \frac{p}{\rho} \right) = 2\Omega^2 y, \quad \frac{\partial}{\partial z}\left(\Omega + \frac{1}{2}v^2 + \frac{p}{\rho} \right) = 0$$

and so

$$\Omega + \frac{1}{2}v^2 + \frac{p}{\rho} = \Omega^2(x^2 + y^2) + C$$

where C is a constant. It follows that

$$gz + \frac{p}{\rho} = \frac{1}{2}\Omega^2(x^2 + y^2) + C.$$

At the free surface the pressure is a constant and equal to the atmospheric pressure p_A and so the equation of the free surface is

$$gz + \frac{p_A - p_0}{\rho} = \frac{1}{2}\Omega^2(x^2 + y^2)$$

where p_0 is the pressure at the origin. This is a paraboloid of revolution.

The fact that the free surface of a rotating liquid assumes the shape of a paraboloid of revolution has been used to make telescope mirrors.

7.06 Prove that the kinetic energy of an incompressible fluid whose motion is irrotational with velocity potential ϕ can be written

$$T = \frac{1}{2}\rho \int_S \phi \frac{\partial \phi}{\partial n} \, dS \tag{7.9}$$

where ρ is the density of the fluid and $\hat{\mathbf{n}}$ has the direction of the outward normal to the boundary surface S.

- *Solution*

The kinetic energy is given by

$$T = \frac{1}{2}\rho \int_R \mathbf{v}^2 \, d\tau$$

where the region R is bounded by the surface S. Since $\mathbf{v} = -\nabla\phi$ we have

$$T = \frac{1}{2}\rho \int_R (\nabla\phi)^2 \, d\tau$$

But $\nabla.(\phi\nabla\phi) = (\nabla\phi)^2 + \phi\nabla^2\phi$ and for an incompressible fluid the continuity equation is div $\mathbf{v} = 0$ giving $\nabla^2\phi = 0$. Hence

$$T = \frac{1}{2}\rho \int \operatorname{div}(\phi\nabla\phi) \, d\tau$$

and so by Gauss's divergence theorem we get

$$T = \frac{1}{2}\rho \int_S (\phi\nabla\phi).\hat{\mathbf{n}} dS$$

which gives the required result.

7.07 Find the kinetic energy of the incompressible fluid confined between two spherical surfaces of radii a and b with $a > b$ having their common centre at the origin, resulting from the presence of a doublet $\mathbf{M}$ at the origin.

• *Solution*

The velocity potential for a doublet is

$$\phi = \frac{\mathbf{M}.\mathbf{r}}{r^3} = \frac{M\cos\theta}{r^2}$$

taking the origin of the polar coordinates r, θ at the doublet and the polar axis in the direction of the dipole. Then $\partial\phi/\partial r = -2Mr^{-3}\cos\theta$.

At the inner surface $r = b$ the outward normal is $\hat{\mathbf{n}} = -\hat{\mathbf{r}}$ and at the outer surface $\hat{\mathbf{n}} = \hat{\mathbf{r}}$. Hence, using the formula (7.9) obtained in the previous problem, we find that

$$T = -2\pi\rho M^2 \left(\frac{1}{a^3} - \frac{1}{b^3}\right) \int_0^\pi \cos^2\theta \sin\theta \, d\theta = \frac{4\pi}{3}\rho M^2 \left(\frac{a^3 - b^3}{a^3 b^3}\right).$$

7.08 An ideal gas obeying Boyle's Law $p = \kappa\rho$, where p is the pressure and ρ is the density, moves in a straight tube of uniform bore. If u is the speed of the gas at a distance x from a fixed point of the tube at time t show that

$$\frac{\partial^2 \rho}{\partial t^2} = \frac{\partial^2}{\partial x^2}[(u^2 + \kappa)\rho].$$

• Solution

This is a one-dimensional problem for which Euler's equation (7.1) and the equation of continuity (7.2) are respectively

$$\frac{du}{dt} = -\frac{1}{\rho}\frac{\partial p}{\partial x}, \quad \frac{\partial\rho}{\partial t} + \frac{\partial(u\rho)}{\partial x} = 0.$$

Hence

$$\frac{\partial^2 \rho}{\partial t^2} = -\frac{\partial}{\partial x}\frac{\partial(u\rho)}{\partial t}.$$

Now $du/dt = \partial u/\partial t + u\partial u/\partial x$ and thus $\partial u/\partial t = -\rho^{-1}\partial p/\partial x - u\partial u/\partial x$ so that

$$\frac{\partial(u\rho)}{\partial t} = u\frac{\partial \rho}{\partial t} + \rho\frac{\partial u}{\partial t} = -u\frac{\partial(u\rho)}{\partial x} - \frac{\partial p}{\partial x} - u\rho\frac{\partial u}{\partial x}.$$

By Boyle's law $p = \kappa\rho$ it follows that

$$-\frac{\partial(u\rho)}{\partial t} = \frac{\partial(u^2\rho)}{\partial x} + \kappa\frac{\partial \rho}{\partial x}$$

which leads to the required differential equation.

7.09 At time t the radius of an expanding sphere surrounded by an incompressible fluid of density ρ is given by $R(t)$. Show that the fluid pressure at the surface of the sphere is

$$p(t) = p_\infty + \frac{1}{2}\rho\left[3\left(\frac{dR}{dt}\right)^2 + 2R\frac{d^2R}{dt^2}\right]$$

where p_∞ is the pressure at infinity.

• *Solution*

Since the motion of the incompressible fluid is irrotational we have $\omega = 0$ and $\mathbf{v} = -\nabla\phi$ so that Euler's equation (7.6) becomes

$$\nabla\left[\frac{p(t)}{\rho} + \frac{1}{2}v^2 - \frac{\partial \phi}{\partial t}\right] = 0.$$

Hence

$$\frac{p(t)}{\rho} + \frac{1}{2}v^2 - \frac{\partial \phi}{\partial t} = F(t)$$

where F is an arbitrary function of the time t. At large distances away from the expanding sphere $p(t) = p_\infty, v = 0$ and $\partial\phi/\partial t = 0$. Thus we have

$$\frac{p(t) - p_\infty}{\rho} = \frac{\partial \phi}{\partial t} - \frac{1}{2}v^2.$$

Since the fluid is incompressible the continuity equation $\nabla.\mathbf{v} = 0$ gives $\nabla^2\phi = 0$. Now by spherical symmetry $\phi(t) = A(t)/r$ and so

$$\mathbf{v} = -\frac{\partial \phi}{\partial r}\hat{\mathbf{r}} = \frac{A(t)}{r^2}\hat{\mathbf{r}}.$$

At the surface of the expanding sphere $r = R(t)$ we have $v_r = dR/dt$ and so $A(t) = R^2 dR/dt$. Hence

$$\frac{\partial \phi}{\partial t} = \frac{1}{r}\frac{d}{dt}\left(R^2\frac{dR}{dt}\right)$$

and so at the surface of the expanding sphere we have

$$\frac{p(t) - p_\infty}{\rho} = \frac{1}{R}\frac{d}{dt}\left(R^2\frac{dR}{dt}\right) - \frac{1}{2}\left(\frac{dR}{dt}\right)^2 = \frac{3}{2}\left(\frac{dR}{dt}\right)^2 + R\frac{d^2R}{dt^2}$$

giving the required result.

7.10 Find the relation between the pressure p and the speed v of a compressible fluid undergoing steady irrotational motion under no external force, if the equation of state of the fluid has the form $p = \kappa \rho^\gamma$ where ρ is the density and κ and γ are constants.

Obtain the speed with which gas of density ρ_1 contained in a vessel at pressure p_1 passes through a small hole into an atmosphere of density ρ_0 at pressure p_0.

• *Solution*

For steady irrotational motion under no external forces Euler's equation (7.6) takes the form

$$\nabla \left(\frac{1}{2} v^2 + P \right) = 0$$

where $P = \int^p \rho^{-1} dp$. Hence

$$\frac{1}{2} v^2 + P = C$$

where C is a constant. Now using the equation of state $p = \kappa \rho^\gamma$ we get

$$P = \int^p \left(\frac{\kappa}{p} \right)^{1/\gamma} dp = \kappa^{1/\gamma} \frac{p^{1-1/\gamma}}{1 - 1/\gamma} = \frac{\gamma}{\gamma - 1} \frac{p}{\rho}$$

and so the relation between the pressure and the speed is

$$\frac{1}{2} v^2 + \frac{\gamma}{\gamma - 1} \frac{p}{\rho} = C.$$

We then have

$$\frac{2\gamma}{\gamma - 1} \frac{p_1}{\rho_1} = \frac{2\gamma}{\gamma - 1} \frac{p_0}{\rho_0} + u^2$$

and so

$$u = \sqrt{\frac{2\gamma}{\gamma - 1} \left(\frac{p_1}{\rho_1} - \frac{p_0}{\rho_0} \right)}.$$

7.11 Find the stream function for the two dimensional motion due to two equal line sources of strength m situated at the points $x = -a$, $y = 0$ and $x = a$, $y = 0$ respectively, and an equal sink at the origin.

• *Solution*

The complex potential $w = \phi + i\psi$ for a line source of strength m through a point z_0 is

$$w = -m \ln(z - z_0). \tag{7.10}$$

Hence the complex potential for the set of line sources in this problem is

$$w = -m \ln(z - a) - m \ln(z + a) + m \ln z = -m \ln \left(\frac{z^2 - a^2}{z} \right) = -m \ln \left(z - \frac{a^2}{z} \right).$$

Putting $z = r e^{i\theta}$ we get

$$w = -m \ln \left(r e^{i\theta} - \frac{a^2}{r} e^{-i\theta} \right)$$

and if we let $re^{i\theta} - a^2 r^{-1} e^{-i\theta} = \rho e^{i\chi}$ we obtain

$$(r - a^2 r^{-1}) \cos \theta = \rho \cos \chi, \quad (r + a^2 r^{-1}) \sin \theta = \rho \sin \chi.$$

It follows that the stream function, given by the imaginary part ψ of w, is

$$\psi = -m\chi = -m \tan^{-1} \left(\frac{r^2 + a^2}{r^2 - a^2} \tan \theta \right).$$

Since $r^2 = x^2 + y^2$ and $\tan \theta = y/x$ this may be rewritten as

$$\psi = -m \tan^{-1} \left(\frac{x^2 + y^2 + a^2}{x^2 + y^2 - a^2} \frac{y}{x} \right).$$

7.12 Two equal line sources of strengths m are situated at the points $x = a$, $y = 0$ and $x = -a$, $y = 0$. Show that $x = 0$ is a stream line and find the velocity of the fluid along it.

- *Solution*

The complex potential $w = \phi + i\psi$ for this pair of line sources is given by, using (7.10),

$$w = -m \ln(z - a) - m \ln(z + a) = -m \ln(z^2 - a^2) = -m \ln(x^2 - y^2 - a^2 + 2ixy).$$

A stream line is given by a constant value of ψ. On $x = 0$ we see that $\psi = 0$ and so it is a stream line.

Now $dw/dz = -u + iv$ where $u = -\partial\phi/\partial x = -\partial\psi/\partial y$ and $v = -\partial\phi/\partial y = \partial\psi/\partial x$ are the x and y components of the velocity vector respectively, and since

$$\frac{dw}{dz} = -\frac{2mz}{z^2 - a^2} = -\frac{2m(x + iy)}{x^2 - y^2 - a^2 + 2ixy}$$

we have $-u + iv = 2myi/(y^2 + a^2)$ on the stream line $x = 0$.

Hence $u = 0$ and $v = 2my/(a^2 + y^2)$. Thus the velocity of the fluid along the stream line $x = 0$ is $2my/(a^2 + y^2)$.

7.13 Find the equation of the stream lines for the two dimensional flow having complex potential given by $z = a \cos w$.

- *Solution*

Putting $z = x + iy$ and $w = \psi + i\psi$ in $z = a \cos w$ we find

$$x + iy = a \cos(\phi + i\psi) = a(\cos \phi \cos i\psi - \sin \phi \sin i\psi)$$

and since $\cos i\psi = \cosh \psi$ and $\sin i\psi = i \sinh \psi$ we obtain

$$x + iy = a(\cos \phi \cosh \psi - i \sin \phi \sinh \psi).$$

Hence $x = a \cos \phi \cosh \psi$ and $y = -a \sin \phi \sinh \psi$ giving

$$\frac{x^2}{a^2 \cosh^2 \psi} + \frac{y^2}{a^2 \sinh^2 \psi} = 1.$$

The stream lines are given by $\psi =$ constant. Thus they are ellipses with semi-axes $a \cosh \psi$ and $a \sinh \psi$.

7.14 Derive the velocity potential and the stream function corresponding to the complex potential

$$w = Az^{\pi/\alpha}$$

and show that it represents the irrotational motion of an incompressible fluid between two rigid plane boundaries inclined at an angle α.

- *Solution*

Putting $z = re^{i\theta}$ we have $w = \phi + i\psi = Ar^{\pi/\alpha}e^{i\pi\theta/\alpha}$ and so the velocity potential is

$$\phi = Ar^{\pi/\alpha} \cos(\pi\theta/\alpha)$$

and the stream function is

$$\psi = Ar^{\pi/\alpha} \sin(\pi\theta/\alpha).$$

Now $\psi = 0$ for $\theta = 0$ and $\theta = \alpha$. Thus $\theta = 0$ and $\theta = \alpha$ are stream lines corresponding to flow along the rigid plane boundaries inclined at the angle α.

7.15 Show that

$$w = ik \ln \left(\frac{z - ia}{z + ia} \right)$$

is the complex potential for a steady flow of liquid about a circular cylinder on one side of a rigid boundary wall at $y = 0$.

- *Solution*

Putting $w = \phi + i\psi$ we obtain

$$\phi + i\psi = ik \left[\ln(z - ia) - \ln(z + ia) \right]$$

and so the stream function is $\psi = k \ln(r_1/r_2)$ where r_1 and r_2 are the radial distances from the points on the imaginary axis at $y = ia$ and $y = -ia$. Hence

$$\frac{r_1}{r_2} = \exp \frac{\psi}{k}$$

and so any stream line $\psi =$ constant is given by taking $r_1/r_2 =$ constant which corresponds to one of a set of coaxial circles with limit points at $y = a, x = 0$ and $y = -a, x = 0$.
$r_1/r_2 = 1$ corresponds to the plane $y = 0$ while $r_1/r_2 = c$, where $c \neq 1$, corresponds to a cylinder with its axis though a point on the line $x = 0$. Thus we can take the circle $r_1/r_2 = c$ and the line $y = 0$ as stream lines so that the steady flow of liquid corresponds to that about a circular cylinder on one side of a rigid boundary wall.

Using Cartesian coordinates, the stream line $r_1/r_2 = c$ is given by

$$\frac{x^2 + (y - a)^2}{x^2 + (y + a)^2} = c^2.$$

This can be transformed into

$$x^2 + \left(y - a\frac{1 + c^2}{1 - c^2} \right)^2 = a^2 \left[\left(\frac{1 + c^2}{1 - c^2} \right)^2 - 1 \right]$$

which is a circle centred at $x = 0$, $y = a(1 + c^2)/(1 - c^2)$ and having radius $a[(1 + c^2)^2/(1 - c^2)^2 - 1]^{1/2}$. If $c < 1$ the circle is on the side $y > 0$ and if $c > 1$ the circle is on the side $y < 0$.

We shall now consider some problems involving *viscous fluids*.

The equation governing the motion of an incompressible fluid with kinematic viscosity ν is the *Navier-Stokes equation*

$$\frac{d\mathbf{v}}{dt} = \mathbf{F} - \frac{1}{\rho}\nabla p + \nu\nabla^2\mathbf{v}. \tag{7.11}$$

If $\nu = 0$ this reduces to Euler's equation (7.1).

7.16 Show that for an incompressible viscous fluid with kinematic viscosity ν moving with velocity $\mathbf{v}$

$$\frac{d\omega}{dt} = \omega.\nabla\mathbf{v} + \nu\nabla^2\omega + \nabla \times \mathbf{F}$$

where $\omega = \nabla \times \mathbf{v}$ is the vorticity and $\mathbf{F}$ is the force acting on the fluid per unit mass.

- *Solution*

Starting from the Navier-Stokes equation (7.11) in the form

$$\frac{\partial\mathbf{v}}{\partial t} + \omega \times \mathbf{v} = \mathbf{F} - \nabla\left(\frac{1}{2}v^2 + \frac{p}{\rho}\right) + \nu\nabla^2\mathbf{v}$$

and taking the vector product with ∇ we obtain

$$\frac{\partial\omega}{\partial t} + \nabla \times (\omega \times \mathbf{v}) = \nabla \times \mathbf{F} + \nu\nabla^2\omega.$$

Since the continuity equation for an incompressible fluid is $\nabla.\mathbf{v} = 0$, and also $\nabla.\omega = 0$, it follows that $\nabla \times (\omega \times \mathbf{v}) = \omega\nabla.\mathbf{v} + \mathbf{v}.\nabla\omega - \omega.\nabla\mathbf{v} - \mathbf{v}\nabla.\omega = \mathbf{v}.\nabla\omega - \omega.\nabla\mathbf{v}$. Now $d\omega/dt = \partial\omega/\partial t + \mathbf{v}.\nabla\omega$ and so the required result is obtained.

7.17 Show that for the steady flow of a viscous liquid under the action of a conservative force $\mathbf{F} = -\nabla\Omega$ per unit mass of liquid

$$\mathbf{v} \times \omega = \nabla\left(\Omega + \frac{1}{2}v^2 + \frac{p}{\rho}\right) + \nu\nabla \times \omega$$

and deduce that

$$\nabla \times (\mathbf{v} \times \omega) = \omega.\nabla\mathbf{v} - \mathbf{v}.\nabla\omega = -\nu\nabla^2\omega.$$

- *Solution*

Since $\partial\mathbf{v}/\partial t = 0$ for steady motion, and setting $\mathbf{F} = -\nabla\Omega$, the Navier-Stokes equation takes the form

$$\mathbf{v} \times \omega = \nabla\left(\Omega + \frac{1}{2}v^2 + \frac{p}{\rho}\right) - \nu\nabla^2\mathbf{v}.$$

Now $\nabla \times \omega = \nabla \times (\nabla \times \mathbf{v}) = \nabla\nabla.\mathbf{v} - \nabla^2\mathbf{v} = -\nabla^2\mathbf{v}$ since the continuity equation for a liquid is $\nabla.\mathbf{v} = 0$ and so the first result is obtained.

It then follows that

$$\nabla \times (\mathbf{v} \times w) = -\nu \nabla^2 w$$

and so we obtain the second result since we know that $\nabla \times (\mathbf{v} \times w) = w.\nabla\mathbf{v} - \mathbf{v}.\nabla w$ from the previous problem.

7.18 Show that for the steady two-dimensional flow of a viscous incompressible fluid under the action of a conservative force, we have

$$\frac{\partial \nabla^2 \psi}{\partial x} \frac{\partial \psi}{\partial y} - \frac{\partial \nabla^2 \psi}{\partial y} \frac{\partial \psi}{\partial x} = -\nu \nabla^2 \nabla^2 \psi$$

where ψ is the stream function satisfying

$$\mathbf{v} = -\frac{\partial \psi}{\partial y}\mathbf{i} + \frac{\partial \psi}{\partial x}\mathbf{j}$$

- *Solution*

We have seen in the previous problem that $\mathbf{v}.\nabla w - w.\nabla\mathbf{v} = \nu\nabla^2 w$. Now

$$w = \nabla \times \mathbf{v} = \begin{vmatrix} \mathbf{i} & \mathbf{j} & \mathbf{k} \\ \frac{\partial}{\partial x} & \frac{\partial}{\partial y} & \frac{\partial}{\partial z} \\ v_x & v_y & v_z \end{vmatrix} = \mathbf{k}\left(\frac{\partial^2 \psi}{\partial x^2} + \frac{\partial^2 \psi}{\partial y^2}\right) = \mathbf{k}\nabla^2\psi$$

since $v_z = 0$ and $\partial v_x/\partial z = \partial v_y/\partial z = 0$ for two-dimensional flow perpendicular to the z axis, and $v_x = -\partial\psi/\partial y$, $v_y = \partial\psi/\partial x$.
Also $w.\nabla\mathbf{v} = \nabla^2\psi\,\partial\mathbf{v}/\partial z = \mathbf{0}$ and

$$\mathbf{v}.\nabla w = \left(v_x\frac{\partial}{\partial x} + v_y\frac{\partial}{\partial y}\right)\nabla^2\psi\,\mathbf{k} = \left(-\frac{\partial\psi}{\partial y}\frac{\partial\nabla^2\psi}{\partial x} + \frac{\partial\psi}{\partial x}\frac{\partial\nabla^2\psi}{\partial y}\right)\mathbf{k}$$

which leads to the required result.

7.19 An incompressible viscous fluid flows with steady motion through an horizontal pipe of length l having elliptical cross section given by

$$\frac{x^2}{a^2} + \frac{y^2}{b^2} = 1.$$

Show that the velocity of the fluid along the pipe is

$$w(x,y) = \frac{\Delta p\, a^2 b^2}{2l\mu(a^2 + b^2)}\left(1 - \frac{x^2}{a^2} - \frac{y^2}{b^2}\right)$$

where $\mu = \nu\rho$ is the dynamic viscosity and Δp is the pressure difference between the ends of the pipe.
Further show that the total volume of fluid discharged per unit time through the tube is given by

$$Q = \frac{\pi a^3 b^3 \Delta p}{4\mu l(a^2 + b^2)}.$$

- *Solution*

The differential equation for the speed $w(x,y)$ of the flow of an incompressible viscous fluid through a pipe which is perpendicular to the x, y plane is

$$\frac{\partial^2 w}{\partial x^2} + \frac{\partial^2 w}{\partial y^2} = -\frac{\Delta p}{l\mu} \qquad (7.12)$$

where $w(x,y) = 0$ over the inner surface of the pipe since it is assumed that the fluid has zero velocity at this boundary surface.

Separating the variables by putting $w(x,y) = X(x) + Y(y)$ we obtain

$$\frac{d^2 X}{dx^2} = -\frac{d^2 Y}{dy^2} - \frac{\Delta p}{l\mu},$$

and for this to hold each side of the equation must be equal to a constant $-k$. Since $w(x,y) = 0$ at $x^2/a^2 + y^2/b^2 = 1$ we see that we can put

$$w(x,y) = c\left(1 - \frac{x^2}{a^2} - \frac{y^2}{b^2}\right).$$

Now substituting into the differential equation for w gives

$$c = \frac{\Delta p \, a^2 b^2}{2l\mu(a^2 + b^2)}$$

and so the formula for $w(x,y)$ follows at once.

The total volume of fluid discharged per unit time through the tube is given by

$$Q = \int_S w(x,y) \, dS$$

where the integration is carried out over the perpendicular cross section S of the tube. Hence we have

$$Q = \frac{\Delta p \, a^2 b^2}{2l\mu(a^2 + b^2)} \int_S \left(1 - \frac{x^2}{a^2} - \frac{y^2}{b^2}\right) dS.$$

We know that $\int_S dS = \pi ab$ since the area of the ellipse is πab. Further, since we may take $dS = 2ydx$, we have

$$\frac{1}{a^2} \int x^2 dS = \frac{1}{a^2} \int_{-a}^{a} x^2 \, 2b\sqrt{1 - \frac{x^2}{a^2}} \, dx = 2ab \int_{-\pi/2}^{\pi/2} \sin^2\theta \cos^2\theta \, d\theta$$

setting $x = a \sin\theta$. Carrying out the integration over θ then gives $\pi ab/4$.

Similarly

$$\frac{1}{b^2} \int y^2 dS = \frac{\pi ab}{4}$$

and so the required expression for Q is obtained.

8
CLASSICAL DYNAMICS

At this stage of the book we shall examine some more advanced dynamical problems than in Chapter 3.

All problems in dynamics can be solved by using Newton's second law of motion. However for complex problems which have to be set up by using a curvilinear coordinate system such as, for example, the spinning top with the three Euler angles θ, ϕ, ψ, it can be advantageous to use the equations obtained by Joseph Louis Lagrange (1736-1813) involving generalized coordinates q_r, or perhaps Hamilton's equations involving generalized coordinates q_r and momenta p_r, as we shall see in the following sections. This can be rather more convenient than using Newton's second law of motion because all that is needed is to be able to write down the kinetic energy T and the potential energy V to give the Lagrangian function $L = T - V$ or the Hamiltonian function $H = T + V$ for a dynamical system which is subject to conservative forces.

Also dynamical problems can be simplified, although not usually made more tractable, by transforming to a different coordinate system by using a contact transformation. This method leads to the Hamilton-Jacobi equation which enables all problems in classical dynamics to be unified within the Hamilton-Jacobi theory. This does not make the problems easier to solve but it does provide a general framework for the solution of all dynamics problems. Moreover the Hamilton-Jacobi equation is the classical high frequency or small wave length approximation to the Schrödinger equation in quantum theory and is of great significance because of this role in applied mathematics.

We shall also consider some problems on vibrations which we will solve by using Lagrange's equations.

8-1 Lagrange's equations

The first set of problems which we shall investigate make use of *Lagrange's equations*.

We consider a dynamical system which is completely specified by a set of n generalized coordinates $q_1, q_2, ..., q_n$. Then *Lagrange's equations* for the dynamical system are

$$\frac{d}{dt}\left(\frac{\partial T}{\partial \dot{q}_r}\right) - \frac{\partial T}{\partial q_r} = Q_r \quad (r = 1, ..., n) \tag{8.1}$$

where $T(q_r, \dot{q}_r)$ is the kinetic energy of the dynamical system and Q_r is the nth generalized force.

For a conservative system with $Q_r = -\partial V/\partial q_r$ where $V(q_r)$ is the potential energy, these equations may be rewritten in terms of the *Lagrangian function $L = T - V$* in the form

$$\frac{d}{dt}\left(\frac{\partial L}{\partial \dot{q}_r}\right) = \frac{\partial L}{\partial q_r} \quad (r = 1, ..., n). \tag{8.2}$$

The following are an elementary set of problems to illustrate the method.

8.01 Use Lagrange's equations to find the equations of motion for:
(i) a particle of mass m falling freely under gravity,
(ii) a simple pendulum,

(iii) a spherical pendulum,

(iv) a particle moving under the action of a central force with potential $V(r)$,

(v) a particle moving under the action of gravity on the inner smooth surface of a cone of semi-angle α whose axis is vertical and vertex downwards.

- *Solution*

(i) A particle falling freely under gravity.

Let x, y, z be the Cartesian coordinates of the particle referred to an inertial frame of reference with z axis chosen vertically upwards. Then the kinetic energy of the particle having mass m is $T = \frac{1}{2}m(\dot{x}^2 + \dot{y}^2 + \dot{z}^2)$ and the potential energy is $V = mgz$+constant where g is the acceleration of gravity.

Lagrange's equation for the z coordinate is

$$\frac{d}{dt}\left(\frac{\partial T}{\partial \dot{z}}\right) - \frac{\partial T}{\partial z} = -\frac{\partial V}{\partial z}$$

which gives $m\ddot{z} = -mg$, that is $\ddot{z} = -g$, while for the x and y coordinates we get $\ddot{x} = 0$ and $\ddot{y} = 0$.

These equations are, of course, given directly by Newton's second law of motion.

(ii) A simple pendulum.

Suppose the string of the pendulum makes an angle θ with the downward vertical. Then the kinetic energy of the pendulum bob is $T = \frac{1}{2}ml^2\dot{\theta}^2$ where m is the mass of the bob and l is the length of the pendulum string. The potential energy of the pendulum bob is $V = -mgl\cos\theta$+constant.

Lagrange's equation for the generalized coordinate θ is

$$\frac{d}{dt}\left(\frac{\partial T}{\partial \dot{\theta}}\right) - \frac{\partial T}{\partial \theta} = -\frac{\partial V}{\partial \theta}$$

and this gives $ml^2\ddot{\theta} = -mgl\sin\theta$, that is the usual equation $\ddot{\theta} + (g/l)\sin\theta = 0$.

(iii) A spherical pendulum.

Here we have a particle of mass m moving under gravity on the inner smooth surface of a sphere of radius a.

We take a spherical polar system of coordinates with origin at the centre of the sphere and axis vertically downwards. Then θ is the angle which the radius vector to the particle makes with the downward vertical and ϕ is the azimuthal angle about the vertical.

The kinetic energy of the particle is $T = \frac{1}{2}m(a^2\dot{\theta}^2 + a^2\sin^2\theta\,\dot{\phi}^2)$ and the potential energy is $V = -mga\cos\theta$+constant. Lagranges equations are

$$\frac{d}{dt}\left(\frac{\partial T}{\partial \dot{\theta}}\right) - \frac{\partial T}{\partial \theta} = -\frac{\partial V}{\partial \theta}, \quad \frac{d}{dt}\left(\frac{\partial T}{\partial \dot{\phi}}\right) - \frac{\partial T}{\partial \phi} = -\frac{\partial V}{\partial \phi}$$

which give $ma^2\ddot{\theta} - ma^2\sin\theta\cos\theta\,\dot{\phi}^2 = -mga\sin\theta$ and $d(ma^2\sin^2\theta\,\dot{\phi})/dt = 0$ respectively. The second equation corresponds to conservation of angular momentum about the vertical axis. From this equation we get $a^2\sin^2\theta\,\dot{\phi} = h$ where h is a constant, and so

$$\ddot{\theta} - \frac{h^2\cos\theta}{a^4\sin^3\theta} = -\frac{g}{a}\sin\theta.$$

Using $\ddot{\theta} = d\left(\frac{1}{2}\dot{\theta}^2\right)/d\theta$ and integrating with respect to θ yields

$$\dot{\theta}^2 + \frac{h^2}{a^4 \sin^2\theta} = C + \frac{2g}{a}\cos\theta$$

where C is a constant. This is equivalent to conservation of energy $T + V = ma^2 C/2$.

(iv) A particle moving under the action of a central force having potential $V(r)$.
The kinetic energy of a particle having velocity $\mathbf{v}$ is

$$T = \frac{1}{2}m\mathbf{v}^2 = \frac{1}{2}m(\dot{x}^2 + \dot{y}^2 + \dot{z}^2)$$

where $x = r\sin\theta\cos\phi, y = r\sin\theta\sin\phi, z = r\cos\theta$ in spherical polar coordinates r, θ, ϕ so that in these coordinates the kinetic energy is given by

$$T = \frac{1}{2}m(\dot{r}^2 + r^2\dot{\theta}^2 + r^2\sin^2\theta\,\dot{\phi}^2).$$

Hence the Lagrange equation for the coordinate r is

$$\frac{d}{dt}\left(\frac{\partial L}{\partial \dot{r}}\right) = \frac{\partial L}{\partial r}$$

which gives the equation of motion

$$m(\ddot{r} - r\dot{\theta}^2 - r\sin^2\theta\,\dot{\phi}^2) = -\frac{dV}{dr}.$$

Also the Lagrange equation for the coordinate θ is

$$\frac{d}{dt}\left(\frac{\partial L}{\partial \dot{\theta}}\right) = \frac{\partial L}{\partial \theta}$$

which gives

$$\frac{d}{dt}(r^2\dot{\theta}) - r^2\sin\theta\cos\theta\,\dot{\phi}^2 = 0.$$

Further the Lagrange equation for the coordinate ϕ is

$$\frac{d}{dt}\left(\frac{\partial L}{\partial \dot{\phi}}\right) = \frac{\partial L}{\partial \phi}$$

and this gives

$$\frac{d}{dt}(r^2\sin^2\theta\,\dot{\phi}) = 0.$$

(v) A particle moving under gravity on the inner smooth surface of a cone whose axis is vertical and vertex downwards.

If r is the horizontal radial distance of the particle from the vertical axis of the cone, ϕ is the azimuthal angle about the vertical axis and z is the height of the particle above the vertex of the cone, the kinetic energy of the particle of mass m is $T = \frac{1}{2}m(\dot{r}^2 + r^2\dot{\phi}^2 + \dot{z}^2)$ and the potential energy is $V = mgz + \text{constant}$.

If the semi-angle at the vertex of the cone is α, we have $r = z\tan\alpha$ and so $T = \frac{1}{2}m[(1 + \tan^2\alpha)\dot{z}^2 + z^2\tan^2\alpha\,\dot{\phi}^2]$.

Lagrange's equations are

$$\frac{d}{dt}\left(\frac{\partial T}{\partial \dot{z}}\right) - \frac{\partial T}{\partial z} = -\frac{\partial V}{\partial z}, \quad \frac{d}{dt}\left(\frac{\partial T}{\partial \dot{\phi}}\right) - \frac{\partial T}{\partial \phi} = -\frac{\partial V}{\partial \phi}$$

which give $m(1 + \tan^2\alpha)\,\ddot{z} - mz\tan^2\alpha\,\dot{\phi}^2 = -mg$ and $d/dt\,(mz^2\tan^2\alpha\,\dot{\phi}) = 0$ respectively.

From the second equation, we get $z^2\tan^2\alpha\,\dot{\phi} = h$ where h is a constant, and so

$$(1 + \tan^2\alpha)\,\ddot{z} - \frac{h^2}{z^3\tan^2\alpha} + g = 0.$$

Using $\ddot{z} = d\left(\frac{1}{2}\dot{z}^2\right)/dz$ and integrating with respect to z yields

$$(1 + \tan^2\alpha)\,\dot{z}^2 + \frac{h^2}{z^2\tan^2\alpha} + 2gz = C$$

where C is a constant. This is equivalent to conservation of energy $T + V = mC/2$.

8.02 A particle P of mass m moves in a plane under the action of a force of magnitude $m\mu r$ directed towards a fixed point O of the plane, r being the distance OP and μ being a constant. Find the Lagrangian function of the system in terms of generalized coordinates q_1 and q_2 given by $x = c\,\cosh q_1 \cos q_2$ and $y = c\,\sinh q_1 \sin q_2$ where x and y are the rectangular Cartesian coordinates of the particle referred to a frame of reference fixed in the plane of motion with origin at O, and c is a constant. Show that Lagrange's equations of motion are satisfied by

$$q_1 = \alpha, \quad \frac{dq_2}{dt} = \pm\mu^{\frac{1}{2}} \quad (\mu > 0),$$

$$q_2 = \beta, \quad \frac{dq_1}{dt} = \pm(-\mu)^{\frac{1}{2}} \quad (\mu < 0)$$

where α and β are constants. Hence find the orbits described by the particle as functions of x and y.

• *Solution*

The kinetic energy $T = \frac{1}{2}m(\dot{x}^2 + \dot{y}^2)$ where $\dot{x} = c(\sinh q_1 \cos q_2\,\dot{q}_1 - \cosh q_1 \sin q_2\,\dot{q}_2)$ and $\dot{y} = c(\cosh q_1 \sin q_2\,\dot{q}_1 + \sinh q_1 \cos q_2\,\dot{q}_2)$ which gives

$$T = \frac{1}{2}mc^2(\sinh^2 q_1 + \sin^2 q_2)(\dot{q}_1^2 + \dot{q}_2^2)$$

using $\cosh^2 q_1 - \sinh^2 q_1 = 1$.

The potential energy $V = \frac{1}{2}m\mu r^2 = \frac{1}{2}m\mu(x^2 + y^2)$ and so we have

$$V = \frac{1}{2}m\mu c^2(\sinh^2 q_1 + \cos^2 q_2).$$

Hence the Lagrangian function $L = T - V$ is given by

$$L = \frac{1}{2}mc^2[(\sinh^2 q_1 + \sin^2 q_2)(\dot{q}_1^2 + \dot{q}_2^2) - \mu(\sinh^2 q_1 + \cos^2 q_2)]$$

and Lagrange's equations for q_1 and q_2 are

$$\frac{d}{dt}[(\sinh^2 q_1 + \sin^2 q_2)\, \dot{q}_1] - \sinh q_1 \cosh q_1 (\dot{q}_1^2 + \dot{q}_2^2 - \mu) = 0,$$

$$\frac{d}{dt}[(\sinh^2 q_1 + \sin^2 q_2)\, \dot{q}_2] - \sin q_2 \cos q_2 (\dot{q}_1^2 + \dot{q}_2^2 + \mu) = 0$$

respectively, that is

$$(\sinh^2 q_1 + \sin^2 q_2)\, \ddot{q}_1 + \sinh q_1 \cosh q_1 (\dot{q}_1^2 - \dot{q}_2^2 + \mu) + 2\sin q_2 \cos q_2\, \dot{q}_1 \dot{q}_2 = 0,$$
$$(\sinh^2 q_1 + \sin^2 q_2)\, \ddot{q}_2 + \sin q_2 \cos q_2 (\dot{q}_2^2 - \dot{q}_1^2 - \mu) + 2\sinh q_1 \cosh q_1\, \dot{q}_1 \dot{q}_2 = 0.$$

Then if $\dot{q}_1 = 0$ and $\ddot{q}_2 = 0$ the second equation gives $\dot{q}_2^2 = \mu$ so that $q_1 = $ constant and $\dot{q}_2 = \pm\sqrt{\mu}$ with $\mu > 0$; and if $\dot{q}_2 = 0$ and $\ddot{q}_1 = 0$ the first equation gives $\dot{q}_1^2 = -\mu$ so that $q_2 = $ constant and $\dot{q}_1 = \pm\sqrt{-\mu}$ with $\mu < 0$.

If $\mu > 0$ we have bounded motion. Setting $q_1 = \alpha$ the path of the particle is the ellipse

$$\frac{x^2}{c^2 \cosh^2 \alpha} + \frac{y^2}{c^2 \sinh^2 \alpha} = 1.$$

If $\mu < 0$ we have unbounded motion. Setting $q_2 = \beta$ the path of the particle is the hyperbola

$$\frac{x^2}{c^2 \cos^2 \beta} - \frac{y^2}{c^2 \sin^2 \beta} = 1.$$

8.03 A uniform rod of length $2a$ is placed in a vertical position with one end on a smooth horizontal table. It is then slightly displaced from the vertical position and left free to move. Use Lagrange's equation for the angle θ which it makes with the horizontal to find the equation of motion of the rod and show that it is consistent with the principle of conservation of energy. Obtain the angular velocity of the rod when it strikes the table.

• *Solution*

Since the table is smooth there is no force in the horizontal direction and so the centre of mass of the rod has no horizontal component of velocity.

If z is the height of the centre of mass of the rod above the table, the kinetic energy of the rod is

$$T = \frac{1}{2}M\,\dot{z}^2 + \frac{1}{2}I\,\dot{\theta}^2$$

where M is the mass of the rod and $I = \frac{1}{3}Ma^2$ is the moment of inertia of the rod about a perpendicular axis through the centre of mass of the rod. The first term is the translational energy of the rod and the second term is the rotational energy of the rod. Since $z = a \sin\theta$ we see that

$$T = \frac{1}{2}Ma^2 \left(\cos^2\theta + \frac{1}{3}\right)\dot{\theta}^2.$$

The potential energy of the rod is $V = Mga\sin\theta$ and so the Lagrange equation for the angle θ

$$\frac{d}{dt}\left(\frac{\partial T}{\partial \dot{\theta}}\right) - \frac{\partial T}{\partial \theta} = -\frac{\partial V}{\partial \theta}$$

yields the equation of motion

$$\frac{d}{dt}\left[Ma^2\left(\cos^2\theta + \frac{1}{3}\right)\dot{\theta}\right] + Ma^2\cos\theta\sin\theta\,\dot{\theta}^2 = -Mga\cos\theta$$

which can be written as

$$\frac{d}{d\theta}\left[\frac{1}{2}Ma^2\left(\cos^2\theta + \frac{1}{3}\right)\dot\theta^2\right] + Mga\cos\theta = 0$$

since $\ddot\theta = d\left(\frac{1}{2}\dot\theta^2\right)/d\theta$.

Now integrating with respect to θ gives

$$\frac{1}{2}Ma^2\left(\cos^2\theta + \frac{1}{3}\right)\dot\theta^2 + Mga\sin\theta = E$$

where E is a constant. This is just the energy equation $T + V = E$.

Initially $\dot\theta = 0$ and $\theta = \pi/2$ so that $E = Mga$.

The rod strikes the table when $\theta = 0$ and then the angular velocity of the rod is given by $\dot\theta = (3g/2a)^{1/2}$.

8.04 Find the equation of motion of a bead of mass m constrained to move on a smooth parabolic wire given by $x^2 = 4ay$ rotating with constant angular velocity ω about the y axis which is fixed in the upward vertical direction.

• Solution

In this problem energy is not conserved, as in the previous problems, because there are variable constraints.

The kinetic energy of the bead is given by

$$T = \frac{1}{2}m(\dot r^2 + \dot y^2 + r^2\omega^2)$$

where r is the radial distance from the y axis, and the potential energy of the bead is $V = mgy$.

Since $r^2 = 4ay$ we have

$$T = \frac{1}{2}m\left[\left(1+\frac{a}{y}\right)\dot y^2 + 4a\omega^2 y\right]$$

and so the Lagrange equation for the y coordinate

$$\frac{d}{dt}\left(\frac{\partial T}{\partial \dot y}\right) - \frac{\partial T}{\partial y} = -\frac{\partial V}{\partial y}$$

yields

$$\frac{d}{dt}\left[m\left(1+\frac{a}{y}\right)\dot y\right] - \frac{\partial}{\partial y}(T-V) = 0$$

that is

$$m\left(1+\frac{a}{y}\right)\ddot y - \frac{1}{2}m\frac{a}{y^2}\dot y^2 - 2ma\omega^2 + mg = 0.$$

Now $\ddot y = d(\frac{1}{2}\dot y^2)/dy$ and so

$$\frac{d}{dy}\left[\frac{1}{2}m\left(1+\frac{a}{y}\right)\dot y^2 - 2ma\omega^2 y + mgy\right] = 0.$$

Integrating with respect to y then gives

$$\frac{1}{2}m\left(1+\frac{a}{y}\right)\dot y^2 - 2ma\omega^2 y + mgy = C$$

where C is a constant.

Thus energy is not conserved in this case. In fact we have $T_2 - T_0 + V = C$ where $T_2 = m(1 + a/y)\,\dot{y}^2\,/2$ and $T_0 = 2maw^2 y$ are the second-order and zero-order terms in $\dot{y}$ of the kinetic energy T. This is a general formula which holds when there are variable constraints but T does not depend on the time.

8.05 A hollow circular cylinder of radius b is made to rotate about its axis with variable angular velocity ω. A uniform circular cylinder of radius $a(< b)$ rolls without slipping inside the hollow cylinder, the axes of the two cylinders being horizontal and the plane through them making an angle θ with the downward vertical. Show that at any time

$$3(b - a)\frac{d^2\theta}{dt^2} - b\frac{d\omega}{dt} + 2g\sin\theta = 0.$$

Show also that if $d\omega/dt$ is a small constant ϵ and the smaller cylinder is at rest in its lowest position at time $t = 0$, then approximately

$$\theta = \frac{\epsilon b}{g}\sin^2\frac{pt}{2}$$

where $p^2 = 2g/[3(b - a)]$.

• *Solution*

Let ϕ be the angle of rotation of the inner cylinder. If the outer cylinder has moment of inertia I and the inner cylinder has mass M, then the kinetic energy of the system is given by

$$T = \frac{1}{2}I\omega^2 + \frac{1}{2}M(b - a)^2\,\dot{\theta}^2 + \frac{1}{2}\left(\frac{1}{2}Ma^2\right)\dot{\phi}^2$$

where the three terms are, respectively, the kinetic energy of rotation of the outer cylinder, and the translational energy and rotational energy of the inner cylinder.

The speed of the axis of the inner cylinder is given by $(b - a)\,\dot{\theta} = b\omega + a\,\dot{\phi}$ and so we may write

$$T = \frac{1}{2}I\omega^2 + \frac{1}{2}M(b - a)^2\,\dot{\theta}^2 + \frac{1}{4}M[(b - a)\,\dot{\theta} - b\omega]^2.$$

Also the potential energy of the system is $V =$ constant $- Mg(b - a)\cos\theta$.

Now using the Lagrange equation for the angle θ

$$\frac{d}{dt}\left(\frac{\partial T}{\partial\dot{\theta}}\right) - \frac{\partial T}{\partial\theta} = -\frac{\partial V}{\partial\theta}$$

we obtain the required equation of motion:

$$3(b - a)\,\ddot{\theta} - b\,\dot{\omega} + 2g\sin\theta = 0.$$

Now let us suppose that $\dot{\omega} = \epsilon$ where ϵ is a small constant. Initially $\theta = 0$ and so θ will be small at first. Then $3(b - a)\,\ddot{\theta} + 2g\theta = b\epsilon$ which has the general solution $\theta = A\cos(pt + \lambda) + b\epsilon/2g$. Since $\theta = \dot{\theta} = 0$ when $t = 0$ it follows that $A\cos\lambda + b\epsilon/2g = 0$ and $pA\sin\lambda = 0$. Hence $\lambda = 0$, $A = -b\epsilon/2g$ and thus we get $\theta = (b\epsilon/2g)(1 - \cos pt)$ which yields the required result. Since ϵ is small it is clear that θ remains small for all time, its maximum value being $b\epsilon/g$.

The next problem involves impulsive motion. To solve this problem we shall use *Lagrange's impulse equations* which, for a dynamical system with n generalized coordinates q_r, have the form

$$\frac{\partial T}{\partial \dot{q}_r} = J_r \quad (r = 1, ..., n)$$ (8.3)

where J_r is the generalized impulse associated with q_r.

8.06 A chain of $2n$ equal uniform bars of mass m, freely jointed at their ends, is laid out in a straight line and set in motion by an impulse J at right angles to its length at the middle point O. If u_r is the initial speed of the rth joint on either side of O, find the initial kinetic energy of the system. Derive the impulse equations and show that they are satisfied by

$$u_r = a\alpha^r + b\beta^r \quad (r = 0, 1, ..., n)$$

where $\alpha = -2 + \sqrt{3}$, $\beta = -2 - \sqrt{3}$, and find a, b in terms of J, α and β.

- *Solution*

The initial kinetic energy of the rth bar on one side of O is

$$T_r = \frac{1}{2}m\left(\frac{u_r + u_{r-1}}{2}\right)^2 + \frac{1}{2}\frac{ml^2}{3}\left(\frac{u_r - u_{r-1}}{2l}\right)^2 = \frac{1}{6}m(u_r^2 + u_r u_{r-1} + u_{r-1}^2)$$

where the first term is the translational energy and the second term is the rotational energy taking $ml^2/3$ as the moment of inertia of a uniform bar of length $2l$. It follows that the initial kinetic energy of the system is

$$T = 2\sum_{r=1}^{n} T_r = \frac{1}{3}m\sum_{r=1}^{n}(u_r^2 + u_r u_{r-1} + u_{r-1}^2).$$

For this problem the Lagrange impulse equations (8.3) become

$$J = \frac{\partial T}{\partial u_0} = \frac{1}{3}m(2u_0 + u_1),$$

$$0 = \frac{\partial T}{\partial u_r} = \frac{1}{3}m(u_{r-1} + 4u_r + u_{r+1}) \quad (r = 1, .., n-1), \quad 0 = \frac{\partial T}{\partial u_n} = \frac{1}{3}m(u_{n-1} + 2u_n).$$

If we put $u_r = a\alpha^r + b\beta^r$ we get that $0 = a\alpha^{r-1} + b\beta^{r-1} + 4(a\alpha^r + b\beta^r) + a\alpha^{r+1} + b\beta^{r+1}$ for $r \neq 0, n$, that is $\alpha^{r-1}a(1 + 4\alpha + \alpha^2) + \beta^{r-1}b(1 + 4\beta + \beta^2) = 0$. Thus α and β satisfy $x^2 + 4x + 1 + 0$ whose solutions are $-2 \pm \sqrt{3}$. Taking $\alpha = -2 + \sqrt{3}$ and $\beta = -2 - \sqrt{3}$ we see that $r = 0$ gives $J = \frac{1}{3}m[2(a+b) + a\alpha + b\beta]$ and $r = n$ gives $0 = a\alpha^{n-1} + b\beta^{n-1} + 2(a\alpha^n + b\beta^n)$ from which it follows that

$$\frac{a}{b} = \left(\frac{\beta}{\alpha}\right)^n = \left(\frac{2 + \sqrt{3}}{2 - \sqrt{3}}\right)^n$$

and

$$a = \frac{J}{m}\frac{\sqrt{3}}{1 - \lambda^n}, \quad b = a\lambda^n, \quad \lambda = \frac{\alpha}{\beta}.$$

In the next problem we use *Euler's equations* for a rigid body having a fixed point O with moments of inertia A, B, C about its principal axes through O, which is rotating with angular velocity $\mathbf{\Omega} = (\omega_1, \omega_2, \omega_3)$ about these axes and is subject to a force exerting a torque $\mathbf{N} = (N_1, N_2, N_3)$ about O. The Euler equations are

$$A\,\dot{\omega}_1 - (B - C)\omega_2\omega_3 = N_1,$$
$$B\,\dot{\omega}_2 - (C - A)\omega_3\omega_1 = N_2, \tag{8.4}$$
$$C\,\dot{\omega}_3 - (A - B)\omega_1\omega_2 = N_3.$$

8.07 A circular disk rotates with constant angular velocity Ω about an axle through its centre inclined at an angle α to its plane. Show that the disk exerts on its bearings a couple $(C - A)\Omega^2 \sin\alpha\cos\alpha$ in the plane through the axle and the perpendicular to the disk, A, A, C being the principal moments of inertia of the disk.

- *Solution*

If we take the 3-axis perpendicular to the disk, and the axle in the plane of the 1-axis and the 3-axis, Euler's equations for this problem are

$$A\,\dot{\omega}_1 - (A - C)\omega_2\omega_3 = N_1,$$
$$A\,\dot{\omega}_2 - (C - A)\omega_3\omega_1 = N_2,$$
$$C\,\dot{\omega}_3 = N_3.$$

Now $\omega_1 = \Omega\cos\alpha$, $\omega_2 = 0$, $\omega_3 = \Omega\sin\alpha$ and $\dot{\omega}_1 = \dot{\omega}_2 = \dot{\omega}_3 = 0$. Hence, as required, we have

$$N_2 = -(C - A)\Omega^2 \sin\alpha\cos\alpha.$$

8-2 Tops

8.08 The principal moments of inertia of a symmetrical top at the fixed point O of its axis, are each equal to $2Ml^2$ where M is the mass of the top and l is the distance from O to its centre of mass G. The top is started with angular velocity $\sqrt{g/l}$ about its axis OG with this axis moving horizontally with angular velocity $\sqrt{g/l}$ about the vertical through O. If lx is the height of G above O at any subsequent time t, show that

$$x = \tanh^2\left(\frac{t}{2}\sqrt{\frac{g}{l}}\right).$$

- *Solution*

Taking θ, ϕ, ψ to be the Euler angles of the top so that θ is the angle that OG makes with the vertical, ϕ is the azimuthal angle of OG, and ψ is the angle of rotation of the top about OG, the kinetic energy of the top is given by

$$T = \frac{1}{2}A\,\dot{\theta}^2 + \frac{1}{2}B\,\dot{\phi}^2 \sin^2\theta + \frac{1}{2}C(\dot{\psi} + \dot{\phi}\cos\theta)^2, \tag{8.5}$$

where A, B, C are the principal moments of inertia of the top at O, and the potential energy of the top is $V = Mgl\cos\theta + \text{constant}$.

Using Lagrange's equations for the angles ψ and ϕ, and putting $B = A$ since tops are symmetrical, gives

$$\frac{d}{dt}[C(\dot{\psi} + \dot{\phi}\cos\theta)] = 0, \quad \frac{d}{dt}[A\,\dot{\phi}\sin^2\theta + C\cos\theta(\dot{\psi} + \dot{\phi}\cos\theta)] = 0$$

so that

$$\dot{\psi} + \dot{\phi}\cos\theta = n, \quad A\dot{\phi}\sin^2\theta + C\cos\theta(\dot{\psi} + \dot{\phi}\cos\theta) = Ch \qquad (8.6)$$

where n and h are constants. The first equation gives the constant spin n about the axis of the top and the second equation gives the constant angular momentum Ch about the vertical direction.

Also the Lagrange equation for θ gives

$$A(\ddot{\theta} - \dot{\phi}^2\sin\theta\cos\theta) + Cn\,\dot{\phi}\sin\theta - Mgl\sin\theta = 0$$

which can be integrated with respect to θ to yield the energy conservation equation

$$A\dot{\theta}^2 + \frac{C^2(h - n\cos\theta)^2}{A\sin^2\theta} + Cn^2 + 2Mgl\cos\theta = 2E \qquad (8.7)$$

using (8.6).

Now putting $2E = Aq + Cn^2$ we get

$$\dot{\theta}^2 + \frac{r^2(h - n\cos\theta)^2}{\sin^2\theta} + \frac{2Mgl}{A}\cos\theta = q \qquad (8.8)$$

where $r = C/A$.

We next set $x = \cos\theta$ and then the above equation becomes

$$\dot{x}^2 = (q - 2Mglx/A)(1 - x^2) - r^2(h - nx)^2. \qquad (8.9)$$

In the present problem $r = 1$, $A = 2Ml^2$, $n = \sqrt{g/l}$, $h = \sqrt{g/l}$ and $\dot{\theta} = 0$ when $\theta = \pi/2$. Hence $q = h^2 = g/l$ and we have

$$\dot{x}^2 = \frac{g}{l}x(1 - x)^2.$$

If we now put $x = y^2$ we see that

$$\frac{dy}{dt} = \frac{1}{2}\sqrt{\frac{g}{l}}(1 - y^2)$$

and so

$$\frac{1}{2}\sqrt{\frac{g}{l}}t = \int_0^y \frac{dy}{1 - y^2}$$

which gives

$$\frac{1}{2}\sqrt{\frac{g}{l}}t = \tanh^{-1}y$$

and the required result follows.

8.09 The axis of a symmetrical top is initially at rest at an angle α to the vertical, and the spin n is large. Show that the axis will never be inclined to the vertical at an angle greater than $\alpha + \beta$ where

$$\beta = \frac{2MglA}{n^2C^2}\sin\alpha,$$

A, A, C being the principal moments of inertia of the top at the fixed point O of its axis of rotation, M the mass of the top and l the distance of the centre of mass of the top from O.

• *Solution*

If the motion of the top is confined between two horizontal circles given by θ_0 and θ_1 we see from the formula (8.9) obtained in the previous problem that

$$\dot{x}^2 = (x_0 - x)\left[\frac{2Mgl}{A}(1 - x^2) - n^2 r^2(x_0 - x)\right]$$

where $x = \cos\theta$ and $x_0 = \cos\theta_0 = h/n = Aq/2Mgl$ so that $x_1 = \cos\theta_1$ satisfies

$$\frac{2Mgl}{A}(1 - x_1^2) - n^2 r^2(x_0 - x_1) = 0.$$

In the case of a fast top for which the initial kinetic energy of rotation is large compared with the maximum change in the potential energy we have $\frac{1}{2}Cn^2 \gg Mgl$. Hence, provided C is not greatly less than A, we see that $n^2 r^2 \gg 2Mgl/A$ and so

$$x_1 \simeq x_0 - \frac{2Mgl}{An^2 r^2}(1 - x_0^2),$$

that is

$$\cos\theta_1 \simeq \cos\theta_0 - \frac{2Mgl}{An^2 r^2}\sin^2\theta_0.$$

Thus for a fast top

$$\cos\theta_1 \simeq \cos\left(\theta_0 + \frac{2Mgl}{An^2 r^2}\sin\theta_0\right).$$

Putting $\theta_0 = \alpha$ we see that $\theta \le \theta_1 \simeq \alpha + \beta$ where $\beta = [2MglA/(n^2 C^2)]\sin\alpha$ and so the required result is obtained.

8.10 A symmetrical top of mass M and principal moments of inertia A, A, C whose centre of mass G is at a distance c from the vertex, is set spinning on a rough horizontal plane with angular velocity n about its axis which is inclined at an angle α to the vertical. If $b = C^2 n^2/4AMg$, and a, b, c are the sides of a triangle with angle α opposite to the side a, show that G will fall until it is a distance $b - a$ above the plane. If $d = c\cos\alpha$, show further that the time between successive returns of G to the highest point of its motion is

$$\left(\frac{2A}{Mg}\right)^{\frac{1}{2}}\int_{b-a}^{d}\frac{dz}{[(d - z)(a - b + z)(a + b - z)]^{\frac{1}{2}}}.$$

• *Solution*

Initially we have that $\dot{\theta} = 0, \dot{\phi} = 0, \dot{\psi} = n$ and $\theta = \alpha$.
Now, as in problem 8.08,

$$\dot{\psi} + \dot{\phi}\cos\theta = n, \quad \dot{\phi}\sin^2\theta + rn\cos\theta = rh$$

and so $n\cos\alpha = h$.
Also

$$\dot{\theta}^2 + \frac{r^2 n^2(\cos\alpha - \cos\theta)^2}{\sin^2\theta} + \frac{2Mgl}{A}\cos\theta = \frac{2Mgl}{A}\cos\alpha$$

and this gives, as in problem 8.09,

$$\dot{x}^2 = (x_0 - x)\left[\frac{2Mgl}{A}(1 - x^2) - r^2 n^2(x_0 - x)\right]$$

where $x = \cos\theta$ and $x_0 = \cos\alpha$.

In the present problem $c = l$ and $b = C^2n^2/4AMg$. Hence we get

$$\dot{x}^2 = \frac{2Mgc}{A}(x_0 - x)\left[1 - x^2 - \frac{2b}{c}(x_0 - x)\right].$$

Let $z = cx$ and $d = c\cos\alpha = cx_0$. Then we see that

$$\dot{z}^2 = \frac{2Mg}{A}(d - z)[c^2 - z^2 - 2b(d - z)].$$

Now $a^2 = b^2 + c^2 - 2bc\cos\alpha = b^2 + c^2 - 2bd$ giving

$$\dot{z}^2 = \frac{2Mg}{A}(d - z)(a - b + z)(a + b - z).$$

Clearly $\dot{z} = 0$ when $z = d$, or $z = b - a$, or $z = a + b$. Because $a + b > c$ we know that $z = a + b$ is disallowed. Hence the motion is confined between $z = d$ and $z = b - a$. Now $a > b - c\cos\alpha = b - d$ and so $d > b - a$. Thus the centre of mass falls from the height d above the plane to the height $b - a$ and then returns to the height d. It follows that the time between successive returns of the centre of mass to the upper limit of its motion is given by

$$T = 2\sqrt{\frac{A}{2Mg}}\int_{b-a}^{d}\frac{dz}{\sqrt{(d - z)(a - b + z)(a + b - z)}}.$$

8-3 Hamilton's equations

The next pair of problems involve the application of *Hamilton's equations*.

The *Hamiltonian function* $H(q_r, p_r)$ for a dynamical system, which is completely specified by n generalized coordinates q_r and n canonical momenta p_r ($r = 1, ..., n$), is defined by

$$H = \sum_{r=1}^{n} p_r\,\dot{q}_r - L \tag{8.10}$$

where L is the Lagrangian function. For a conservative system with potential energy V and kinetic energy T we have $L = T - V$ and $H = T + V$. It is important to note that H must be expressed as a function of the coordinates q_r and the momenta p_r by eliminating the velocities $\dot{q}_r$ using the definition $p_r = \partial L/\partial\dot{q}_r$.

Hamilton's equations are the first-order differential equations

$$\dot{q}_r = \frac{\partial H}{\partial p_r}, \quad \dot{p}_r = -\frac{\partial H}{\partial q_r} \quad (r = 1, ..., n). \tag{8.11}$$

8.11 Use Hamilton's equations to find the equations of motion for:
(i) a particle of mass m falling freely under gravity,
(ii) a simple pendulum,
(iii) a spherical pendulum,
(iv) a particle moving under the action of a central force with potential $V(r)$.

• *Solution*

(i) A particle falling freely under gravity.

In Cartesian coordinates with the z axis in the upward vertical direction, the kinetic energy is given by $T = \frac{1}{2}m(\dot{x}^2 + \dot{y}^2 + \dot{z}^2)$ and the potential energy is $V = mgz$.

We have $p_x = \partial L/\partial \dot{x} = m\,\dot{x}$ and similarly $p_y = m\,\dot{y}$, $p_z = m\,\dot{z}$. Hence

$$H = T + V = \frac{1}{2m}\left(p_x^2 + p_y^2 + p_z^2\right) + mgz.$$

The first Hamilton equation gives $\dot{x} = \partial H/\partial p_x = p_x/m$ and similarly $\dot{y} = p_y/m$, $\dot{z} = p_z/m$.

The second Hamilton equation gives $\dot{p}_x = -\partial H/\partial x = 0$, $\dot{p}_y = -\partial H/\partial y = 0$, $\dot{p}_z = -\partial H/\partial z = -mg$ and so $\ddot{x} = 0$, $\ddot{y} = 0$, $\ddot{z} = -g$ which are just Newton's second law of motion equations.

(ii) A simple pendulum.

If θ is the angle of inclination of the pendulum string with the downward vertical and l is the length of the string, the kinetic energy of the pendulum bob is $T = \frac{1}{2}ml^2\,\dot{\theta}^2$ and its potential energy is $V = -mgl\cos\theta$ where m is the mass of the bob. Then $p_\theta = \partial L/\partial \dot{\theta} = \partial T/\partial \dot{\theta} = ml^2\,\dot{\theta}$. Hence $T = p_\theta^2/(2ml^2)$ and so

$$H = \frac{p_\theta^2}{2ml^2} - mgl\cos\theta.$$

Hamilton's equations give $\dot{\theta} = \partial H/\partial p_\theta = p_\theta/(ml^2)$ and $\dot{p}_\theta = -\partial H/\partial \theta = -mgl\sin\theta$ so that we get $ml^2\,\ddot{\theta} = -mgl\sin\theta$ which is the same as Newton's equation of motion for a simple pendulum.

(iii) A spherical pendulum.

If θ is the angle which the radius from the centre of the sphere to the particle makes with the downward vertical, and ϕ is the azimuthal angle about the vertical diameter, the kinetic energy is $T = \frac{1}{2}m(a^2\,\dot{\theta}^2 + a^2\sin^2\theta\,\dot{\phi}^2)$ and the potential energy is $V = -mga\cos\theta$. Then $p_\theta = \partial L/\partial \dot{\theta} = \partial T/\partial \dot{\theta} = ma^2\,\dot{\theta}$, $p_\phi = \partial L/\partial \dot{\phi} = ma^2\sin^2\theta\,\dot{\phi}$. Hence

$$H = \frac{1}{2ma^2}\left(p_\theta^2 + \frac{p_\phi^2}{\sin^2\theta}\right) - mga\cos\theta.$$

Hamilton's equations give $\dot{\theta} = \partial H/\partial p_\theta = p_\theta/(ma^2)$, $\dot{\phi} = \partial H/\partial p_\phi = p_\phi/(ma^2\sin^2\theta)$ and $\dot{p}_\theta = -\partial H/\partial \theta = p_\phi^2\cos\theta/ma^2\sin^3\theta - mga\sin\theta$, $\dot{p}_\phi = -\partial H/\partial \phi = 0$. It follows from these that

$$ma^2\,\ddot{\theta} = ma^2\sin\theta\cos\theta\,\dot{\phi}^2 - mga\sin\theta, \quad \frac{d}{dt}(ma^2\sin^2\theta\,\dot{\phi}) = 0$$

which are the same as we obtained using Lagrange's equations in problem 8.01.

(iv) A particle moving under the action of a central force with potential $V(r)$.

Using the formula for the kinetic energy T given in problem 8.01 (iv) we see that

$$p_r = \partial L/\partial \dot{r} = m\,\dot{r}, \quad p_\theta = \partial L/\partial \dot{\theta} = mr^2\,\dot{\theta}, \quad p_\phi = \partial L/\partial \dot{\phi} = mr^2\sin^2\theta\,\dot{\phi}$$

and so we may write the Hamiltonian $H = T + V$ in the form

$$H = \frac{1}{2m}\left(p_r^2 + \frac{p_\theta^2}{r^2} + \frac{p_\phi^2}{r^2\sin^2\theta}\right) + V(r).$$

Then the Hamilton equations (8.11) give

$$\dot{p_r} = -\frac{\partial H}{\partial r} = \frac{1}{mr^3}\left(p_\theta^2 + \frac{p_\phi^2}{\sin^2\theta}\right) - \frac{dV}{dr},$$

$$\dot{p_\theta} = -\frac{\partial H}{\partial \theta} = \frac{p_\phi^2 \cos\theta}{mr^2 \sin^3\theta}$$

and

$$\dot{p_\phi} = -\frac{\partial H}{\partial \phi} = 0.$$

These give

$$m(\ddot{r} - r\dot{\theta}^2 - r\sin^2\theta\,\dot{\phi}^2) = -\frac{dV}{dr}, \quad \frac{d}{dt}(r^2\dot{\theta}) - r^2\sin\theta\cos\theta\,\dot{\phi}^2 = 0, \quad \frac{d}{dt}(r^2\sin^2\theta\,\dot{\phi}) = 0$$

which were obtained previously in 8.01 (iv) using Lagrange's equations.

The other Hamilton's equations

$$\dot{r} = \frac{\partial H}{\partial p_r} = \frac{p_r}{m}, \quad \dot{\theta} = \frac{\partial H}{\partial p_\theta} = \frac{p_\theta}{mr^2}, \quad \dot{\phi} = \frac{\partial H}{\partial p_\phi} = \frac{p_\phi}{mr^2\sin^2\theta}$$

just yield the definitions of the generalized momenta.

8.12 A particle of unit mass moves in a plane under the action of a uniform field of force F parallel to the x axis of a fixed rectangular frame of reference $O(x, y)$ in the plane of motion. If $x + iy = \frac{1}{2}(u + iv)^2$ find the generalized momenta p_u and p_v and derive Hamilton's equations. Find u as a function of the time t, given that v is a constant of the motion and that $u = 0$ initially.

• *Solution*

The kinetic energy is $T = \frac{1}{2}(\dot{x}^2 + \dot{y}^2)$ and the potential energy is $V = -Fx$. Since $x + iy = \frac{1}{2}(u+iv)^2$ we have $x = \frac{1}{2}(u^2 - v^2)$ and $\dot{x} + i\dot{y} = (u+iv)(\dot{u} + i\dot{v})$. It follows that

$$T = \frac{1}{2}(u^2 + v^2)(\dot{u}^2 + \dot{v}^2)$$

and so $p_u = \partial L/\partial \dot{u} = \partial T/\partial \dot{u} = (u^2 + v^2)\dot{u}$, $p_v = \partial L/\partial \dot{v} = \partial T/\partial \dot{v} = (u^2 + v^2)\dot{v}$ which enables us to write

$$T = \frac{1}{2(u^2 + v^2)}(p_u^2 + p_v^2).$$

Since $V = -F(u^2 - v^2)/2$ the Hamiltonian is given by

$$H = T + V = \frac{1}{2(u^2 + v^2)}(p_u^2 + p_v^2) - \frac{1}{2}F(u^2 - v^2).$$

Hence Hamilton's equations are

$$\dot{u} = \partial H/\partial p_u = p_u/(u^2 + v^2), \quad \dot{v} = \partial H/\partial p_v = p_v/(u^2 + v^2)$$

and

$$\dot{p_u} = -\partial H/\partial u = \frac{u(p_u^2 + p_v^2)}{(u^2 + v^2)^2} + Fu, \quad \dot{p_v} = -\partial H/\partial v = \frac{v(p_u^2 + p_v^2)}{(u^2 + v^2)^2} - Fv.$$

If $v = $constant we have $\dot{v} = \ddot{v} = 0$. Hence $p_v = \dot{p_v} = 0$ and so $p_u^2/(u^2 + v^2)^2 = F$, that is $\dot{u}^2 = F$. Hence $\dot{u} = \pm\sqrt{F}$ giving $u = \pm\sqrt{F}t$ since $u = 0$ when $t = 0$.

8-4 Contact transformations

8.13 Show that the following are *contact transformations:*
(i) $P = \frac{1}{2}(p^2 + q^2), \quad Q = \tan^{-1}(q/p)$;
(ii) $P = 2(1 + q^{1/2}\cos p)q^{1/2}\sin p, \quad Q = \ln(1 + q^{1/2}\cos p)$;
(iii) $Q = \ln(q^{-1}\sin p), \quad P = q\cot p$.

• *Solution*

(i) We see that

$$dQ = \frac{pdq - qdp}{p^2 + q^2}$$

and so

$$pdq - PdQ = \frac{1}{2}(pdq + qdp) = d\left(\frac{1}{2}qp\right).$$

Thus we have a contact transformation with *generating function* $W = \frac{1}{2}qp$ since $pdq - PdQ = dW$.

(ii) We have

$$dQ = \frac{-q^{1/2}\sin p\, dp + \frac{1}{2}q^{-1/2}\cos p\, dq}{1 + q^{1/2}\cos p}$$

and so

$$pdq - PdQ = pdq + 2q\sin^2 p\, dp - \sin p\cos p\, dq = d(pq - q\sin p\cos p).$$

Hence this is a contact transformation with generating function $W = q(p - \sin p\cos p)$.

(iii) Now

$$dQ = \frac{q}{\sin p}\left(-\frac{\sin p\, dq}{q^2} + \frac{\cos p\, dp}{q}\right)$$

and so

$$pdq - PdQ = pdq + \frac{\cos p\, dq}{\sin p} - \frac{\cos^2 p}{\sin^2 p}qdp = d\left(q\frac{\cos p}{\sin p} + qp\right).$$

Hence we have a contact transformation with generating function $W = q(\cot p + p)$.

8.14 (a) Obtain the Hamiltonian function for a harmonic oscillator with angular frequency ω.
(b) Prove that the transformation

$$q = \sqrt{\frac{2P}{m\omega}}\sin Q, \quad p = \sqrt{2m\omega P}\cos Q$$

is a contact transformation and determine its generating function.
(c) Express the Hamiltonian in terms of P and Q showing that Q is cyclic.
(d) Hence obtain the solution of the harmonic oscillator problem.

• *Solution*

(a) The kinetic energy of a particle of mass m is $T = \frac{1}{2}m\,\dot{q}^2 = p^2/2m$ since the momentum of the particle is given by $p = m\,\dot{q}$ where q is the coordinate specifying the position of the particle. The potential energy of a harmonic oscillator is $V = \frac{1}{2}m\omega^2 q^2$ and so the Hamiltonian function is

$$H = \frac{1}{2m}p^2 + \frac{1}{2}m\omega^2 q^2.$$

(b) We have $pdq = \sin Q \cos Q \, dP + 2P\cos^2 Q \, dQ$ and so

$$pdq - PdQ = \sin Q \cos Q \, dP + P(\cos^2 Q - \sin^2 Q)dQ = d(P\sin Q\cos Q).$$

Hence we have a contact transformation with generating function
$W = P\sin Q\cos Q = \frac{1}{2}m\omega q^2 \cot Q$.

(c) The Hamiltonian function can now be expressed in terms of the new variables Q, P in the form

$$K = \omega P\cos^2 Q + \omega P\sin^2 Q = \omega P$$

and thus Q is a *cyclic* coordinate since K is independent of Q.

(d) Hamilton's equations for Q, P and the Hamiltonian K are

$$\dot{Q} = \frac{\partial K}{\partial P}, \quad \dot{P} = -\frac{\partial K}{\partial Q}$$

and these give $\dot{Q} = \omega$ so that $Q = \omega t + \epsilon$, and $\dot{P} = 0$ so that $P = \beta$ where ϵ and β are constants. Hence, as expected,

$$q = A\sin(\omega t + \epsilon), \quad p = m\omega A\cos(\omega t + \epsilon)$$

where the constant $A = \sqrt{2\beta/m\omega}$.

8-5 Hamilton-Jacobi equation

8.15 Use the *Hamilton-Jacobi equation* to solve the problem of the harmonic oscillator.

● *Solution*

To obtain the Hamilton-Jacobi equation we put $p = \partial S/\partial q$ in the Hamiltonian function for the harmonic oscillator found in the previous problem. Then we get

$$\frac{1}{2m}\left(\frac{\partial S}{\partial q}\right)^2 + \frac{1}{2}m\omega^2 q^2 = \alpha_1$$

where α_1 is a constant. Integrating with respect to q gives

$$S(q) = \int \sqrt{2m\alpha_1 - m^2\omega^2 q^2} \, dq$$

and so

$$\frac{\partial S}{\partial \alpha_1} = \int \frac{m\,dq}{\sqrt{2m\alpha_1 - m^2\omega^2 q^2}} = t + \beta_1$$

according to Hamilton-Jacobi theory. Carrying out the integration we obtain

$$t + \beta_1 = \frac{1}{\omega}\sin^{-1}\frac{q}{\sqrt{2\alpha_1/m\omega^2}}$$

giving

$$q = \sqrt{\frac{2\alpha_1}{m\omega^2}}\sin[\omega(t+\beta_1)].$$

8.16 Show that the solution of the Hamilton-Jacobi equation for the Hamiltonian $H = \frac{1}{2}p^2 - aq^{-1}$, where a is a constant, can be put in the form

$$W = \frac{at}{b} + \sqrt{2ab}\sin^{-1}\sqrt{\frac{q}{b}} + \sqrt{\frac{2aq(b-q)}{b}}$$

where b is a constant.

• *Solution*

To get the Hamilton-Jacobi equation we put $p = \partial W/\partial q$ in

$$\frac{\partial W}{\partial t} + H(q,p) = 0$$

which gives in the present problem

$$\frac{\partial W}{\partial t} + \frac{1}{2}\left(\frac{\partial W}{\partial q}\right)^2 - \frac{a}{q} = 0.$$

To solve this equation we separate the variables by putting

$$W = f(t) + g(q).$$

Then we obtain

$$\frac{df}{dt} = \frac{a}{q} - \frac{1}{2}\left(\frac{dg}{dq}\right)^2.$$

Each side of this equation must be a constant which we take to be a/b. Then we get $f(t) = at/b$ and

$$\left(\frac{dg}{dq}\right)^2 = 2a\left(\frac{1}{q} - \frac{1}{b}\right)$$

giving

$$g(q) = \int\sqrt{\frac{2a(b-q)}{bq}}dq = \sqrt{\frac{2aq(b-q)}{b}} - \int q\frac{d}{dq}\sqrt{\frac{2a(b-q)}{bq}}dq.$$

The integral can be evaluated by putting $q = x^2$. Then we get

$$-\int q\frac{d}{dq}\sqrt{\frac{2a(b-q)}{bq}}dq = \sqrt{2ab}\int\frac{dx}{\sqrt{b-x^2}} = \sqrt{2ab}\sin^{-1}\sqrt{\frac{q}{b}}$$

and the required result follows directly.

8.17 A particle of mass m moves in a straight line under the action of an attractive force $m\omega^2 x$ where x is the distance of the particle from a fixed point on the line. Show that Hamilton's principle function is

$$W = \frac{m\omega}{2\sin\omega t}[(x^2 + x_0^2)\cos\omega t - 2xx_0]$$

where t is the time taken by the particle to travel from the point x_0 to the point x.
Verify that

$$\frac{\partial W}{\partial x} = p, \quad \frac{\partial W}{\partial t} + H = 0$$

where $p = m\,\dot{x}$ is the momentum of the particle and H is the Hamiltonian function.

• *Solution*

Hamilton's principal function is defined to be $W = \int_0^t L\,dt$ where L is the Lagrangian function.

In the present problem the kinetic energy is $T = \frac{1}{2}m\,\dot{x}^2$ and the potential energy is $V = \frac{1}{2}m\omega^2x^2$. Since we are dealing with the simple harmonic oscillator problem, we know that we can write the solution in the form $x = a\sin\omega t + x_0\cos\omega t$, where x_0 is the initial value of x, and so $\dot{x} = \omega(a\cos\omega t - x_0\sin\omega t)$. Hence we have

$$W = \int_0^t (T - V)dt = \frac{m}{2}\int_0^t (\dot{x}^2 - \omega^2x^2)dt = \frac{m\omega^2}{2}\int_0^t [(a^2 - x_0^2)\cos 2\omega t - 2ax_0\sin 2\omega t]dt$$

which can be easily integrated to give the first result.
Further

$$\frac{\partial W}{\partial x} = \frac{m\omega}{\sin\omega t}(x\cos\omega t - x_0) = m\,\dot{x}$$

as required, and

$$\frac{\partial W}{\partial t} = -\frac{m\omega^2\cos\omega t}{2\sin^2\omega t}[(x^2 + x_0^2)\cos\omega t - 2xx_0] - \frac{m\omega^2}{2}(x^2 + x_0^2) = -\frac{m\omega^2}{2}(x_0^2 + a^2)$$

Now

$$H = \frac{1}{2}m\,\dot{x}^2 + \frac{1}{2}m\omega^2x^2 = \frac{1}{2}m\omega^2(a^2 + x_0^2)$$

and so $\partial W/\partial t = -H$ as required.

8.18 Show that Hamilton's principal function for a particle of mass m moving under the influence of a uniform gravitational field with potential $V = mgz$ is

$$W = \frac{1}{2}m\left[\frac{(x - x_0)^2 + (y - y_0)^2 + (z - z_0)^2}{t}\right] - \frac{1}{2}mg(z + z_0)t - \frac{1}{24}mg^2t^3$$

where x_0, y_0, z_0 are the rectangular Cartesian coordinates of the particle at time $t = 0$.
Verify that

$$\frac{\partial W}{\partial z} = p, \quad \frac{\partial W}{\partial t} + H = 0$$

where $p = m\,\dot{z}$ is the momentum of the particle and H is the Hamiltonian function.

• *Solution*

Let u, v, w be the components of the velocity at time $t = 0$. Then the Lagrangian function is $L = \frac{1}{2}m(\dot{x}^2 + \dot{y}^2 + \dot{z}^2) - mgz$ where $\dot{x} = u, \dot{y} = v, \dot{z} = w - gt$ and $z = z_0 + wt - \frac{1}{2}gt^2$. Hence Hamilton's principal function is given by

$$W = \int_0^t L\,dt = \frac{1}{2}m\left(u^2 + v^2 + w^2\right)t - mgwt^2 + \frac{1}{3}mg^2t^3 - mgz_0t.$$

Substituting $u = (x - x_0)/t$, $v = (y - y_0)/t$ and $w = (z - z_0)/t + \frac{1}{2}gt$ into the above equation then leads to the required formula for W.

We now see that

$$\frac{\partial W}{\partial z} = m\frac{z - z_0}{t} - \frac{1}{2}mgt = m(w - gt) = m\,\dot{z}$$

and

$$\frac{\partial W}{\partial t} = -\frac{1}{2}m\left[\frac{(x - x_0)^2 + (y - y_0)^2 + (z - z_0)^2}{t^2}\right] - \frac{1}{2}mg(z + z_0) - \frac{1}{8}mg^2t^2.$$

But $H = \frac{1}{2}m(\dot{x}^2 + \dot{y}^2 + \dot{z}^2) + mgz = \frac{1}{2}m(u^2 + v^2) + \frac{1}{2}m(w - gt)^2 + mg\left(z_0 + wt - \frac{1}{2}gt^2\right)$ and so we have $H = \frac{1}{2}m(u^2 + v^2 + w^2) + mgz_0$ from which it follows that $\partial W/\partial t = -H$ as required.

8-6 Vibrational motion

8.19 A double pendulum is composed of two particles P_1 and P_2 each having mass m. The particle P_1 is attached to a fixed point O by a string of length l and the particle P_2 is attached to the particle P_1 by another string of length l. Find the normal modes of small vibrations of the double pendulum.

• *Solution*

Let θ_1 and θ_2 be the angles which the strings OP_1 and P_1P_2 make with the downward vertical, respectively.

Assuming that the angles are small, the kinetic energy of the double pendulum is

$$T = \frac{1}{2}ml^2\,\dot{\theta_1}^2 + \frac{1}{2}ml^2(\dot{\theta}_1 + \dot{\theta}_2)^2$$

and its potential energy is

$$V = c - mgl\cos\theta_1 - mgl(\cos\theta_1 + \cos\theta_2)$$

which may be approximated by

$$V = \frac{1}{2}mgl(2\theta_1^2 + \theta_2^2) + c'$$

where c and c' are constants.

Using Lagrange's equations for the angles θ_1 and θ_2 and putting $n^2 = g/l$ we get the coupled equations

$$2\,\ddot{\theta}_1 + \ddot{\theta}_2 + 2n^2\theta_1 = 0,$$
$$\ddot{\theta}_1 + \ddot{\theta}_2 + n^2\theta_2 = 0.$$

Now we let $\theta_1 = a_1\sin(pt + \epsilon)$ and $\theta_2 = a_2\sin(pt + \epsilon)$ and this leads to

$$2(n^2 - p^2)a_1 - p^2a_2 = 0,$$
$$-p^2a_1 + (n^2 - p^2)a_2 = 0.$$

This pair of algebraic equations are consistent if $2(n^2 - p^2)^2 - p^4 = 0$, that is if $p^2 = \mp\sqrt{2}(n^2 - p^2)$ whose solutions are $p_1^2 = (2 - \sqrt{2})n^2$ and $p_2^2 = (2 + \sqrt{2})n^2$. For p_1 we have $a_2 = \sqrt{2}a_1$ and for p_2 we have $a_2 = -\sqrt{2}a_1$.

Thus the normal modes are

$$\theta_1 = a_1 \sin(p_1 t + \epsilon_1), \quad \theta_2 = \sqrt{2} a_1 \sin(p_1 t + \epsilon_1),$$

and

$$\theta_1 = a_1 \sin(p_2 t + \epsilon_2), \quad \theta_2 = -\sqrt{2} a_1 \sin(p_2 t + \epsilon_2).$$

8.20 Three equal springs OP_1, $P_1 P_2$, $P_2 P_3$ are of natural length a and modulus of elasticity λ. Particles of equal mass $2m$ join OP_1 to $P_1 P_2$ at P_1, and $P_1 P_2$ to $P_2 P_3$ at P_2, and a particle of mass m is at P_3. If O is fixed and the springs are put in a straight line on a smooth horizontal table, find the normal modes of vibration along the straight line.

• *Solution*

Suppose q_1, q_2, q_3 are the coordinates of the particles P_1, P_2, P_3 referred to their equilibrium positions. Then the kinetic energy is

$$T = \frac{1}{2}(2m \, \dot{q_1}^2 + 2m \, \dot{q_2}^2 + m \, \dot{q_3}^2)$$

and the potential energy is

$$V = \frac{1}{2}\frac{\lambda}{a}[q_1^2 + (q_2 - q_1)^2 + (q_3 - q_2)^2]$$

since the potential energy of a spring of modulus λ which has been extended by an amount x from its natural length a is $\lambda x^2 / 2a$.

Using Lagrange's equations for q_1, q_2, q_3 we obtain

$$2m \, \ddot{q_1} + \lambda q_1 / a - \lambda(q_2 - q_1)/a = 0,$$
$$2m \, \ddot{q_2} + \lambda(q_2 - q_1)/a - \lambda(q_3 - q_2)/a = 0,$$
$$m \, \ddot{q_3} + \lambda(q_3 - q_2)/a = 0.$$

Putting $q_i = a_i \sin(pt + \epsilon)$ and setting $\lambda/2ma = n^2$ we get

$$(2n^2 - p^2)a_1 - n^2 a_2 = 0,$$
$$-n^2 a_1 + (2n^2 - p^2)a_2 - n^2 a_3 = 0,$$
$$-2n^2 a_2 + (2n^2 - p^2)a_3 = 0,$$

and these three algebraic equations are consistent if

$$\begin{vmatrix} 2-k & -1 & 0 \\ -1 & 2-k & -1 \\ 0 & -2 & 2-k \end{vmatrix} = 0$$

where we have put $p^2/n^2 = k$. Then we have $(2-k)^3 - 3(2-k) = 0$, that is $(2-k)(k^2 - 4k + 1) = 0$ giving $k = 2$ or $2 \pm \sqrt{3}$. Hence p^2 has the values

$$p_1^2 = (2 - \sqrt{3})n^2, \quad p_2^2 = 2, \quad p_3^2 = (2 + \sqrt{3})n^2.$$

For p_1, that is $k = 2 - \sqrt{3}$, we have $a_2 = \sqrt{3}a_1, a_3 = 2a_1$; for p_2, that is $k = 2$, we have $a_2 = 0, a_3 = -a_1$; and for p_3, that is $k = 2 + \sqrt{3}$, we have $a_2 = -\sqrt{3}a_1, a_3 = 2a_1$.

Hence the normal modes are

$$q_1 = a_1 \sin(p_1 t + \epsilon_1), \quad q_2 = \sqrt{3}a_1 \sin(p_1 t + \epsilon_1), \quad q_3 = 2a_1 \sin(p_1 t + \epsilon_1);$$
$$q_1 = a_1 \sin(p_2 t + \epsilon_2), \quad q_2 = 0, \quad q_3 = -a_1 \sin(p_2 t + \epsilon_2);$$
$$q_1 = a_1 \sin(p_3 t + \epsilon_3), \quad q_2 = -\sqrt{3}a_1 \sin(p_3 t + \epsilon_3), \quad q_3 = 2a_1 \sin(p_3 t + \epsilon_3).$$

8.21 Three particles of equal masses m are attached to a light string of length $4a$ so as to divide the string into four equal parts of length a. If the string is held stationary at its ends and stretched at tension P on a smooth horizontal table, displaced horizontally and then released from rest, find the normal modes of vibration of the string perpendicular to its equilibrium position.

- *Solution*

If y_1, y_2, y_3 are the displacements of the particles from their equilibrium positions, in order along the length of the string, the kinetic energy of the system is

$$T = \frac{1}{2}m(\dot{y_1}^2 + \dot{y_2}^2 + \dot{y_3}^2)$$

and the potential energy is

$$V = \frac{1}{2}\frac{P}{a}[y_1^2 + (y_2 - y_1)^2 + (y_3 - y_2)^2 + y_3^2].$$

Lagrange's equations for the coordinates y_1, y_2, y_3 are

$$m\,\ddot{y_1} + P[y_1 - (y_2 - y_1)]/a = 0,$$
$$m\,\ddot{y_2} + P[y_2 - y_1 - (y_3 - y_2)]/a = 0,$$
$$m\,\ddot{y_3} + P[(y_3 - y_2) + y_3]/a = 0.$$

Putting $y_i = a_i \sin(pt + \epsilon)$ and $n^2 = P/ma$ we obtain

$$(2n^2 - p^2)a_1 - n^2 a_2 = 0,$$
$$-n^2 a_1 + (2n^2 - p^2)a_2 - n^2 a_3 = 0,$$
$$-n^2 a_2 + (2n^2 - p^2)a_3 = 0,$$

and these three algebraic equations are consistent if

$$\begin{vmatrix} 2-k & -1 & 0 \\ -1 & 2-k & -1 \\ 0 & -1 & 2-k \end{vmatrix} = 0$$

where $k = p^2/n^2$. This gives $(2-k)^3 - 2(2-k) = 0$, that is $(2-k)(k^2 - 4k + 2) = 0$ so that $k = 2$ or $k = 2 \pm \sqrt{2}$. Hence p^2 has the values $p_1^2 = (2-\sqrt{2})n^2, p_2^2 = 2n^2, p_3^2 = (2+\sqrt{2})n^2$.
For p_1, that is $k = 2 - \sqrt{2}$, we have $a_2 = \sqrt{2}a_1, a_3 = a_1$; for p_2, that is $k = 2$, we have $a_2 = 0, a_3 = -a_1$; and for p_3, that is $k = 2 + \sqrt{2}$, we have $a_2 = -\sqrt{2}a_1, a_3 = a_1$.
Hence the normal modes are

$$y_1 = a_1 \sin(p_1 t + \epsilon_1), \quad y_2 = \sqrt{2}a_1 \sin(p_1 t + \epsilon_1), \quad y_3 = a_1 \sin(p_1 t + \epsilon_1);$$
$$y_1 = a_1 \sin(p_2 t + \epsilon_2), \quad y_2 = 0, \quad y_3 = -a_1 \sin(p_2 t + \epsilon_2);$$
$$y_1 = a_1 \sin(p_3 t + \epsilon_3), \quad y_2 = -\sqrt{2}a_1 \sin(p_3 t + \epsilon_3), \quad y_3 = a_1 \sin(p_3 t + \epsilon_3).$$

9

FOURIER SERIES, FOURIER AND LAPLACE TRANSFORMS

The subjects of *Fourier series, Fourier transforms* and *Laplace transforms* are essential for the study of wave motion and heat conduction to be discussed in Chapters 11 and 12. Fourier series and transforms were first developed by Jean Baptiste Joseph Fourier (1758-1830) in connection with his study of heat conduction in 1822, and Laplace transforms were introduced in 1782 by Pierre Simon Laplace (1749-1827) for the solution of linear difference equations and differential equations which arise in various physical problems.

9-1 Fourier series

The importance of the Fourier series, which is composed of the complete set of functions $\sin(n\pi x/L), \cos(n\pi x/L)$ $(n = 0, 1, 2, ...)$ defined in the range $(-L \leq x \leq L)$, is that any integrable square function, or more generally a function of bounded variation, may be expanded in terms of these trigonometric functions and thus may be used to solve the equations of applied mathematics.

Consider the following examples:

9.01 Find the values of A_1, A_2, A_3 such that the function

$$A_1 \sin \frac{\pi x}{L} + A_2 \sin \frac{2\pi x}{L} + A_3 \sin \frac{3\pi x}{L}$$

is the best mean square approximation to the function $f(x) = C$ over the interval $(0, L)$.

- *Solution*

The best mean square approximation to $f(x)$ using the Fourier expansion

$$\sum_{n=1}^{m} A_n \sin \frac{n\pi x}{L}$$

occurs for

$$A_n = \frac{2}{L} \int_0^L f(x) \sin \frac{n\pi x}{L} \, dx.$$

In the present problem $f(x) = C, m = 3$ and so we get

$$A_1 = \frac{2}{L} C \int_0^L \sin \frac{\pi x}{L} \, dx = \frac{4C}{\pi}, \quad A_2 = \frac{2}{L} \int_0^L \sin \frac{2\pi x}{L} \, dx = 0, \quad A_3 = \frac{2}{L} \int_0^L \sin \frac{3\pi x}{L} \, dx = \frac{4C}{3\pi}.$$

9.02 Find the Fourier series for the following functions $f(x)$ defined for $-\pi < x < \pi$:

$$(i) \ \ f(x) = x, \quad (ii) \ \ f(x) = e^x.$$

● *Solution*

(i) $f(x) = x$ is an odd function and so we may expand it using the *Fourier sine series*

$$f(x) = \sum_{n=1}^{\infty} b_n \sin nx \tag{9.1}$$

where

$$b_n = \frac{2}{\pi} \int_0^{\pi} f(x) \sin nx\, dx. \tag{9.2}$$

Then we have

$$b_n = \frac{2}{\pi} \int_0^{\pi} x \sin nx\, dx = -\frac{2}{n} \cos n\pi = -\frac{2}{n}(-1)^n$$

giving

$$x = \sum_{n=1}^{\infty} \frac{2(-1)^{n+1}}{n} \sin nx.$$

(ii) $f(x) = e^x$ is neither an even nor an odd function and so we must expand it using the full *Fourier series*

$$f(x) = \frac{a_0}{2} + \sum_{n=1}^{\infty} (a_n \cos nx + b_n \sin nx) \tag{9.3}$$

where

$$a_0 = \frac{1}{\pi} \int_{-\pi}^{\pi} f(x)\, dx, \quad a_n = \frac{1}{\pi} \int_{-\pi}^{\pi} f(x) \cos nx\, dx, \quad b_n = \frac{1}{\pi} \int_{-\pi}^{\pi} f(x) \sin nx\, dx. \tag{9.4}$$

Then we have

$$a_0 = \frac{1}{\pi} \int_{-\pi}^{\pi} e^x\, dx = \frac{1}{\pi}(e^{\pi} - e^{-\pi}) = \frac{2}{\pi} \sinh \pi,$$

$$a_n = \frac{1}{\pi} \int_{-\pi}^{\pi} e^x \cos nx\, dx = \frac{1}{2\pi} \int_{-\pi}^{\pi} [e^{(1+in)x} + e^{(1-in)x}]\, dx = \frac{2}{\pi} \frac{(-1)^n \sinh \pi}{1 + n^2},$$

$$b_n = \frac{1}{\pi} \int_{-\pi}^{\pi} e^x \sin nx\, dx = \frac{1}{2\pi i} \int_{-\pi}^{\pi} [e^{(1+in)x} - e^{(1-in)x}]\, dx = -\frac{2n}{\pi} \frac{(-1)^n \sinh \pi}{1 + n^2}$$

giving

$$e^x = \frac{1}{\pi} \sinh \pi \left[1 + 2 \sum_{n=1}^{\infty} \frac{(-1)^n}{1 + n^2} (\cos nx - n \sin nx) \right].$$

9.03 Find (a) the Fourier sine series and (b) the Fourier cosine series for the following functions $f(x)$ defined for $0 < x < \pi$:

$$(i)\ \ f(x) = x, \quad (ii)\ \ f(x) = \cos x.$$

● *Solution*

(i) $f(x) = x$

(a) Fourier sine series:
We have already found the solution for this in problem 9.02.

(b) Fourier cosine series:
We put

$$f(x) = \frac{a_0}{2} + \sum_{n=1}^{\infty} a_n \cos nx \qquad (9.5)$$

where

$$a_0 = \frac{2}{\pi} \int_0^{\pi} f(x)\, dx, \quad a_n = \frac{2}{\pi} \int_0^{\pi} f(x) \cos nx\, dx. \qquad (9.6)$$

Then we get

$$a_0 = \frac{2}{\pi} \int_0^{\pi} x\, dx = \pi,$$

$$a_n = \frac{2}{\pi} \int_0^{\pi} x \cos nx\, dx = \frac{2}{\pi} \frac{\cos n\pi - 1}{n^2} = \frac{2}{\pi} \frac{(-1)^n - 1}{n^2}$$

giving

$$f(x) = \frac{\pi}{2} + \frac{2}{\pi} \sum_{n=1}^{\infty} \frac{(-1)^n - 1}{n^2} \cos nx$$

which may be rewritten as

$$f(x) = \frac{\pi}{2} - \frac{4}{\pi} \sum_{n=0}^{\infty} \frac{\cos(2n+1)x}{(2n+1)^2}.$$

If we put $x = 0$ it can be readily verified that formula (9.7) is obtained.

(ii) $f(x) = \cos x$

(a) Fourier sine series:
We put

$$f(x) = \sum_{n=1}^{\infty} b_n \sin nx$$

where we have

$$b_n = \frac{2}{\pi} \int_0^{\pi} \cos x \sin nx\, dx = \frac{1}{\pi} \int_0^{\pi} [\sin(n+1)x + \sin(n-1)x]\, dx = \frac{1}{\pi}[1 - (-1)^{n+1}]\frac{2n}{n^2 - 1}$$

giving

$$\cos x = \frac{2}{\pi} \sum_{n=1}^{\infty} [1 + (-1)^n]\frac{n}{n^2 - 1} \sin nx$$

which may be rewritten as

$$\cos x = \frac{8}{\pi} \sum_{n=1}^{\infty} \frac{n}{4n^2 - 1} \sin 2nx.$$

(b) Fourier cosine series:
This is trivial since the solution is just $\cos x$.

9.04 Find the Fourier series for the function $f(x) = |x|$ defined for $-L < x < L$. Hence show that

$$\sum_{n=0}^{\infty} \frac{1}{(2n+1)^2} = \frac{\pi^2}{8}. \qquad (9.7)$$

• *Solution*

This is an even function and so we may expand in terms of the Fourier cosine series

$$f(x) = \frac{a_0}{2} + \sum_{n=1}^{\infty} a_n \cos \frac{n\pi x}{L}$$

where

$$a_0 = \frac{2}{L} \int_0^L f(x)\, dx, \quad a_n = \frac{2}{L} \int_0^L f(x) \cos \frac{n\pi x}{L}\, dx.$$

Then we get

$$a_0 = \frac{2}{L} \int_0^L x\, dx = L,$$

$$a_n = \frac{2}{L} \int_0^L x \cos \frac{n\pi x}{L}\, dx = \frac{2L}{n^2\pi^2}(\cos n\pi - 1) = \frac{2L}{n^2\pi^2}[(-1)^n - 1]$$

giving

$$|x| = \frac{L}{2} + \frac{2L}{\pi^2} \sum_{n=1}^{\infty} \frac{(-1)^n - 1}{n^2} \cos \frac{n\pi x}{L}$$

which may be rewritten as

$$|x| = \frac{L}{2} - \frac{4L}{\pi^2} \sum_{n=0}^{\infty} \frac{1}{(2n+1)^2} \cos \frac{(2n+1)\pi x}{L}.$$

Now put $x = L$. Then we obtain

$$L = \frac{L}{2} + \frac{4L}{\pi^2} \sum_{n=0}^{\infty} \frac{1}{(2n+1)^2}$$

which yields the required result.

9.05 Show that

$$x^2 = \frac{\pi^2}{3} + 4 \sum_{n=1}^{\infty} \frac{(-1)^n}{n^2} \cos nx$$

and hence evaluate

$$\sum_{n=1}^{\infty} \frac{(-1)^{n+1}}{n^2}, \quad \sum_{n=1}^{\infty} \frac{1}{n^2}.$$

Further, by using Parseval's theorem, obtain the formulae

$$\sum_{n=1}^{\infty} \frac{1}{n^4} = \frac{\pi^4}{90}, \quad \sum_{n=0}^{\infty} \frac{1}{(2n+1)^4} = \frac{\pi^4}{96}. \qquad (9.8)$$

• *Solution*

Since x^2 is an even function we may use the cosine series (9.5) given in problem 9.03. We obtain

$$a_0 = \frac{2}{\pi} \int_0^\pi x^2\, dx = \frac{2}{3}\pi^2,$$

$$a_n = \frac{2}{\pi} \int_0^\pi x^2 \cos nx\, dx = \frac{4}{n^2} \cos n\pi = \frac{4(-1)^n}{n^2}$$

and so required cosine series for x^2 follows.

Now setting $x = 0$ and $x = \pi$ respectively yields

$$\sum_{n=1}^{\infty} \frac{(-1)^{n+1}}{n^2} = \frac{\pi^2}{12}, \quad \sum_{n=1}^{\infty} \frac{1}{n^2} = \frac{\pi^2}{6}. \tag{9.9}$$

To evaluate the third and fourth sums we use Parseval's formula for the scalar product of two square integrable functions $f(x)$ and $g(x)$ in the form

$$\frac{1}{\pi} \int_{-\pi}^{\pi} f(x)g(x) \; dx = \frac{1}{2}a_0 a_0' + \sum_{n=1}^{\infty} (a_n a_n' + b_n b_n')$$

where a_n, b_n and a_n', b_n' are the Fourier coefficients of $f(x)$ and $g(x)$ respectively.

Putting $f(x) = g(x) = x^2$ we get $a_0 = a_0' = 2\pi^2/3$, $a_n = a_n' = 4(-1)^n n^{-2}$ and $b_n = b_n' = 0$. Now $\pi^{-1} \int_{-\pi}^{\pi} x^4 \; dx = 2\pi^4/5$ and so Parseval's formula gives $2\pi^4/5 = 2\pi^4/9 + 16 \sum_{n=1}^{\infty} n^{-4}$ which yields the required result for $\sum_{n=1}^{\infty} n^{-4}$.

Next putting $f(x) = x, g(x) = x^2$ we have $a_0 = \pi, a_n = (2/\pi)[(-1)^n - 1]n^{-2}$ and so Parseval's formula gives $\pi^3/2 = \pi^3/3 + (8/\pi) \sum_{n=1}^{\infty}[1 - (-1)^n]n^{-4}$ which gives the required result for $\sum_{n=0}^{\infty}(2n + 1)^{-4}$.

9.06 Obtain the Fourier series for the square wave with period T given by

$$f(t) = \begin{cases} 1 & (0 < t < T/2) \\ 0 & (T/2 < t < T) \end{cases}$$

- *Solution*

The Fourier series is given by

$$f(t) = \frac{a_0}{2} + \sum_{n=1}^{\infty} \left(a_n \cos \frac{2n\pi t}{T} + b_n \sin \frac{2n\pi t}{T} \right)$$

where

$$a_n = \frac{2}{T} \int_{-T/2}^{T/2} f(t) \cos \frac{2n\pi t}{T} \; dt, \quad b_n = \frac{2}{T} \int_{-T/2}^{T/2} f(t) \sin \frac{2n\pi t}{T} \; dt.$$

Hence we have for the square wave

$$a_0 = 1, \quad a_n = \frac{2}{T} \int_0^{T/2} \cos \frac{2n\pi t}{T} \; dt = \frac{1}{n\pi} \sin n\pi = 0 \quad (n \neq 0),$$

$$b_n = \frac{2}{T} \int_0^{T/2} \sin \frac{2n\pi t}{T} \; dt = \frac{1}{n\pi}(1 - \cos n\pi) = \frac{1}{n\pi}[1 - (-1)^n]$$

giving

$$f(t) = \frac{1}{2} + \sum_{n=1}^{\infty} \frac{1 - (-1)^n}{n\pi} \sin \frac{2n\pi t}{T}.$$

We obtain a sine series because $f(t) - \frac{1}{2}$ is an odd function about $t = T/2$.

9.07 Obtain the Fourier series for the function with period T given by

$$f(t) = \left| \cos \left(\frac{2\pi t}{T} \right) \right|.$$

- *Solution*

Since $f(t)$ is an even function with period $T/2$ we use the Fourier cosine series

$$f(t) = \frac{a_0}{2} + \sum_{n=1}^{\infty} a_n \cos \frac{4\pi n t}{T}$$

where

$$a_n = \frac{8}{T} \int_0^{T/4} f(t) \cos \frac{4\pi n t}{T} \, dt.$$

So we have

$$a_0 = \frac{8}{T} \int_0^{T/4} \cos \frac{2\pi t}{T} \, dt = \frac{4}{\pi},$$

$$a_n = \frac{8}{T} \int_0^{T/4} \cos \frac{2\pi t}{T} \cos \frac{4\pi n t}{T} \, dt = \frac{4}{T} \int_0^{T/4} \left[\cos \frac{2\pi(2n+1)t}{T} + \cos \frac{2\pi(2n-1)t}{T} \right] dt$$

and so

$$a_n = \frac{2}{\pi} \left[\frac{1}{2n+1} \sin \frac{\pi(2n+1)}{2} + \frac{1}{2n-1} \sin \frac{\pi(2n-1)}{2} \right] = \frac{4}{\pi} \frac{(-1)^{n+1}}{4n^2 - 1}.$$

Hence

$$f(t) = \frac{2}{\pi} + \frac{4}{\pi} \sum_{n=1}^{\infty} \frac{(-1)^{n+1}}{4n^2 - 1} \cos \frac{4\pi n t}{T}.$$

9.08 Show that

$$\exp x = \frac{\sinh \pi}{\pi} \sum_{n=-\infty}^{\infty} \frac{(-1)^n}{1 - in} \exp inx \quad (-\pi < x < \pi).$$

• *Solution*

We use the exponential form of the Fourier series

$$f(x) = \sum_{n=-\infty}^{\infty} c_n \exp \left(\frac{in\pi x}{L} \right) \tag{9.10}$$

where

$$c_n = \frac{1}{2L} \int_{-L}^{L} f(x) \exp \left(-\frac{in\pi x}{L} \right) dx. \tag{9.11}$$

Then, putting $L = \pi$ and $f(x) = \exp x$, we have

$$c_n = \frac{1}{2\pi} \int_{-\pi}^{\pi} \exp(1-in)x \, dx = \frac{1}{2\pi} \frac{1}{1 - in} \{ \exp[(1 - in)\pi] - \exp[-(1 - in)\pi] \} = \frac{\sinh \pi}{\pi} \frac{(-1)^n}{1 - in}$$

and this gives the required result.

9-2 Fourier transforms

In 1823 Fourier found the following *reciprocal* formulae:
(i) the *sine transform*

$$F(s) = \sqrt{\frac{2}{\pi}} \int_0^{\infty} \sin sx \, f(x) \, dx, \quad f(x) = \sqrt{\frac{2}{\pi}} \int_0^{\infty} \sin sx \, F(s) \, ds, \tag{9.12}$$

(ii) the *cosine transform*

$$F(s) = \sqrt{\frac{2}{\pi}} \int_0^\infty \cos sx \, f(x) \, dx, \quad f(x) = \sqrt{\frac{2}{\pi}} \int_0^\infty \cos sx \, F(s) \, ds. \qquad (9.13)$$

Also we have (iii) the *exponential transform*

$$F(u) = \frac{1}{\sqrt{2\pi}} \int_{-\infty}^\infty e^{iux} f(x) \, dx, \quad f(x) = \frac{1}{\sqrt{2\pi}} \int_{-\infty}^\infty e^{-iux} F(u) \, du. \qquad (9.14)$$

Fourier transforms can be used for solving heat conduction problems for semi-infinite and infinite solids since the transformed equations are more readily solved than the original equations, as we shall see in Chapter 12.

They can also be used, for example, in quantum theory where it may be convenient to transform from the coordinate representation to the momentum representation by means of a Fourier transform.

Meanwhile we shall use these reciprocal formulae in the following problems.

9.09 Find the function which satisfies the integral equation

$$\exp(-s) = \int_0^\infty f(x) \sin sx \, dx.$$

• *Solution*

We use the Fourier sine transform (9.12) here. We have

$$f(x) = \frac{2}{\pi} \int_0^\infty \exp(-s) \sin sx \, ds = \frac{2}{\pi} \int_0^\infty \frac{\exp\{-(1 - ix)s\} - \exp\{-(1 + ix)s\}}{2i} \, ds$$

which gives as the solution to the integral equation

$$f(x) = \frac{1}{\pi i} \left[\frac{1}{1 - ix} - \frac{1}{1 + ix} \right] = \frac{2}{\pi} \frac{x}{1 + x^2}.$$

9.10 If

$$f(x) = \begin{cases} 1 & (|x| < 1) \\ 0 & (|x| > 1) \end{cases}$$

and $f(1) = f(-1) = \frac{1}{2}$, show that

$$f(x) = \frac{2}{\pi} \int_0^\infty \frac{\sin s \cos(sx)}{s} \, ds \qquad (-\infty < x < \infty).$$

• *Solution*

Since $f(x)$ is even we use the Fourier cosine transformation (9.13) and put

$$f(x) = \frac{2}{\pi} \int_0^\infty F(s) \cos sx \, ds$$

where

$$F(s) = \int_0^\infty f(x) \cos sx \, dx = \int_0^1 \cos sx \, dx = \frac{\sin s}{s}$$

which proves the result.

9.11 If $f(-x) = f(x)$ and

$$f(x) = \begin{cases} 1 - x & (0 \le x \le 1) \\ 0 & (x > 1) \end{cases}$$

find the function $F(x)$ which satisfies

$$f(x) = \int_0^\infty F(s) \cos(sx) \, ds.$$

• *Solution*

Since $f(x)$ is an even function we can use the Fourier cosine transform (9.13) which gives

$$F(s) = \frac{2}{\pi} \int_0^\infty f(x) \cos sx \, dx = \frac{2}{\pi} \int_0^1 (1 - x) \cos sx \, dx = \frac{2}{\pi} \frac{1 - \cos s}{s^2}.$$

9.12 If $f(0) = \frac{1}{2}$ and

$$f(x) = \begin{cases} 0 & (x < 0) \\ e^{-x} & (x > 0) \end{cases}$$

show that

$$f(x) = \frac{1}{\pi} \int_0^\infty \frac{\cos(ux) + u \sin(ux)}{1 + u^2} \, du \qquad (-\infty < x < \infty).$$

• *Solution*

Here we use the Fourier exponential transform (9.14):

$$F(u) = \frac{1}{\sqrt{2\pi}} \int_{-\infty}^\infty e^{iux} f(x) \, dx = \frac{1}{\sqrt{2\pi}} \int_0^\infty e^{-(1-iu)x} \, dx = \frac{1}{\sqrt{2\pi}} \frac{1}{1 - iu}$$

so that we have

$$f(x) = \frac{1}{\sqrt{2\pi}} \int_{-\infty}^\infty e^{-iux} F(u) \, du = \frac{1}{2\pi} \int_{-\infty}^\infty \frac{e^{-iux}}{1 - iu} \, du$$

which gives, remembering that $f(x)$ is real,

$$f(x) = \frac{1}{2\pi} \int_{-\infty}^\infty \frac{\cos ux + u \sin ux}{1 + u^2} \, du$$

leading to the required result.

9.13 Show that

$$e^{-|x|} = \frac{2}{\pi} \int_0^\infty \frac{\cos(sx)}{1 + s^2} \, ds \quad (-\infty < x < \infty).$$

• *Solution*

Since $e^{-|x|}$ is an even function we use the cosine transform (9.13):

$$F(s) = \sqrt{\frac{2}{\pi}} \int_0^\infty f(x) \cos sx \, dx = \sqrt{\frac{2}{\pi}} \int_0^\infty e^{-x} \cos sx \, dx = \sqrt{\frac{2}{\pi}} \frac{1}{1 + s^2}$$

and so

$$f(x) = e^{-|x|} = \sqrt{\frac{2}{\pi}} \int_0^\infty F(s) \cos sx \, ds = \frac{2}{\pi} \int_0^\infty \frac{\cos(sx)}{1+s^2} \, ds.$$

9.14 Show that if

$$f(t) = \begin{cases} \cos at & (-T/2 < t < T/2) \\ 0 & (t \geq |T/2|) \end{cases}$$

then

$$f(t) = \frac{1}{\pi} \int_0^\infty \left\{ \frac{\sin[T(a-s)/2]}{a-s} + \frac{\sin[T(a+s)/2]}{a+s} \right\} \cos(st) \, ds.$$

• *Solution*

Since $f(t)$ is an even function we use the cosine transform (9.13).
We have

$$F(s) = \sqrt{\frac{2}{\pi}} \int_0^\infty f(t) \cos st \, dt = \sqrt{\frac{2}{\pi}} \int_0^{T/2} \cos at \cos st \, dt$$

which gives

$$F(s) = \sqrt{\frac{2}{\pi}} \int_0^{T/2} \frac{1}{2} [\cos(a+s)t + \cos(a-s)t] \, dt = \frac{1}{\sqrt{2\pi}} \left\{ \frac{\sin[(a+s)T/2]}{a+s} + \frac{\sin[(a-s)T/2]}{a-s} \right\}$$

and so the result follows from the reciprocal formula for $f(t)$.

9.15 If $f(x), f'(x), f''(x)$ are continuous for $x \geq 0$, and if f and $f' \to 0$ as $x \to \infty$, find the Fourier cosine transform of $f''(x)$ in terms of the Fourier cosine transform of $f(x)$ and $f'(0)$.
Hence evaluate

$$\int_0^\infty \frac{\cos sx}{a^2 + s^2} \, ds$$

by setting $f(x) = e^{-ax}$.

• *Solution*

Since

$$\int_0^\infty f''(x) \cos sx \, dx = [f'(x) \cos sx + s f(x) \sin sx]_0^\infty - s^2 \int_0^\infty f(x) \cos sx \, dx$$

we obtain

$$F''(s) = -\sqrt{2/\pi} f'(0) - s^2 F(s)$$

where $F(s)$ and $F''(s)$ are the Fourier cosine transforms of $f(x)$ and $f''(x)$.
We now set $f(x) = e^{-ax}$. Then $f'(0) = -a$ and so $F''(s) = \sqrt{2/\pi}a - s^2 F(s)$.
But $F''(s) = \sqrt{2/\pi} \int_0^\infty f''(x) \cos sx \, dx = a^2 F(s)$ and hence

$$F(s) = \sqrt{\frac{2}{\pi}} \frac{a}{a^2 + s^2}. \tag{9.15}$$

Since

$$f(x) = \sqrt{\frac{2}{\pi}} \int_0^\infty F(s) \cos sx \, ds$$

it follows that

$$e^{-ax} = \frac{2a}{\pi} \int_0^\infty \frac{\cos sx}{a^2 + s^2} \, ds.$$

9.16 If $F_1(s)$ and $F_2(s)$ are the Fourier cosine transforms of $f_1(u)$ and $f_2(u)$ respectively, show that the Fourier cosine transform of

$$\alpha(x) = \frac{1}{\sqrt{2\pi}} \int_0^\infty f_1(u)[f_2(|x - u|) + f_2(x + u)] \, du$$

is $\beta(s) = F_1(s)F_2(s)$.

Hence show that

$$\int_0^\infty f_1(u)f_2(u) \, du = \int_0^\infty F_1(s)F_2(s) \, ds.$$

By setting $f_1(u) = e^{-au}$ and $f_2(u) = e^{-bu}$ evaluate the integral

$$\int_0^\infty \frac{ds}{(a^2 + s^2)(b^2 + s^2)}.$$

• *Solution*

We have that

$$\beta(s) = \sqrt{\frac{2}{\pi}} \int_0^\infty \alpha(x) \cos sx \, dx = \frac{1}{\pi} \int_0^\infty \cos sx \, dx \int_0^\infty f_1(u)[f_2(|x - u|) + f_2(x + u)] \, du$$

and this gives

$$\beta(s) = \frac{1}{\pi} \int_0^\infty f_1(u) \, du \left\{ \int_{-u}^\infty \cos[s(u + t)]f_2(|t|) \, dt + \int_u^\infty \cos[s(t - u)]f_2(t) \, dt \right\}$$

which can be rewritten as

$$\beta(s) = \frac{2}{\pi} \int_0^\infty f_1(u) \cos su \, du \int_0^\infty f_2(t) \cos st \, dt = F_1(s)F_2(s).$$

Hence, using the reciprocal cosine formula for $\alpha(x)$ we get

$$\int_0^\infty f_1(u)[f_2(|x - u|) + f_2(x + u)] \, du = 2 \int_0^\infty F_1(s)F_2(s) \cos xs \, ds$$

and then putting $x = 0$ we obtain

$$\int_0^\infty f_1(u)f_2(u) \, du = \int_0^\infty F_1(s)F_2(s) \, ds.$$

Now we set $f_1(u) = e^{-au}$ and $f_2(u) = e^{-bu}$. Then, using (9.15), we have

$$F_1(s) = \sqrt{\frac{2}{\pi}} \frac{a}{a^2 + s^2}$$

from which we see that

$$\int_0^\infty F_1(s)F_2(s) \, ds = \frac{2ab}{\pi} \int_0^\infty \frac{ds}{(a^2 + s^2)(b^2 + s^2)},$$

and also we have

$$\int_0^\infty f_1(u)f_2(u) \, du = \int_0^\infty e^{-(a+b)u} \, du = \frac{1}{a + b}$$

giving as the final result

$$\int_0^\infty \frac{ds}{(a^2 + s^2)(b^2 + s^2)} = \frac{\pi}{2ab(a + b)}.$$

9-3 Laplace transforms

The *Laplace transform* of a function $f(t)$ is given by

$$F(s) = L\{f(t)\} = \int_0^\infty f(t)e^{-st}\,dt. \tag{9.16}$$

It can be used to evaluate integrals and to solve certain linear differential equations, for example those which arise in electric circuit theory.

9.17 Find the Laplace transform of e^{at}.

• *Solution*

We have

$$L[e^{at}] = \int_0^\infty e^{-(s-a)t}\,dt = \frac{1}{s-a} \tag{9.17}$$

if $s > a$.

9.18 Find the Laplace transform of $e^{at} - e^{bt}$.

• *Solution*

We have

$$L(e^{at} - e^{bt}) = \int_0^\infty (e^{at} - e^{bt})e^{-st}\,dt = \frac{1}{s-a} - \frac{1}{s-b}$$

if $s > a$ and b. Hence

$$L\{e^{at} - e^{bt}\} = \frac{a-b}{(s-a)(s-b)}. \tag{9.18}$$

9.19 If

$$I = \int_0^\infty e^{-x^2}\,dx$$

evaluate

$$I^2 = \int_0^\infty \int_0^\infty e^{-(x^2+y^2)}\,dx\,dy$$

using polar coordinates. Hence show that for $s > 0$

$$L\{t^{-\frac{1}{2}}\} = \left(\frac{\pi}{s}\right)^{\frac{1}{2}}. \tag{9.19}$$

• *Solution*

Transforming to polar coordinates r, θ we obtain

$$I^2 = \int_0^\infty e^{-x^2}\,dx \int_0^\infty e^{-y^2}\,dy = \int_0^\infty e^{-r^2} r\,dr \int_0^{\pi/2} d\theta$$

since $r^2 = x^2 + y^2$ and $dx\,dy = r\,dr\,d\theta$. Hence

$$I^2 = \frac{\pi}{2} \int_0^\infty e^{-r^2} r\,dr = \frac{\pi}{4}$$

and so $I = \sqrt{\pi}/2$.

Now

$$L\{t^{-1/2}\} = \int_0^\infty t^{-1/2} e^{-st}\, dt$$

and putting $x^2 = st$ we obtain

$$L\{t^{-1/2}\} = \frac{2}{\sqrt{s}} \int_0^\infty e^{-x^2}\, dx = \sqrt{\frac{\pi}{s}}.$$

9.20 Obtain the Laplace transform of $t^n e^{-at}$.

- *Solution*

We have

$$L\{t^n e^{-at}\} = \int_0^\infty t^n e^{-(s+a)t}\, dt.$$

Now

$$\int_0^\infty x^n e^{-\alpha x}\, dx = \frac{n!}{\alpha^{n+1}}$$

and so it follows that

$$L\{t^n e^{-at}\} = \frac{n!}{(s+a)^{n+1}} \quad (s > -a). \tag{9.20}$$

9.21 Find the inverse transforms of

$$(i) \quad f(s) = \frac{s}{(s-a)(s-b)},$$

$$(ii) \quad f(s) = \frac{a^n}{s(s+a)^n}.$$

- *Solution*

(i) Denoting the inverse transform of $f(s)$ by $L^{-1}\{f(s)\}$ and noting that

$$f(s) = \frac{1}{a-b}\left(\frac{a}{s-a} - \frac{b}{s-b}\right)$$

we get

$$L^{-1}\{f(s)\} = \frac{1}{a-b}\left[aL^{-1}\{(s-a)^{-1}\} - bL^{-1}\{(s-b)^{-1}\}\right].$$

Now using (9.17) we obtain

$$L^{-1}\{f(s)\} = \frac{ae^{at} - be^{bt}}{a-b}.$$

(ii) To solve this problem we use the expansion

$$\frac{a^n}{s(s+a)^n} = \frac{1}{s} - \frac{1}{s+a} - \frac{a}{(s+a)^2} - \cdots - \frac{a^{n-1}}{(s+a)^n}.$$

This gives

$$L^{-1}\left\{\frac{a^n}{s(s+a)^n}\right\} = L^{-1}\left\{\frac{1}{s} - \frac{1}{s+a} - \frac{a}{(s+a)^2} - \cdots - \frac{a^{n-1}}{(s+a)^n}\right\}$$

and so, using the result (9.20), we obtain

$$L^{-1}\left\{\frac{a^n}{s(s+a)^n}\right\} = 1 - e^{-at} - ate^{-at} - \cdots - \frac{(at)^{n-1}}{(n-1)!}e^{-at} = 1 - e^{-at}\sum_{r=0}^{n-1}\frac{(at)^r}{r!}.$$

9.22 Using Laplace transforms solve the differential equations
(i) $f''(t) - (a+b)f'(t) + abf(t) = 0$ for $f(t)$ with $f(0) = 0, f'(0) = 1$,
(ii) $tf''(t) + 2f'(t) - (t-2)f(t) = 2e^t$ for $f(t)$ with $f(0) = 0, f'(0) = 1$.

- *Solution*

Let $L\{f(t)\} = F(s)$. Then

$$L\{f'(t)\} = \int_0^\infty f'(t)e^{-st}dt = [f(t)e^{-st}]_0^\infty + s\int_0^\infty f(t)e^{-st}dt = sF(s) - f(0) \quad (9.21)$$

and so

$$L\{f''(t)\} = \int_0^\infty f''(t)e^{-st}dt = [f'(t)e^{-st}]_0^\infty + s\int_0^\infty f'(t)e^{-st}dt = s^2F(s) - sf(0) - f'(0).$$
$$(9.22)$$

(i) Applying the Laplace transform to the first differential equation and using the above results gives

$$s^2F(s) - 1 - (a+b)sF(s) + abF(s) = 0$$

and so

$$F(s) = \frac{1}{s^2 - (a+b)s + ab} = \frac{1}{(s-a)(s-b)}.$$

Hence using the result (9.18) we get

$$f(t) = \frac{e^{at} - e^{bt}}{a - b}.$$

(ii) We have that

$$L\{tf(t)\} = -\frac{d}{ds}L\{f(t)\} = -F'(s),$$

and using (9.22) we find that

$$L\{tf''(t)\} = -\frac{d}{ds}L\{f''(t)\} = -\frac{d}{ds}[s^2F(s) - sf(0) - f'(0)] = -2sF(s) - s^2F'(s) + f(0).$$
$$(9.23)$$

Then, since $L\{e^t\} = (s-1)^{-1}$, if we apply the Laplace transform to the second differential equation and use (9.21) and (9.23), we get

$$-2sF - s^2F' + 2sF + F' + 2F = \frac{2}{s-1}$$

so that F satisfies the first order differential equation

$$(1 - s^2)F' + 2F = \frac{2}{s-1}.$$

This can be written

$$\frac{d}{ds}\left(\frac{1+s}{1-s}F\right) = \frac{2}{(s-1)^3}$$

and so

$$F = \frac{1}{s^2 - 1} = \frac{1}{2}\left(\frac{1}{s-1} - \frac{1}{s+1}\right).$$

Hence

$$f(t) = \frac{1}{2}(e^t - e^{-t}) = \sinh t.$$

9.23 If $F(s) = L\{f(t)\}$ show that

$$L\{f^{(n)}(t)\} = s^nF(s) - s^{n-1}f(0) - s^{n-2}f'(0) - \dots - f^{(n-1)}(0)$$

for $s > \lambda$, where $f(t)$ and its derivatives $f'(t), \dots, f^{(n-1)}(t)$ are of order $e^{\lambda t}$ as $t \to \infty$.

- *Solution*

We shall prove this theorem by using the principle of induction.
The result is true for $n = 1$, as we have seen in the previous problem.
Let us suppose the result is true for $n - 1$. Then we have for $s > \lambda$

$$L\{f^{(n-1)}(t)\} = s^{n-1}F(s) - s^{n-2}f(0) - s^{n-3}f'(0) - \dots - f^{(n-2)}(0).$$

Now

$$L\left\{\frac{d}{dt}f^{(n-1)}(t)\right\} = sL\{f^{(n-1)}(t)\} - f^{(n-1)}(0)$$

and so the result is true for n. Thus, by the principle of induction, the theorem follows.

9.24 If $F(s) = L\{f(t)\}$ show that

$$F(bs - a) = L\left\{\frac{1}{b}\exp\left(\frac{at}{b}\right)f\left(\frac{t}{b}\right)\right\}. \tag{9.24}$$

- *Solution*

We have, putting $bu = t$, that

$$F(bs - a) = \int_0^\infty e^{-(bs-a)u}f(u)\,du = \frac{1}{b}\int_0^\infty e^{-st}\exp\left(\frac{at}{b}\right)f\left(\frac{t}{b}\right)dt$$

and this yields the required result.

9.25 Use the convolution theorem to show that

$$L^{-1}\left\{\frac{1}{s^{\frac{1}{2}}(s-1)}\right\} = e^t\,\mathrm{erf}\ t^{\frac{1}{2}}$$

where the *error function* is defined by

$$\mathrm{erf}\ x = \frac{2}{\sqrt{\pi}}\int_0^x e^{-r^2}dr.$$

Hence evaluate $L\{\mathrm{erf}\ t^{\frac{1}{2}}\}$.

- *Solution*

If $F(s)$ and $G(s)$ are the Laplace transforms of $f(t)$ and $g(t)$ respectively, the *convolution theorem* states that $F(s)G(s)$ is the Laplace transform of

$$\int_0^t f(t - u)g(u)\,du.$$

To verify that this is correct, without any attempt at a rigorous argument, we note that

$$\int_0^\infty e^{-st}dt\int_0^t f(t - u)g(u)\,du = \int_0^\infty e^{-sv}f(v)\,dv\int_0^\infty e^{-su}g(u)\,du = F(s)G(s)$$

on setting $v = t - u$ and taking $f(t)$ to vanish for $t < 0$.

Now we know from (9.19) that $s^{-1/2} = \pi^{-1/2}L\{t^{-1/2}\}$. Also $(s-1)^{-1} = L\{e^t\}$ and so we see that

$$L^{-1}\left\{\frac{1}{s^{1/2}(s-1)}\right\} = \frac{1}{\sqrt{\pi}}\int_0^t \frac{e^{t-u}}{u^{1/2}}\,du = \frac{2e^t}{\sqrt{\pi}}\int_0^{t^{1/2}} e^{-r^2}\,dr = e^t\,\mathrm{erf}\;t^{1/2}$$

as required.

Hence

$$L^{-1}\left\{\frac{1}{s(s+1)^{1/2}}\right\} = \mathrm{erf}\;t^{1/2}$$

since $F(s-a) = L\{e^{at}f(t)\}$, putting $b=1$ in (9.24). It follows that

$$L\{\mathrm{erf}\;t^{1/2}\} = \frac{1}{s(s+1)^{1/2}}.$$

9.26 Show that

$$L\left\{\frac{e^{-at}-e^{-bt}}{t}\right\} = \ln\frac{s+b}{s+a}.$$

• *Solution*

We have $L\{e^{-at}\} = (s+a)^{-1}$ and

$$\frac{\partial}{\partial s}L\left\{\frac{e^{-at}-e^{-bt}}{t}\right\} = -L\left\{e^{-at}-e^{-bt}\right\}.$$

Hence

$$L\left\{\frac{e^{-at}-e^{-bt}}{t}\right\} = \int_s^\infty L\{e^{-at}-e^{-bt}\}\,ds = \int_s^\infty\left(\frac{1}{x+a}-\frac{1}{x+b}\right)dx = [\ln(x+a)-\ln(x+b)]_s^\infty$$

and this gives the required formula.

9.27 Show that

$$L\{-Ei(-t)\} = s^{-1}\ln(1+s)$$

where

$$E_i(t) = \int_{-\infty}^t \frac{e^x}{x}\,dx$$

is the exponential integral.

• *Solution*

We may write

$$-E_i(-t) = \int_{-t}^{-\infty}\frac{e^x}{x}\,dx = \int_t^\infty \frac{e^{-y}}{y}\,dy = \int_1^\infty \frac{e^{-tu}}{u}\,du$$

and so

$$L\{-E_i(-t)\} = \int_0^\infty e^{-ts}\,dt\int_1^\infty \frac{e^{-tu}}{u}\,du = \int_1^\infty \frac{du}{u}\int_0^\infty e^{-(s+u)t}\,dt = \int_1^\infty \frac{du}{u(s+u)}$$

which gives

$$L\{-E_i(-t)\} = \frac{1}{s}\int_1^\infty\left(\frac{1}{u}-\frac{1}{s+u}\right)du = \frac{1}{s}[\ln u - \ln(s+u)]_1^\infty = \frac{1}{s}\ln(1+s).$$

9.28 Show that the Laplace transform of the triangular wave defined for all t by

$$f(t) = \begin{cases} t & (0 < t < T/2) \\ T - t & (T/2 < t < T) \end{cases}$$

where $f(t + T) = f(t)$, is

$$\frac{1}{s^2} \tanh \frac{sT}{4}.$$

● *Solution*

We have that the Laplace transform of $f(t)$ is given by

$$F(s) = \int_0^\infty e^{-st} f(t) \, dt = \sum_{n=0}^\infty \int_{nT}^{(n+1)T} e^{-st} f(t) \, dt.$$

Next we set $u = t - nT$ and use $f(u + nT) = f(u)$. Then we get

$$F(s) = \sum_{n=0}^\infty e^{-nTs} \int_0^T e^{-su} f(u) \, du = (1 - e^{-Ts})^{-1} \int_0^T e^{-su} f(u) \, du \quad (s > 0)$$

since the sum of the geometric series which occurs here is $\sum_{n=0}^\infty e^{-nTs} = (1 - e^{-Ts})^{-1}$.
Now

$$\int_0^T e^{-su} f(u) \, du = \int_0^{T/2} e^{-su} u \, du + \int_{T/2}^T e^{-su}(T - u) \, du = \frac{1}{s^2}(1 - 2e^{-sT/2} + e^{-sT})$$

and so

$$F(s) = \frac{1}{s^2} \frac{(1 - e^{-sT/2})^2}{1 - e^{-sT}} = \frac{1}{s^2} \frac{1 - e^{-sT/2}}{1 + e^{-sT/2}} = \frac{1}{s^2} \frac{e^{sT/4} - e^{-sT/4}}{e^{sT/4} + e^{-sT/4}}$$

which gives the required result.

9.29 Show that the Laplace transform of the half wave rectification $f(t)$ of the function $\sin at$ given by

$$f(t) = \begin{cases} \sin at & (0 < t < \pi/a) \\ 0 & (\pi/a < t < 2\pi/a) \end{cases}$$

where $f(t + 2\pi/a) = f(t)$, is

$$F(s) = \frac{a}{s^2 + a^2} \frac{1}{1 - e^{-\pi s/a}}.$$

● *Solution*

The Laplace transform of $f(t)$ is given by

$$F(s) = \int_0^\infty e^{-st} f(t) \, dt = \sum_{n=0}^\infty \int_{2\pi n/a}^{2\pi(n+1)/a} e^{-st} f(t) \, dt.$$

Setting $u = t - 2\pi n/a$ and using $f(u + 2\pi n/a) = f(u)$ we obtain

$$F(s) = \sum_{n=0}^\infty e^{-2\pi ns/a} \int_0^{2\pi/a} e^{-su} f(u) \, du = (1 - e^{-2\pi s/a})^{-1} \int_0^{2\pi/a} e^{-su} f(u) \, du.$$

Now

$$\int_0^{2\pi/a} e^{-su} f(u)\, du = \int_0^{\pi/a} e^{-su} \sin au\, du = \frac{1}{2i}\int_0^{\pi/a} [e^{(ia-s)u} - e^{-(ia+s)u}]\, du = \frac{a(1 + e^{-s\pi/a})}{s^2 + a^2}$$

and the required result follows.

9.30 An electric circuit is composed of an inductance L, a resistance R and a capacity C in series with an electromotive force $E_0\delta(t)$ where δ is the Dirac δ function given by

$$\delta(t) = \lim_{h\to 0} I(h, t)$$

and I is the impulse function defined as

$$I(h, t) = \begin{cases} h^{-1} & (0 \leq t \leq h \\ 0 & (t < 0, t > h). \end{cases}$$

If $L/C > R^2/4$ find the charge on the condenser at any time $t > 0$ given that the initial charge and the initial current are both zero.

• *Solution*

The differential equation for the circuit is

$$L\frac{di}{dt} + Ri + \frac{q}{C} = E_0\delta(t)$$

where q is the charge on the condenser and $i = dq/dt$ is the current in the circuit, and so we have

$$L\frac{d^2q}{dt^2} + R\frac{dq}{dt} + \frac{q}{C} = E_0\delta(t). \tag{9.25}$$

Now the Laplace transform of the impulse function is

$$L\{I(h, t)\} = \frac{1}{h}\int_0^h e^{-st}\, dt = \frac{1}{sh}(1 - e^{-sh})$$

and so the Laplace transform of the δ function is

$$L\{\delta(t)\} = \lim_{h\to 0} \frac{1}{sh}(1 - e^{-sh}) = 1.$$

Hence, taking Laplace transforms and using (9.21) and (9.22), we obtain

$$s^2 QL + sQR + \frac{Q}{C} = E_0$$

where Q is the Laplace transform of q, and so

$$Q = \frac{E_0}{s^2 L + sR + 1/C} = \frac{E_0}{L}\left\{\frac{1}{[s + R/(2L)]^2 + 1/LC - R^2/4L^2}\right\}.$$

Since $L\{e^{-bt} \sin at\} = \int_0^\infty e^{-(s+b)t} \sin at\, dt = (2i)^{-1}\int_0^\infty [e^{-(s+b-ia)t} - e^{-(s+b+ia)t}]\, dt$ it follows that $L\{e^{-bt} \sin at\} = a/[(s + b)^2 + a^2]$ and so we find that

$$q(t) = \frac{E_0}{L}\left(\frac{1}{LC} - \frac{R^2}{4L^2}\right)^{-1/2} \exp\left(-\frac{Rt}{2L}\right) \sin\left[\left(\frac{1}{LC} - \frac{R^2}{4L^2}\right)^{1/2} t\right]$$

remembering that $L/C > R^2/4$.

This satisfies the initial condition $q(0) = 0$. There is an instantaneous jump $[i] = E_0/L$ in the current i at $t = 0$.

We can see that this is correct by integrating both sides of (9.25) over the interval $[-\epsilon, \epsilon]$ and then letting $\epsilon \to 0$. Thus the term on the right hand side yields E_0 while the first term on the left hand side yields $L[i]$ while the remaining terms give a zero contribution.

By using the Laplace transform in the above electric circuit problem, we have been able to convert a second-order differential equation into an easily solvable equation in algebra which, on using the inverse Laplace transform, yields the solution to the original problem.

10
INTEGRAL EQUATIONS

The history of integral equations goes back to Laplace who in 1782 used the integral transform associated with his name, discussed in the previous chapter, to solve difference and differential equations. However, the name *integral equation* for any equation with the unknown function $\phi(x)$ under the integral sign was introduced by du Bois-Reymond in 1888.

Linear integral equations are named after Vito Volterra (1860-1940) and Erik Ivar Fredholm (1866-1927) in the following way:

The *Fredholm equation of the first kind* has the form

$$f(x) = \int_a^b K(x,s)\phi(s)\,ds \quad (a \le x \le b) \tag{10.1}$$

and the *Fredholm equation of the second kind* has the form

$$\phi(x) = f(x) + \int_a^b K(x,s)\phi(s)\,ds \quad (a \le x \le b) \tag{10.2}$$

while its corresponding *homogeneous equation* is

$$\phi(x) = \int_a^b K(x,s)\phi(s)\,ds \quad (a \le x \le b). \tag{10.3}$$

The *Volterra equation of the first kind* has the form

$$f(x) = \int_a^x K(x,s)\phi(s)\,ds \quad (a \le x \le b) \tag{10.4}$$

and the *Volterra equation of the second kind* has the form

$$\phi(x) = f(x) + \int_a^x K(x,s)\phi(s)\,ds \quad (a \le x \le b). \tag{10.5}$$

In integral equations the boundary conditions are automatically incorporated within the formulation whereas in the case of differential equations the boundary conditions have to be imposed externally.

This can be seen in the following example:

10.01 Transform the Volterra integral equation of the first kind

$$x = \int_0^x (e^x + e^s)\phi(s)\,ds$$

into a Volterra equation of the second kind. Show that the solution ϕ satisfies a first-order differential equation and hence solve the integral equation.

- *Solution*

By differentiating the Volterra integral equation of the first kind we get

$$1 = 2e^x \phi(x) + e^x \int_0^x \phi(s)\, ds$$

and hence

$$\phi(x) = \frac{1}{2}e^{-x} - \frac{1}{2}\int_0^x \phi(s)\, ds$$

which is a Volterra equation of the second kind.

Differentiating once more then gives the first order differential equation $\phi'(x) + \frac{1}{2}\phi(x) = -\frac{1}{2}e^{-x}$ which has the particular integral e^{-x} and complementary function $Ce^{-x/2}$. Since $\phi(0) = \frac{1}{2}$ it follows that the solution to the integral equation is

$$\phi(x) = e^{-x} - \frac{1}{2}e^{-x/2}.$$

10.02 Solve the Fredholm equation

$$\phi(x) = 1 + \lambda \int_0^1 x^n s^m \phi(s)\, ds$$

and show that the characteristic value of the associated homogeneous equation is $\lambda_1 = m + n + 1$.

- *Solution*

This is an example of a Fredholm equation of the second kind (10.2) with *separable kernel* $K(x,s) = \lambda u(x)\overline{v}(s)$:

$$\phi(x) = f(x) + \lambda u(x) \int_a^b \overline{v}(s)\phi(s)\, ds \tag{10.6}$$

whose solution can be obtained by multiplying across by $\overline{v}(x)$ and integrating. This gives

$$\int_a^b \overline{v}(x)\phi(x)\, dx = \frac{\int_a^b \overline{v}(x)f(x)\, dx}{1 - \lambda \int_a^b \overline{v}(x)u(x)\, dx}$$

and so we obtain the solution

$$\phi(x) = f(x) + \lambda \int_a^b R(x,s;\lambda)f(s)\, ds \tag{10.7}$$

where

$$R(x,s;\lambda) = \frac{u(x)\overline{v}(s)}{1 - \lambda \int_a^b \overline{v}(t)u(t)\, dt} \tag{10.8}$$

is called the *resolvent* kernel.

In the present problem $f(x) = 1, u(x) = x^n, v(s) = s^m$ and so the resolvent kernel is

$$R(x,s;\lambda) = \frac{x^n s^m}{1 - \lambda \alpha}$$

where $\alpha = \int_0^1 t^{m+n}\, dt = (m+n+1)^{-1}$. Hence

$$\phi(x) = 1 + \lambda x^n \left(1 - \frac{\lambda}{m+n+1}\right)^{-1} \int_0^1 s^m\, ds = 1 + \frac{\lambda(m+n+1)x^n}{(m+1)(m+n+1-\lambda)} \qquad (\lambda \neq m+n+1).$$

The homogeneous equation is

$$\phi(x) = \lambda x^n \int_0^1 s^m \phi(s)\, ds.$$

The values of λ for which the homogeneous equation can be solved are called the *characteristic values*. This equation has just one characteristic value λ_1 which satisfies

$$1 = \lambda_1 \int_0^1 x^{m+n}\, ds = \frac{\lambda_1}{m+n+1}$$

giving $\lambda_1 = m + n + 1 = \alpha^{-1}$.

10.03 Solve the Fredholm equation

$$\phi(x) = x + \lambda \int_0^\pi \sin nx \sin ns\ \phi(s)\, ds$$

where n is an integer and show that the characteristic value of the associated homogeneous equation is $\lambda_1 = 2/\pi$.

• *Solution*

For this example of a Fredholm equation of the second kind (10.6) we have $f(x) = x, u(x) = \sin nx, v(s) = \sin ns$. Then the resolvent kernel is

$$R(x, s; \lambda) = \frac{\sin nx \sin ns}{1 - \lambda \alpha}$$

where $\alpha = \int_0^\pi \sin^2 ns\, ds = \pi/2$.

Hence the solution of the Fredholm equation is

$$\phi(x) = x + \frac{\lambda \sin nx}{1 - \lambda \pi/2} \int_0^\pi s \sin ns\, ds = x + \frac{2\lambda \sin nx}{2 - \lambda \pi}\left(\frac{-\pi \cos n\pi}{n}\right) = x - \frac{2(-1)^n \lambda \pi \sin nx}{n(2 - \lambda \pi)}.$$

The homogeneous equation is

$$\phi(x) = \lambda \sin nx \int_0^\pi \sin ns\ \phi(s)\, ds$$

and its characteristic value λ_1 satisfies $1 = \lambda_1 \int_0^\pi \sin^2 x\, dx = \lambda_1 \pi/2$ giving $\lambda_1 = 2/\pi = \alpha^{-1}$.

10.04 Verify that the *singular integral equation*

$$\phi(x) = \lambda \sqrt{\frac{2}{\pi}} \int_0^\infty \sin xs\ \phi(s)\, ds$$

has the characteristic solutions

$$\phi_\pm(x) = \sqrt{\frac{\pi}{2}} e^{-ax} \pm \frac{x}{a^2 + x^2} \qquad (x > 0)$$

associated with characteristic values $\lambda = \pm 1$ for all $a > 0$.

• *Solution*

To solve this equation we use the Fourier sine transform

$$\Phi(x) = \sqrt{\frac{2}{\pi}} \int_0^\infty \sin xs \ \phi(s) \, ds$$

which gives $\phi(x) = \lambda \Phi(x)$ and so, using the reciprocal Fourier sine transform given by (9.12)

$$\phi(x) = \sqrt{\frac{2}{\pi}} \int_0^\infty \sin xs \ \Phi(s) \, ds,$$

we obtain

$$\lambda\phi(x) = \sqrt{\frac{2}{\pi}} \int_0^\infty \sin xs \ \phi(s) \, ds = \Phi(x)$$

Hence $\phi(x) = \lambda^2 \phi(x)$ giving $\lambda^2 = 1$ so that $\lambda = \pm 1$.
Now

$$\int_0^\infty e^{-as} \sin xs \, ds = \frac{1}{2i} \int_0^\infty [e^{-(a-ix)s} - e^{-(a+ix)s}] \, ds = \frac{x}{a^2 + x^2} \quad (a > 0)$$

and so by the reciprocal Fourier sine transform we have

$$\int_0^\infty \frac{x}{a^2 + x^2} \sin xs \, dx = \frac{\pi}{2} e^{-as} \quad (a > 0).$$

Hence we see that

$$\sqrt{\frac{\pi}{2}} e^{-ax} \pm \frac{x}{a^2 + x^2} = \pm \sqrt{\frac{2}{\pi}} \int_0^\infty \sin xs \ \left(\sqrt{\frac{\pi}{2}} e^{-as} \pm \frac{s}{a^2 + s^2} \right) ds \quad (x > 0)$$

and this proves the result.

10.05 Solve the *non-linear* integral equation

$$\phi(x) = 1 + \lambda \int_0^1 [\phi(s)]^2 \, ds$$

and show that $\lambda = \frac{1}{4}$ is a *bifurcation point* while $\lambda = 0$ is a *singular point* of the spectrum of characteristic values.

- *Solution*

We put

$$\int_0^1 [\phi(s)]^2 \, ds = \alpha$$

and then we have $\phi(x) = 1 + \lambda\alpha$ so that $(1 + \lambda\alpha)^2 = \alpha$, that is $\lambda^2\alpha^2 + (2\lambda - 1)\alpha + 1 = 0$.
This gives

$$\alpha = \frac{1 - 2\lambda \pm \sqrt{1 - 4\lambda}}{2\lambda^2}$$

and hence

$$\phi(x) = \frac{1 \pm \sqrt{1 - 4\lambda}}{2\lambda}.$$

There are two real solutions if $\lambda < \frac{1}{4}$, one real solution if $\lambda = \frac{1}{4}$ and no real solutions if $\lambda > \frac{1}{4}$. Thus $\lambda = \frac{1}{4}$ is a bifurcation point.
$\lambda = 0$ is a singular point. One solution is $\phi(x) = 1$ while the other solution is infinite since for small λ the associated solution behaves as $\phi(x) \sim \lambda^{-1}$.

10.06 Solve the Volterra equation of the second kind

$$\phi(x) = f(x) + \lambda \int_0^x \sin(x - s)\, \phi(s)\, ds$$

for (i) $f(x) = x$ and $\lambda = 1$,   (ii) $f(x) = e^{-x}$ and $\lambda = 2$.

• *Solution*

We solve this integral equation by transforming it into a differential equation. We see that

$$\phi'(x) = f'(x) + \lambda \int_0^x \cos(x - s)\, \phi(s)\, ds$$

and

$$\phi''(x) = f''(x) + \lambda\phi(x) - \lambda \int_0^x \sin(x - s)\, \phi(s)\, ds$$

giving

$$\phi''(x) + (1 - \lambda)\phi(x) = f(x) + f''(x).$$

(i) Since we have $f(x) = x$ and $\lambda = 1$, it follows that $\phi''(x) = x$ and so $\phi'(x) = \frac{1}{2}x^2 + 1$ since $\phi'(0) = 1$. Hence

$$\phi(x) = \frac{1}{6}x^3 + x$$

since $\phi(0) = 0$.

(ii) Here $f(x) = e^{-x}$, $\lambda = 2$ and hence $\phi''(x) - \phi(x) = 2e^{-x}$. This differential equation has a particular integral $-xe^{-x}$ and the complementary function $A \sinh x + B \cosh x$ where A and B are arbitrary constants. Now $\phi(0) = 1$ and so $B = 1$. Also $\phi'(0) = -1$ and so $A = 0$. Hence the solution to the integral equation is

$$\phi(x) = \cosh x - xe^{-x}.$$

10.07 Show that the Volterra equation of the first kind

$$f(x) = \int_0^x \frac{(x - s)^{n-1}}{(n - 1)!} \phi(s)\, ds$$

has the continuous solution $\phi(s) = f^{(n)}(s)$ when $f(x)$ has continuous derivatives $f'(x)$, $f''(x), ..., f^{(n)}(x)$, and $f(x)$ and its first $n - 1$ derivatives vanish at $x = 0$.

• *Solution*

We have

$$f'(x) = \int_0^x \frac{(x - s)^{n-2}}{(n - 2)!} \phi(s)\, ds,$$

$$f''(x) = \int_0^x \frac{(x - s)^{n-3}}{(n - 3)!} \phi(s)\, ds,$$

and so on, until we reach

$$f^{(n-1)}(x) = \int_0^x \phi(s)\, ds$$

giving

$$f^{(n)}(x) = \phi(x)$$

as the continuous solution of the integral equation.

10.08 Show that the solution of the differential equation

$$\frac{d^2\phi}{dx^2} + \omega^2\phi = F[x, \phi(x)] \qquad (0 \le x \le l)$$

obeying the end conditions $\phi(0) = \phi(l) = 0$, satisfies the integral equation

$$\phi(x) = \int_0^l G(x, s)F[s, \phi(s)] \, ds$$

where

$$G(x, s) = \begin{cases} -(\omega \sin \omega l)^{-1} \sin \omega x \sin \omega(l - s) & (x \le s) \\ -(\omega \sin \omega l)^{-1} \sin \omega(l - x) \sin \omega s & (x \ge s) \end{cases}$$

is the Green's function for the linear differential operator $d^2/dx^2 + \omega^2$.

• *Solution*

The solution $v(x)$ of the second order linear differential equation

$$\frac{d}{dx}\left[p(x)\frac{dv(x)}{dx}\right] + q(x)v(x) = r(x) \quad (a \le x \le b)$$

satisfying homogeneous boundary conditions at $x = a$ and $x = b$ may be expressed in the form

$$v(x) = \int_a^b G(x, s)r(s) \, ds$$

where $G(x, s)$ is the *Green's function*.

If $v_1(x)$ and $v_2(x)$ are two linearly independent solutions of the homogeneous equation

$$\frac{d}{dx}\left[p(x)\frac{dv_i(x)}{dx}\right] + q(x)v_i(x) = 0 \quad (a \le x \le b)$$

satisfying the same boundary conditions at $x = a$ and $x = b$ respectively, we have for the Green's function

$$G(x, s) = \begin{cases} A^{-1}v_1(x)v_2(s) & (x \le s) \\ A^{-1}v_2(x)v_1(s) & (x \ge s) \end{cases}$$

where

$$A = p(x)[v_1(x)v_2'(x) - v_2(x)v_1'(x)].$$

In the present problem we have the linearly independent solutions $v_1(x) = \sin \omega x$ and $v_2(x) = \sin \omega(l - x)$ satisfying

$$\frac{d^2 v_i}{dx^2} + \omega^2 v_i(x) = 0$$

with the boundary conditions $v_1(0) = 0$ and $v_2(l) = 0$ respectively. Hence

$$G(x, s) = \begin{cases} A^{-1} \sin \omega x \sin \omega(l - s) & (x \le s) \\ A^{-1} \sin \omega(l - x) \sin \omega s & (x \ge s) \end{cases}$$

where $A = v_1(x)v_2'(x) - v_2(x)v_1'(x) = -\omega[\sin \omega x \cos \omega(l - x) + \sin \omega(l - x) \cos \omega x]$ giving $A = -\omega \sin \omega l$ as required.

10.09 Show that

$$f(x) = \int_{-\infty}^{\infty} e^{-|x-s|}\phi(s) \, ds \qquad (-\infty < x < \infty)$$

has the solution $\phi(x) = \frac{1}{2}[f(x) - f''(x)]$ provided f, $f' \to 0$ as $x \to \pm\infty$.

- *Solution*

This is an example of an *integral equation of the convolution type*

$$f(x) = \int_{-\infty}^{\infty} k(x-s)\phi(s)\, ds \qquad (10.9)$$

which can be solved by using the *convolution theorem* for Fourier transforms which states that $\sqrt{2\pi}K(u)\Phi(u)$ is the exponential Fourier transform of $\int_{-\infty}^{\infty} k(x-s)\phi(s)\, ds$, where $K(u)$ and $\Phi(u)$ are the exponential Fourier transforms of $k(x)$ and $\phi(x)$ respectively.

To verify that this is correct we note that

$$\frac{1}{\sqrt{2\pi}} \int_{-\infty}^{\infty} e^{iux}\, dx \int_{-\infty}^{\infty} k(x-s)\phi(s)\, ds = \frac{1}{\sqrt{2\pi}} \int_{-\infty}^{\infty} e^{iuv}k(v)\, dv \int_{-\infty}^{\infty} e^{ius}\phi(s)\, ds$$

on setting $v = x - s$.

Hence, if $F(u)$ is the exponential Fourier transform (9.14) of $f(x)$, we have

$$F(u) = \sqrt{2\pi}K(u)\Phi(u)$$

and so, using the reciprocal exponential Fourier transform, we obtain the solution

$$\phi(x) = \frac{1}{2\pi} \int_{-\infty}^{\infty} e^{-ixu}\frac{F(u)}{K(u)}\, du.$$

In the present problem $k(x) = e^{-|x|}$ for which the exponential Fourier transform is given by

$$K(u) = \frac{1}{\sqrt{2\pi}} \int_{-\infty}^{\infty} e^{iux}e^{-|x|}\, dx = \sqrt{\frac{2}{\pi}} \int_{0}^{\infty} e^{-x}\cos ux\, dx = \sqrt{\frac{2}{\pi}}\frac{1}{1+u^2}.$$

Thus

$$\phi(x) = \frac{1}{2\sqrt{2\pi}} \int_{-\infty}^{\infty} e^{-ixu}(1+u^2)F(u)\, du.$$

Now

$$\int_{-\infty}^{\infty} e^{iux}f''(x)\, dx = \left[e^{iux}\{f'(x) - iuf(x)\}\right]_{-\infty}^{\infty} - u^2 \int_{-\infty}^{\infty} e^{iux}f(x)\, dx$$

and since $f, f' \to 0$ as $x \to \pm\infty$ it follows that $-u^2 F(u)$ is the exponential Fourier transform of $f''(x)$.

Hence the solution of the integral equation is $\phi(x) = \frac{1}{2}[f(x) - f''(x)]$ as required.

10.10 If $F(u)$ is the Laplace transform of $f(x)$ show that the transform of $f'(x)$ is $uF(u) - f(0)$ for $u > \alpha$ provided $f(x)e^{-\alpha x} \to 0$ as $x \to \infty$. Hence find the solution of the Volterra equation of the first kind

$$f(x) = \int_{0}^{x} e^{x-s}\phi(s)\, ds$$

where $f(0) = 0$.

- *Solution*

From problem 9.22 we know that $L\{f'(x)\} = uF(u) - f(0)$ and from problem 9.17 we know that $L\{e^x\} = (u-1)^{-1}$.

Hence using the convolution theorem for Laplace transforms given in problem 9.25 we see that

$$\phi(x) = L^{-1}\{(u-1)F(u)\} = f'(x) - f(x)$$

since $f(0) = 0$.

This result can also be obtained by direct differentiation of both sides of the integral equation:

$$f'(x) = \phi(x) + \int_0^x e^{x-s}\phi(s)\,ds = \phi(x) + f(x).$$

10.11 Show that the solution of *Fox's integral equation*

$$\phi(x) = f(x) + \lambda\sqrt{\frac{2}{\pi}}\int_0^\infty \sin xs\,\phi(s)\,ds \qquad (0 < x < \infty)$$

is given by

$$\phi(x) = \frac{f(x)}{1-\lambda^2} + \frac{\lambda}{1-\lambda^2}\sqrt{\frac{2}{\pi}}\int_0^\infty \sin xs\ f(s)\,ds \qquad (0 < x < \infty).$$

• *Solution*

We have that

$$\phi(x) = f(x) + \lambda\Phi(x)$$

where $\Phi(x)$ is the Fourier sine transform of $\phi(s)$.

Now taking the Fourier sine transform of both sides of the above we get

$$\Phi(s) = F(s) + \lambda\phi(s)$$

where

$$F(s) = \sqrt{\frac{2}{\pi}}\int_0^\infty \sin xs\ f(x)\,dx$$

and so

$$\phi(x) = f(x) + \lambda[F(x) + \lambda\phi(x)]$$

which gives

$$\phi(x) = \frac{f(x) + \lambda F(x)}{1-\lambda^2}$$

and so the required solution is obtained.

10.12 Use the *method of successive approximations* to solve the Fredholm equation of the second kind

$$\phi(x) = x + \lambda\int_a^b u(x)\overline{v}(s)\phi(s)\,ds$$

with separable kernel $K(x,s) = u(x)\overline{v}(s)$.

• *Solution*

The Fredholm integral equation of the second kind

$$\phi(x) = f(x) + \int_a^b K(x,s)\phi(s)\,ds \quad (a \le x \le b)$$

can be solved by an iterative method which yields a sequence of approximations producing an infinite series solution called the *Neumann series*. The Nth approximation is the sum

$$\phi_N(x) = \sum_{n=0}^{N} \lambda^n \phi^{(n)}(x) \quad (N = 0, 1, 2, ...)$$

where $\phi^{(0)}(x) = f(x), \phi^{(1)}(x) = \int_a^b K(x,s)\phi^{(0)}(s)\,ds, ...$, and in general

$$\phi^{(n)}(x) = \int_a^b K(x,s)\phi^{(n-1)}(s)\,ds \quad (n \ge 1).$$

If $f(x)$ and $K(x,s)$ are continuous functions we may write $|f(x)| \le m$ for $a \le x \le b$, and $|K(x,s)| \le M$ for $a \le x \le b$ and $a \le s \le b$ where m and M are positive constants. Then

$$\phi(x) = \sum_{n=0}^{\infty} \lambda^n \phi^{(n)}(x)$$

is a continuous solution of the Fredholm integral equation if $|\lambda| < [M(b-a)]^{-1}$.

This can be written in the form

$$\phi(x) = f(x) + \lambda \int_a^b R(x,s;\lambda)f(s)\,ds$$

where the resolvent kernel

$$R(x,s;\lambda) = \sum_{n=0}^{\infty} \lambda^n K_{n+1}(x,s) \tag{10.10}$$

is given in terms of the *iterated kernels* $K_{n+1}(x,s)$ where $K_1(x,s) = K(x,s)$ and in general $K_{n+1}(x,s) = \int_a^b K(x,t)K_n(t,s)\,dt$.

In the present problem we have $K_1(x,s) = K(x,s) = u(x)\overline{v}(s)$ and $K_2(x,s) = \int_a^b K_1(x,t)K_1(t,s)\,dt = \alpha u(x)\overline{v}(s)$ where $\alpha = \int_a^b \overline{v}(t)u(t)\,dt$, and in general $K_{n+1}(x,s) = \alpha^n u(x)\overline{v}(s)$.

Hence the resolvent kernel is

$$R(x,s;\lambda) = u(x)\overline{v}(s) \sum_{n=0}^{\infty} (\lambda\alpha)^n = \frac{u(x)\overline{v}(s)}{1 - \lambda\alpha}$$

provided $|\lambda\alpha| < 1$, which is the same as (10.8).

10.13 Use the method of successive approximations to solve

$$\phi(x) = x + \lambda \int_0^\pi \sin nx \sin ns\, \phi(s)\,ds$$

where n is an integer and verify that the solution agrees with the solution to problem 10.03 for $\lambda < 2/\pi$.

- *Solution*

In the present problem $K(x,s) = \sin nx \sin ns$, $u(x) = \sin nx$, $v(s) = \sin ns$ and $K_{n+1}(x,s) = \alpha^n \sin nx \sin ns$ where

$$\alpha = \int_0^\pi \overline{v}(t)u(t)\,dt = \int_0^\pi \sin^2 nt\,dt = \frac{\pi}{2}.$$

Hence $K_{n+1}(x,s) = (\pi/2)^n \sin nx \sin ns$ and so the resolvent kernel is

$$R(x,s;\lambda) = \sin nx \sin ns \sum_{n=0}^{\infty} \left(\frac{\lambda\pi}{2}\right)^n = \frac{\sin nx \sin ns}{1 - \lambda\pi/2}$$

if $\lambda < 2/\pi$. This is the same as the solution obtained in problem 10.03.

10.14 Obtain the solution of

$$\phi(x) = f(x) + \int_0^{2\pi} K(x,s)\phi(s)\,ds$$

where

$$K(x,s) = \sum_{i=1}^{\infty} a_i \sin ix \cos is$$

and

$$\sum_{i=1}^{\infty} |a_i| < \infty,$$

using the method of successive approximations.

• *Solution*

We have $K_1(x,s) = K(x,s) = \sum_{i=1}^{\infty} a_i \sin ix \cos is$ and

$$K_2(x,s) = \int_0^{2\pi} K(x,t)K(t,s)\,dt = \sum_{i=1}^{\infty}\sum_{j=1}^{\infty} a_i a_j \sin ix \cos js \int_0^{2\pi} \cos it \sin jt\,dt = 0$$

since $\int_0^{2\pi} \cos it \sin jt\,dt = 0$. Hence $K_{n+1}(x,s) = 0$ for $n \geq 1$. Thus from (10.10) $R(x,s;\lambda) = K(x,s)$ and so the solution is

$$\phi(x) = f(x) + \int_0^{2\pi} K(x,s)f(s)\,ds.$$

10.15 Use the method of successive approximations to obtain the resolvent kernel for the Volterra integral equation

$$\phi(x) = 1 + \lambda \int_0^x xs\phi(s)\,ds$$

and hence find the solution to this equation.

• *Solution*

This equation is a Volterra equation of the second kind (10.5) . Volterra kernels satisfy $K(x,s) = 0$ for $x < s$ and the Neumann series is convergent for all λ when the Volterra kernel is continuous.
We have

$$K_1(x,s) = K(x,s) = xs \quad (x > s)$$

and

$$K_2(x,s) = \int_s^x K(x,t)K(t,s)\,dt = xs\int_s^x t^2\,dt = \frac{1}{3}xs(x^3 - s^3) \quad (x > s).$$

Let us assume that

$$K_n(x,s) = \frac{xs(x^3 - s^3)^{n-1}}{3^{n-1}(n-1)!} \quad (x > s).$$

Then

$$K_{n+1}(x,s) = \int_s^x K(x,t)K_n(t,s)\,dt = \frac{xs}{3^{n-1}(n-1)!}\int_s^x t^2(t^3 - s^3)^{n-1}\,dt = \frac{xs}{3^n n!}(x^3 - s^3)^n$$

and so the formula for $K_n(x,s)$ is established by induction.

Hence, using (10.10) we have

$$R(x,s;\lambda) = \sum_{n=0}^{\infty} \lambda^n K_{n+1}(x,s) = xs \sum_{n=0}^{\infty} \left(\frac{\lambda}{3}\right)^n \frac{(x^3 - s^3)^n}{n!}$$

and so

$$\phi(x) = 1 + \lambda \int_0^1 \sum_{n=0}^{\infty} \left(\frac{\lambda}{3}\right)^n \frac{x^{3n+3}}{n!}(1 - t^3)^n t\,dt$$

where we have put $t = s/x$.

Now, integrating by parts n times, we get

$$\int_0^1 (1 - t^3)^n t\,dt = \frac{3^n n!}{2.5...(3n + 2)}$$

and so the solution to the Volterra integral equation of the second kind is

$$\phi(x) = 1 + \lambda \sum_{n=0}^{\infty} \frac{\lambda^n x^{3(n+1)}}{2.5...(3n + 2)}.$$

10.16 Evaluate

$$\frac{1}{\pi} P \int_{-\infty}^{\infty} \frac{e^{is}}{s - x}\,ds$$

using Cauchy's residue theorem, and hence find the solutions of the pair of reciprocal Hilbert transforms

$$(i) \quad \sin x = -\frac{1}{\pi} P \int_{-\infty}^{\infty} \frac{f(s)}{s - x}\,ds,$$

$$(ii) \quad \cos x = \frac{1}{\pi} P \int_{-\infty}^{\infty} \frac{g(s)}{s - x}\,ds$$

where P denotes the *principal part* of the integral.

• *Solution*

We apply Cauchy's residue theorem to a closed contour C composed of a large semi-circle Γ of radius R centred at the origin and taken in the anticlockwise sense, the straight line along the real axis from $-R$ to $x - r$, the small semi-circle γ of radius r centred at the point x on the real axis taken in the clockwise sense, and the straight line along the real axis from $x + r$ to R. Since this contour contains no poles we have

$$\oint_C \frac{e^{iz}}{z - x}\,dz = 0.$$

The contribution from the semi-circle Γ vanishes as $R \to \infty$ and since

$$\oint_\gamma \frac{e^{iz}}{z - x} \, dz = -\pi i e^{ix}$$

we obtain

$$P \int_{-\infty}^\infty \frac{e^{is}}{s - x} \, ds = \pi i e^{ix} = -\pi \sin x + \pi i \cos x.$$

Hence we have (i) $f(s) = \cos x$ and (ii) $g(s) = \sin s$. Thus $\sin s$ is the Hilbert transform of $\cos s$ and (i) and (ii) are a pair of reciprocal *Hilbert transforms* given by

$$g = Hf, \quad f = -Hg \tag{10.11}$$

where H is the integral operator

$$H = -\frac{1}{\pi} P \int_{-\infty}^\infty \frac{ds}{s - x}. \tag{10.12}$$

10.17 Find the solution of

$$\frac{1}{1 + x^2} = \frac{1}{\pi} P \int_{-\infty}^\infty \frac{f(s)}{s - x} \, ds$$

• *Solution*

Using the reciprocal Hilbert transform given by (10.11) we obtain

$$f(x) = -P \frac{1}{\pi} \int_{-\infty}^\infty \frac{1}{1 + s^2} \frac{1}{s - x} \, ds.$$

Then applying Cauchy's residue theorem to the closed contour C defined in the previous problem we get

$$\oint_C \frac{dz}{(1 + z^2)(z - x)} = -\frac{\pi}{x - i}$$

since the contour C encloses the pole at $z = i$.

Also we have

$$\oint_\gamma \frac{dz}{(1 + z^2)(z - x)} = -\frac{\pi i}{1 + x^2}$$

and so

$$P \int_{-\infty}^\infty \frac{ds}{(1 + s^2)(s - x)} - \frac{\pi i}{1 + x^2} = -\frac{\pi}{x - i}.$$

Hence

$$f(x) = \frac{x}{1 + x^2}.$$

10.18 If the linearly independent functions $\phi_1, \phi_2, ..., \phi_n$; $\psi_1, \psi_2, ..., \psi_n$ form a bi-orthogonal set of functions so that the inner product $(\phi_i, \psi_j) = \int_a^b \overline{\psi}_j(x) \phi_i(x) \, dx = \delta_{ij}$ $(i, j = 1, 2, ..., n)$, show that the kernel

$$K(x, s) = \sum_{i=1}^n a_i \phi_i(x) \overline{\psi}_i(s) \quad (a \le x \le b, \ a \le s \le b)$$

has the characteristic functions $\phi_i(x)$ with characteristic values a_i^{-1} $(i = 1, 2, ..., n)$.

Further, use the Neumann expansion to show that the resolvent kernel of $K(x, s)$ converges to

$$R(x, s; \lambda) = \sum_{i=1}^n \frac{a_i \phi_i(x) \overline{\psi}_i(s)}{1 - a_i \lambda}$$

if $|a_i \lambda| < 1$ for all i.

• *Solution*

We have that

$$\int_a^b K(x,s)\phi_i(s)\,ds = a_i\phi_i(x)$$

since $\int_a^b \overline{\psi}_j(s)\phi_i(s)\,ds = \delta_{ij}$. Hence the characteristic functions are ϕ_i with characteristic values a_i^{-1} $(i = 1,...,n)$.

Now $K_1(x,s) = K(x,s)$ and

$$K_2(x,s) = \int_a^b K(x,t)K(t,s)\,dt = \sum_{i=1}^n a_i^2 \phi_i(x)\overline{\psi}_i(s).$$

In general, if we assume that

$$K_m(x,s) = \sum_{i=1}^n a_i^m \phi_i(x)\overline{\psi}_i(s)$$

we obtain

$$K_{m+1}(x,s) = \int_a^b K(x,t)K_m(t,s)\,dt = \sum_{i=1}^n a_i^{m+1}\phi_i(x)\overline{\psi}_i(s)$$

and so the formula for $K_{m+1}(x,s)$ follows by induction. Hence the resolvent kernel is given by

$$R(x,s;\lambda) = \sum_{m=0}^\infty \lambda^m K_{m+1}(x,s) = \sum_{i=1}^n \phi_i(x)\overline{\psi}_i(s)\sum_{m=0}^\infty \lambda^m a_i^{m+1}$$

and the required formula for the resolvent kernel follows since $\sum_{m=0}^\infty \lambda^m a_i^{m+1} = a_i(1 - \lambda a_i)^{-1}$ if $|a_i\lambda| < 1$ for all i.

10.19 Show that the kernel

$$K(x,s) = \sum_{\nu=0}^\infty a_\nu \cos \nu x \cos \nu s \quad (0 \le x \le \pi, 0 \le s \le \pi)$$

where $\sum_{\nu=0}^\infty |a_\nu| < \infty$, has the characteristic functions $1/\sqrt{\pi}$, $\sqrt{2/\pi}\cos \nu x$ $(\nu = 1,2,...)$ with characteristic values $1/\pi a_0$, $2/\pi a_\nu$ respectively.

Use the Neumann expansion to show that for sufficiently small λ the resolvent kernel for $K(x,s)$ is given by

$$R(x,s;\lambda) = \frac{a_0}{1 - \pi a_0 \lambda} + \sum_{\nu=1}^\infty \frac{a_\nu \cos \nu x \cos \nu s}{1 - \pi a_\nu \lambda/2}.$$

• *Solution*

We have

$$\int_0^\pi K(x,s)\cos \nu s\,ds = a_\nu \cos \nu x \int_0^\pi \cos^2 \nu s\,ds = \frac{\pi}{2}a_\nu \cos \nu x$$

and so the characteristic function $\sqrt{2/\pi}\cos \nu x$ has the characteristic value $2/\pi a_\nu$. Also $\int_0^\pi K(x,s)\,ds = \pi a_0$ and so the characteristic function $1/\sqrt{\pi}$ has the characteristic value $1/\pi a_0$.

Now $K_1(x, s) = K(x, s)$ and

$$K_2(x, s) = \int_0^\pi K(x, t) K(t, s)\, dt = \sum_{\nu=0}^\infty a_\nu^2 \cos \nu x \cos \nu s \int_0^\pi \cos^2 \nu t\, dt$$

which leads to

$$K_2(x, s) = \pi a_0^2 + \sum_{\nu=1}^\infty \frac{\pi}{2} a_\nu^2 \cos \nu x \cos \nu s.$$

Then assuming that

$$K_n(x, s) = (\pi a_0)^{n-1} a_0 + \sum_{\nu=1}^\infty \left(\frac{\pi}{2} a_\nu\right)^{n-1} a_\nu \cos \nu x \cos \nu s$$

we obtain

$$K_{n+1}(x, s) = \int_0^\pi K(x, t) K_n(t, s)\, dt = (\pi a_0)^n a_0 + \sum_{\nu=1}^\infty \left(\frac{\pi}{2} a_\nu\right)^n a_\nu \cos \nu x \cos \nu s$$

and so the formula for $K_n(x, s)$ follows by induction. Then we have

$$R(x, s; \lambda) = \sum_{n=0}^\infty \lambda^n K_{n+1}(x, s) = a_0 \sum_{n=0}^\infty (\pi a_0 \lambda)^n + \sum_{\nu=1}^\infty a_\nu \cos \nu x \cos \nu s \sum_{n=0}^\infty \left(\frac{\pi a_\nu \lambda}{2}\right)^n$$

which gives the required formula for the resolvent kernel for sufficiently small λ.

10.20 Show that the Poisson kernel

$$K(x, s) = \frac{1}{2\pi} \frac{1 - a^2}{1 + a^2 - 2a \cos(x - s)} \qquad (0 \le x \le 2\pi, 0 \le s \le 2\pi)$$

where $|a| < 1$, can be expressed as the Fourier series

$$\frac{1}{2\pi} + \frac{1}{\pi} \sum_{\nu=1}^\infty a^\nu \cos \nu(x - s).$$

Obtain the Neumann expansion for the resolvent kernel of $K(x, s)$ and show that it converges to

$$R(x, s; \lambda) = \frac{1}{2\pi} \frac{1}{1 - \lambda} + \frac{1}{\pi} \sum_{\nu=1}^\infty \frac{a^\nu}{1 - a^\nu \lambda} \cos \nu(x - s)$$

if $|\lambda| < 1$.

- *Solution*

We have

$$\frac{1}{2} + \sum_{\nu=1}^\infty a^\nu e^{i\nu x} = \frac{1}{2} + \frac{ae^{ix}}{1 - ae^{ix}} = \frac{1}{2} \frac{1 + ae^{ix}}{1 - ae^{ix}} = \frac{1}{2} \frac{1 - a^2 + 2ia \sin x}{1 + a^2 - 2a \cos x}$$

if $|a| < 1$, which yields

$$\frac{1}{2} + \sum_{\nu=1}^\infty a^\nu \cos \nu x = \frac{1}{2} \frac{1 - a^2}{1 + a^2 - 2a \cos x}.$$

Hence the Poisson kernel can be expressed in the form

$$K(x, s) = \frac{1}{2\pi} + \frac{1}{\pi} \sum_{\nu=1}^{\infty} a^{\nu} \cos \nu(x - s).$$

Using $\cos \nu(x - s) = \cos \nu x \cos \nu s + \sin \nu x \sin \nu s$ we get

$$K_2(x, s) = \int_0^{2\pi} K(x, t)K(t, s)\, dt = \frac{1}{2\pi} + \frac{1}{\pi} \sum_{\nu=1}^{\infty} a^{2\nu}(\cos \nu x \cos \nu s + \sin \nu x \sin \nu s)$$

and using the principle of induction, as in the previous two problems, it can be shown that

$$K_{n+1}(x, s) = \frac{1}{2\pi} + \frac{1}{\pi} \sum_{\nu=1}^{\infty} a^{(n+1)\nu}(\cos \nu x \cos \nu s + \sin \nu x \sin \nu s).$$

Hence the resolvent kernel is given by

$$R(x, s; \lambda) = \sum_{n=0}^{\infty} \lambda^n K_{n+1}(x, s) = \frac{1}{2\pi} \sum_{n=0}^{\infty} \lambda^n + \frac{1}{\pi} \sum_{\nu=1}^{\infty} a^{\nu} \cos \nu(x - s) \sum_{n=0}^{\infty} (a^{\nu}\lambda)^n$$

and the required formula is obtained if $|\lambda| < 1$.

10.21 Obtain the resolvent kernel for the symmetric kernel

$$K(x, s) = 1 + xs \quad (0 \leq x \leq 1,\ 0 \leq s \leq 1)$$

and verify that its characteristic values are real numbers.

• *Solution*

This is an example of a *degenerate kernel* having the separable form

$$K(x, s) = \sum_{i=1}^{n} u_i(x)\bar{v}_i(s) \quad (a \leq x \leq b,\ a \leq s \leq b) \tag{10.13}$$

where $u_1(x), ..., u_n(x)$ and $v_1(x), ..., v_n(x)$ are two sets of linearly independent square integrable functions. The least integer n for which a degenerate kernel can be expressed in this form is called the *rank* of the kernel.

The corresponding Fredholm linear integral equation of the second kind has the form

$$\phi(x) = f(x) + \lambda \sum_{i=1}^{n} u_i(x) \int_a^b \bar{v}_i(s)\phi(s)\, ds \tag{10.14}$$

and its resolvent kernel is given by

$$R(x, s; \lambda) = \frac{1}{d(\lambda)} \sum_{i=1}^{n} \sum_{j=1}^{n} u_i(x)d_{ij}(\lambda)\bar{v}_j(s) \tag{10.15}$$

where $d(\lambda)$ is the determinant of $I - \lambda A$, I being the unit matrix and A the matrix with elements $a_{ij} = \int_a^b \bar{v}_j(s)u_i(s)ds$, and (d_{ij}) are the elements of the adjugate matrix D of the matrix $I - \lambda A$, that is the transposed matrix of its cofactors.

The characteristic values satisfy $d(\lambda) = 0$.

In the present problem $u_1(x) = 1$, $u_2(x) = x$, $v_1(s) = 1$, $v_2(s) = s$ and $a_{11} = 1$, $a_{22} = 1/3$, $a_{12} = a_{21} = 1/2$ and so

$$d(\lambda) = \begin{vmatrix} 1 - \lambda & -\lambda/2 \\ -\lambda/2 & 1 - \lambda/3 \end{vmatrix} = 1 - \frac{4}{3}\lambda + \frac{1}{12}\lambda^2.$$

The characteristic values satisfy $d(\lambda) = 0$, that is $\lambda^2 - 16\lambda + 12 = 0$. They are the real numbers $\lambda = 8 \pm 2\sqrt{13}$.

Now $d_{11} = 1 - \lambda/3$, $d_{22} = 1 - \lambda$, $d_{12} = d_{21} = \lambda/2$ and so

$$R(x, s; \lambda) = \frac{1 - \lambda/3 + (1 - \lambda)xs + \lambda(x + s)/2}{1 - 4\lambda/3 + \lambda^2/12}.$$

10.22 Obtain the resolvent kernel for the skew-symmetric kernel

$$K(x, s) = \sin(x - s) \quad (0 \le x \le 2\pi, \ 0 \le s \le 2\pi)$$

and verify that its characteristic values are pure imaginary numbers.

• *Solution*

We use the same theory as in the previous problem. In the present problem $K(x, s) = \sin x \cos s - \cos x \sin s$ $(0 \le x \le 2\pi, \ 0 \le s \le 2\pi)$ and so $u_1(x) = \sin x$, $u_2(x) = \cos x$, $v_1(s) = \cos s$, $v_2(s) = -\sin s$. Then we have $a_{11} = 0$, $a_{22} = 0$, $a_{12} = -\pi$, $a_{21} = \pi$ and so

$$d(\lambda) = \begin{vmatrix} 1 & \pi\lambda \\ -\pi\lambda & 1 \end{vmatrix} = 1 + \pi^2\lambda^2.$$

The characteristic values are given by $1 + \pi^2\lambda^2 = 0$ and so are the pure imaginary numbers $\lambda = \pm i/\pi$.

Now $d_{11} = 1$, $d_{22} = 1$, $d_{12} = -\pi\lambda$, $d_{21} = \pi\lambda$ and so the resolvent kernel is

$$R(x, s; \lambda) = \frac{\sin x \cos s - \cos x \sin s + \pi\lambda(\sin x \sin s + \cos x \cos s)}{1 + \pi^2\lambda^2}$$

that is

$$R(x, s; \lambda) = \frac{\sin(x - s) + \pi\lambda\cos(x - s)}{1 + \pi^2\lambda^2}.$$

10.23 Show that the non-symmetric kernel

$$K(x, s) = \sin x \cos s \quad (0 \le x \le \pi, \ 0 \le s \le \pi)$$

has no characteristic values.

• *Solution*

In this problem we have a kernel of rank 1 and $u_1(x) = \sin x$, $v_1(s) = \cos s$. The only element of the matrix A is $a_{11} = \int_0^\pi \cos s \sin s \, ds = 0$. Hence $d(\lambda) = 1$ and so no solution of $d(\lambda) = 0$ is possible. Thus no characteristic values exist for this kernel and the only solution of the homogeneous equation

$$\phi(x) = \lambda \int_a^b K(x, s)\phi(s) \, ds$$

is the trivial solution $\phi(x) = 0$.

10.24 Show that the homogeneous equation

$$\phi(x) = \lambda \int_0^1 K(x, s)\phi(s) \, ds$$

with the symmetric kernel

$$K(x, s) = \begin{cases} x(1 - s) & (0 \le x \le s \le 1) \\ s(1 - x) & (0 \le s \le x \le 1) \end{cases}$$

is equivalent to the differential equation $\phi'' + \lambda\phi = 0$ with $\phi(0) = \phi(1) = 0$.

Hence show that the kernel has characteristic functions $\sqrt{2}\sin\nu\pi x$ $(\nu = 1, 2, ...)$ with characteristic values $(\nu\pi)^2$ respectively. Further show that the kernel $K(x, s)$ is given by the uniformly convergent series

$$2\sum_{\nu=1}^{\infty} \frac{\sin\nu\pi x \sin\nu\pi s}{\nu^2\pi^2}$$

and also obtain an expansion for the resolvent kernel.

- *Solution*

This homogeneous equation can be written as

$$\phi(x) = \lambda \int_0^x s(1 - x)\phi(s) \, ds + \lambda \int_x^1 x(1 - s)\phi(s) \, ds.$$

We see that the end conditions $\phi(0) = 0$ and $\phi(1) = 0$ are satisfied. Differentiation gives

$$\phi'(x) = \lambda x(1 - x)\phi(x) - \lambda \int_0^x s\phi(s) \, ds - \lambda x(1 - x)\phi(x) + \lambda \int_x^1 (1 - s)\phi(s) \, ds,$$

that is

$$\phi'(x) = -\lambda \int_0^1 s\phi(s) \, ds + \lambda \int_x^1 \phi(s) \, ds$$

and so

$$\phi''(x) = -\lambda\phi(x).$$

The normalized solution of this which vanishes at $x = 0$ is $\phi(x) = \sqrt{2}\sin\sqrt{\lambda}x$ and this also vanishes at $x = 1$ if $\sqrt{\lambda} = \nu\pi$ $(\nu = 1, 2, ...)$. Hence the characteristic functions are $\phi_\nu(x) = \sqrt{2}\sin\nu\pi x$ with characteristic values $\lambda_\nu = \nu^2\pi^2$ as required.

Since $K(x, s)$ is a continuous positive kernel, the *expansion theorem* tells us that it is given by the uniformly convergent series

$$K(x, s) = \sum_{\nu=1}^{\infty} \frac{\phi_\nu(x)\overline{\phi}_\nu(s)}{\lambda_\nu} = 2\sum_{\nu=1}^{\infty} \frac{\sin\nu\pi x \sin\nu\pi s}{\nu^2\pi^2}$$

and that the resolvent kernel is given by the uniformly convergent series

$$R(x, s; \lambda) = \sum_{\nu=1}^{\infty} \frac{\phi_\nu(x)\overline{\phi}_\nu(s)}{\lambda_\nu - \lambda} = 2\sum_{\nu=1}^{\infty} \frac{\sin\nu\pi x \sin\nu\pi s}{\nu^2\pi^2 - \lambda}.$$

10.25 Show that the homogeneous equation with the symmetric kernel

$$K(x,s) = \begin{cases} x & (0 \le x \le s \le 1) \\ s & (0 \le s \le x \le 1) \end{cases}$$

is equivalent to the differential equation $\phi'' + \lambda\phi = 0$ with the end conditions $\phi(0) = 0$ and $\phi'(1) = 0$.

Hence show that the kernel has characteristic functions $\sqrt{2}\sin[(2\nu - 1)\pi x/2]$ ($\nu = 1, 2, ...$) with characteristic values $[(2\nu - 1)/2]^2\pi^2$. Further obtain uniformly convergent series for $K(x,s)$ and the resolvent kernel.

- *Solution*

This homogeneous integral equation can be expressed as

$$\phi(x) = \lambda \int_0^x s\phi(s)\,ds + \lambda x \int_x^1 \phi(s)\,ds$$

and satisfies the end condition $\phi(0) = 0$. Differentiation gives

$$\phi'(x) = \lambda x\phi(x) + \lambda \int_x^1 \phi(s)\,ds - \lambda x\phi(x) = \lambda \int_x^1 \phi(s)\,ds$$

and so the second end condition $\phi'(1) = 0$ is satisfied. Differentiating again we get

$$\phi''(x) = -\lambda\phi(x)$$

whose normalized solution which satisfies the end condition $\phi(0) = 0$ is $\phi(x) = \sqrt{2}\sin\sqrt{\lambda}x$. The other end condition $\phi'(1) = 0$ is satisfied if $\sqrt{\lambda} = (2\nu - 1)\pi/2$ ($\nu = 1, 2, ...$). Thus the characteristic functions are $\phi_\nu(x) = \sqrt{2}\sin[(2\nu - 1)\pi x/2]$ with characteristic values $\lambda_\nu = [(2\nu - 1)/2]^2\pi^2$.

Since the kernel $K(x,s)$ is continuous and positive, the expansion theorem gives

$$K(x,s) = \sum_{\nu=1}^{\infty} \frac{\phi_\nu(x)\overline{\phi}_\nu(s)}{\lambda_\nu} = 2\sum_{\nu=1}^{\infty} \frac{\sin[(2\nu - 1)\pi x/2]\sin[(2\nu - 1)\pi s/2]}{[(2\nu - 1)/2]^2\pi^2}$$

and for the resolvent kernel

$$R(x,s;\lambda) = \sum_{\nu=1}^{\infty} \frac{\phi_\nu(x)\overline{\phi}_\nu(s)}{\lambda_\nu - \lambda} = 2\sum_{\nu=1}^{\infty} \frac{\sin[(2\nu - 1)\pi x/2]\sin[(2\nu - 1)\pi s/2]}{[(2\nu - 1)/2]^2\pi^2 - \lambda},$$

both being uniformly convergent series.

Integral equations occur in various branches of applied mathematics. In particular they are found in quantum theory where, for example, the energies and the scattering amplitudes can be expressed in terms of integrals whose integrands contain the unknown solution of the Schrödinger equation. Then by using trial functions to represent the wave function solution to the Schrödinger equation, approximate values for the energies, and scattering amplitudes and phase shifts, can be obtained. This approach yields very accurate values for the energies of the ground states of quantum mechanical systems where other procedures do considerably less well.

Perturbation expansions involving the method of successive approximations can also be used and produce a Neumann series for the scattering amplitudes and phase shifts known as the Born series. The first term of this expansion is called the Born approximation and an example of this is given in problem 15.34.

11
WAVE MOTION

The subject of *wave motion* was first discussed in detail by *Lord Rayleigh* (1842-1919) in his treatise *The Theory of Sound* published in 1877.

In this chapter we shall be concerned with the solutions of the wave equations for *vibrating strings*, *sound waves* and *waves in water*.

These solutions are generally obtained by expanding in a Fourier series.

11-1 Vibrations of strings

The wave equation for a flexible string making small transverse vibrations is

$$\frac{\partial^2 y}{\partial x^2} = \frac{1}{c^2}\frac{\partial^2 y}{\partial t^2} \tag{11.1}$$

where the small displacement of the string at the point x at time t is denoted by $y(x,t)$ and c is the phase velocity of the wave motion.

By separating the x and t variables, a solution of the wave equation can be written in the form

$$y(x,t) = (a\cos kx + b\sin kx)(A\cos ckt + B\sin ckt) \tag{11.2}$$

where a, b, A, B and k are constants. The angular frequency is given by $\omega = ck$ and the periodic time of the wave motion is $T = 2\pi/\omega$.

In the following problems the string is of finite length and is fixed at its ends.

11.01 A string of uniform line density ρ is stretched to a tension $P = \rho c^2$, the ends of the string being fixed at the points $x = 0$ and $x = l$. The string is released from rest in the shape of an arc of the parabola $y = \alpha x(l - x)$ where αl^2 is small. Show that at time t its form is given by

$$y = \frac{8\alpha l^2}{\pi^3}\sum_{n=0}^{\infty}\frac{1}{(2n+1)^3}\sin\frac{(2n+1)\pi x}{l}\cos\frac{(2n+1)\pi ct}{l}.$$

- *Solution*

We expand the displacement of the string y in the form of a Fourier series

$$y(x,t) = \sum_{s=1}^{\infty}\sin\frac{s\pi x}{l}\left[A_s\cos\frac{s\pi ct}{l} + B_s\sin\frac{s\pi ct}{l}\right]$$

since this is a solution of the wave equation for the string which vanishes at $x = 0$ and $x = l$.

Since at time $t = 0$ we have $y = \alpha x(l - x)$ and $\partial y/\partial t = 0$, it follows that

$$A_s = \frac{2}{l}\int_0^l \alpha x(l - x)\sin\frac{s\pi x}{l}\,dx, \quad B_s = 0.$$

Now

$$\int_0^l x \sin \frac{s\pi x}{l}\, dx = -\frac{l^2}{s\pi}(-1)^s$$

and

$$\int_0^l x^2 \sin \frac{s\pi x}{l}\, dx = -\frac{l^3}{s\pi}(-1)^s + \frac{2l^3}{s^3\pi^3}[(-1)^s - 1].$$

Hence

$$\int_0^l x(l - x) \sin \frac{s\pi x}{l}\, dx = \frac{2l^3}{s^3\pi^3}[1 - (-1)^s]$$

which vanishes if s is even and is $4l^3/s^3\pi^3$ if s is odd. Putting $s = 2n + 1$ and summing from $n = 0$ to ∞ then yields the required series formula for the small displacement of the string.

11.02 A string of uniform line density ρ is stretched at tension ρc^2 with its ends held fixed at the points $x = 0$ and $x = l$. The string is hit with a hammer so as to give it the initial velocity distribution

$$v(x) = \begin{cases} 0 & (0 \le x < a - d) \\ V & (a - d \le x \le a + d) \\ 0 & (a + d < x \le l) \end{cases}$$

where V is a constant. Find the displacement of the string at any subsequent time t.

• *Solution*

Since the string is held fixed at its ends $x = 0$ and $x = l$, and $y = 0$ when $t = 0$, the solution takes the form

$$y(x, t) = \sum_{n=1}^{\infty} B_n \sin \frac{n\pi x}{l} \sin \frac{n\pi ct}{l} \quad (0 \le x \le l).$$

Now $\partial y/\partial t = v(x)$ when $t = 0$ and so we have

$$v(x) = \sum_{n=1}^{\infty} \frac{n\pi c}{l} B_n \sin \frac{n\pi x}{l}$$

which gives

$$B_n = \frac{2}{l}\int_0^l \frac{l}{n\pi c} v(x) \sin \frac{n\pi x}{l}\, dx = \frac{2V}{n\pi c}\int_{a-d}^{a+d} \sin \frac{n\pi x}{l}\, dx = \frac{4lV}{n^2\pi^2 c} \sin \frac{n\pi a}{l} \sin \frac{n\pi d}{l}$$

and so

$$y(x, t) = \frac{4lV}{\pi^2 c}\sum_{n=1}^{\infty} \frac{1}{n^2} \sin \frac{n\pi a}{l} \sin \frac{n\pi d}{l} \sin \frac{n\pi x}{l} \sin \frac{n\pi ct}{l}.$$

11.03 A stretched string of length l has one end fixed and the other end attached to a light ring which can slide on a smooth rod which is perpendicular to the string in its equilibrium position. If the ring is displaced a small distance a from the position of equilibrium and the system starts from rest, show that the displacement of the string at time t of a point of the string at distance x from the fixed end is

$$\frac{8a}{\pi^2}\sum_{n=0}^{\infty} \frac{(-1)^n}{(2n + 1)^2} \sin \frac{(2n + 1)\pi x}{2l} \cos \frac{(2n + 1)\pi ct}{2l}.$$

• *Solution*

Since the string is fixed at the end $x = 0$ we may write the displacement of the string in the form

$$y(x, t) = \sin kx \ (A \cos ckt + B \sin ckt).$$

Since it starts from rest so that $\partial y / \partial t = 0$ at time $t = 0$, we have $B = 0$.

At $x = l$ the string is attached to a ring of negligible mass which can slide on a smooth rod and so we have $\partial y / \partial x = 0$ at $x = l$ since there is no force acting on the string along the rod. Hence $\cos kl = 0$ so that $kl = (2n + 1)\pi / 2 \quad (n = 0, 1, 2, ...)$.

It follows that we can expand the displacement of the string in the form of the Fourier series

$$y(x, t) = \sum_{n=0}^{\infty} A_n \sin \frac{(2n + 1)\pi x}{2l} \cos \frac{(2n + 1)\pi ct}{2l}.$$

Now $y = ax/l$ when $t = 0$ and so

$$\frac{ax}{l} = \sum_{n=0}^{\infty} A_n \sin \frac{(2n + 1)\pi x}{2l}$$

and hence

$$A_n = \frac{2a}{l^2} \int_0^l x \sin \frac{(2n + 1)\pi x}{2l} \, dx = \frac{8a(-1)^n}{\pi^2 (2n + 1)^2}$$

which leads to the required result.

11.04 A uniform string is stretched between two points distant l apart, and a bead of mass equal to that of the string is attached to one of the points of trisection. Prove that the angular frequency ω of the small transverse vibrations of the system satisfies

$$\frac{\omega l}{c} \sin \frac{2\omega l}{3c} = 1 + 2 \cos \frac{2\omega l}{3c}.$$

• *Solution*

Suppose the bead is at the point $x = a$ and that $y_1(x, t)$ and $y_2(x, t)$ denote the displacements of the string for $0 \le x \le a$ and $a \le x \le l$ respectively. If the displacement of the bead is $y(a, t) = \alpha \sin \omega t$ then, since $y_1(0, t) = 0$ and $y_2(l, t) = 0$, we have

$$y_1(x, t) = \alpha \frac{\sin(\omega x/c)}{\sin(\omega a/c)} \sin \omega t, \quad y_2(x, t) = \alpha \frac{\sin[\omega(l - x)/c]}{\sin[\omega(l - a)/c]} \sin \omega t. \tag{11.3}$$

Now the equation of motion of the bead is

$$m \frac{d^2 y(a, t)}{dt^2} = P \left(\frac{\partial y_2}{\partial x} - \frac{\partial y_1}{\partial x} \right)_{x=a} \tag{11.4}$$

where m is the mass of the bead and P is the tension in the string.

Hence we get

$$-m\omega^2 \alpha \sin \omega t = -\frac{P\omega}{c} \alpha \sin \omega t \left[\cot \frac{\omega(l - a)}{c} + \cot \frac{\omega a}{c} \right]$$

and so we have

$$\frac{m\omega c}{P} = \frac{\sin(\omega l/c)}{\sin[\omega(l - a)/c] \sin(\omega a/c)}. \tag{11.5}$$

If ρ is the line density of the uniform string we have $P = \rho c^2$. In the present problem $m = \rho l$ and $a = l/3$. Hence (11.5) gives

$$\frac{\omega l}{c} \sin(2\omega l/3c) = \frac{\sin(\omega l/c)}{\sin(\omega l/3c)}.$$

Now $\sin 3\theta = \sin \theta \cos 2\theta + \cos \theta \sin 2\theta = \sin \theta (\cos 2\theta + 2 \cos^2 \theta) = \sin \theta (1 + 2 \cos 2\theta)$ and so the required result follows by putting $\theta = \omega l/3c$.

11.05 Two uniform strings of densities ρ_1 and ρ_2 and of equal lengths are fastened together at one end and the other two ends are tied to two fixed points at a distance $2l$ apart. If the tension of the combined string is P show that the periods of vibration are given by $2\pi/\omega$ where

$$c_1 \tan \frac{\omega l}{c_1} + c_2 \tan \frac{\omega l}{c_2} = 0$$

and $c_1^2 \rho_1 = c_2^2 \rho_2 = P$.

- *Solution*

Suppose the strings are joined at the point $x = 0$ and the fixed ends of the two strings are at $x = -l$ and $x = l$. If y_1 and y_2 are the displacements of the strings we have $y_1 = y_2$ and $\partial y_1/\partial x = \partial y_2/\partial x$ at $x = 0$. Now we may write

$$y_1(x,t) = \left(a_1 \cos \frac{\omega x}{c_1} + b_1 \sin \frac{\omega x}{c_1} \right) \sin \omega t, \quad y_2(x,t) = \left(a_2 \cos \frac{\omega x}{c_2} + b_2 \sin \frac{\omega x}{c_2} \right) \sin \omega t$$

where $c_1^2 \rho_1 = c_2^2 \rho_2$ is the tension in the strings. Putting $x = -l$ we get

$$a_1 \cos(\omega l/c_1) - b_1 \sin(\omega l/c_1) = 0$$

and putting $x = l$ we get

$$a_2 \cos(\omega l/c_2) + b_2 \sin(\omega l/c_2) = 0.$$

But $a_1 = a_2$ since $y_1 = y_2$ at $x = 0$ and so we obtain

$$b_1 \tan \frac{\omega l}{c_1} + b_2 \tan \frac{\omega l}{c_2} = 0.$$

The required result now follows since $b_1/c_1 = b_2/c_2$, obtained by putting $\partial y_1/\partial x = \partial y_2/\partial x$ at $x = 0$.

11.06 A ring of mass M slides on a horizontal rod, its motion being determined by a spring so that its period of oscillation is $2\pi/\omega_0$. The ring is tied to one end of a string of length l whose other end is fixed, and in the equilibrium position the string is perpendicular to the rod. Show that for small oscillations of the system

$$M(\omega_0^2 - \omega^2) \tan \frac{\omega l}{c} + \frac{P\omega}{c} = 0$$

where P is the tension in the string.

- *Solution*

We may write the displacement of the string in the form

$$y(x, t) = \left(a \cos \frac{\omega x}{c} + b \sin \frac{\omega x}{c} \right) \sin \omega t.$$

The equation of motion of the ring, assumed to be at $x = 0$, is

$$M \left(\frac{\partial^2 y}{\partial t^2} + \omega_0^2 y \right) = P \frac{\partial y}{\partial x}$$

and so we have

$$M(-\omega^2 + \omega_0^2) a = P \frac{b\omega}{c}.$$

The string is fixed at $x = l$ and so

$$a \cos \frac{\omega l}{c} + b \sin \frac{\omega l}{c} = 0.$$

Hence

$$M(\omega_0^2 - \omega^2) \sin \frac{\omega l}{c} = -\frac{P\omega}{c} \cos \frac{\omega l}{c}$$

and the result follows.

11.07 The two ends of a stretched string of length $2l$ are connected to rings of mass M which slide on parallel horizontal rods perpendicular to the string. If P is the tension of the string, prove that for small oscillations the angular frequency ω satisfies

$$P \tan \frac{\omega l}{c} + M\omega c = 0.$$

Hence show that if M is very large, the effect of the slight motion of the ends of the string is to increase the pitch of any normal mode of oscillation, the increase being greater the lower the frequency of oscillation.

• *Solution*

The equations of motion of the rings, taken to be at $x = -l$ and $x = l$, are

$$M \frac{\partial^2 y}{\partial t^2} = P \frac{\partial y}{\partial x} \quad (x = -l), \quad M \frac{\partial^2 y}{\partial t^2} = -P \frac{\partial y}{\partial x} \quad (x = l).$$

Now the displacement of the string may be written in the form

$$y(x, t) = \left(a \cos \frac{\omega x}{c} + b \sin \frac{\omega x}{c} \right) \sin \omega t$$

and so we have at $x = -l$

$$-M\omega^2 \left(a \cos \frac{\omega l}{c} - b \sin \frac{\omega l}{c} \right) = P \frac{\omega}{c} \left(a \sin \frac{\omega l}{c} + b \cos \frac{\omega l}{c} \right)$$

and at $x = l$

$$-M\omega^2 \left(a \cos \frac{\omega l}{c} + b \sin \frac{\omega l}{c} \right) = -P \frac{\omega}{c} \left(-a \sin \frac{\omega l}{c} + b \cos \frac{\omega l}{c} \right)$$

giving

$$a \left(\frac{P}{c} \sin \frac{\omega l}{c} + M\omega \cos \frac{\omega l}{c} \right) = b \left(M\omega \sin \frac{\omega l}{c} - \frac{P}{c} \cos \frac{\omega l}{c} \right),$$

$$a\left(\frac{P}{c}\sin\frac{\omega l}{c} + M\omega\cos\frac{\omega l}{c}\right) = b\left(\frac{P}{c}\cos\frac{\omega l}{c} - M\omega\sin\frac{\omega l}{c}\right).$$

This pair of equations can be satisfied by taking $b = 0$ and

$$\frac{P}{c}\sin\frac{\omega l}{c} + M\omega\cos\frac{\omega l}{c} = 0$$

corresponding to the required result.

If we let $M \to \infty$ the problem tends to the case of a string of length $2l$ with fixed ends for which the angular frequencies are $\omega = (2n+1)c\pi/2l$ where n is an integer. Hence we write $\omega = (2n+1)c\pi/2l + \epsilon$ and this leads to

$$\frac{P}{Mc} = \omega\tan\frac{l\epsilon}{c}.$$

Since M is large this gives $P/Mc \simeq \omega l\epsilon/c \simeq (2n+1)\pi\epsilon/2$ and so the frequency of the normal mode is increased, this increase becoming larger as the value of n is made smaller.

11.08 A uniform string having mass ρ per unit length is stretched at tension P between two points at a distance l apart. Initially the string is at rest and has the shape $y = a\sin(n\pi x/l)$ where n is a positive integer. Assuming that the resistance of the air is $\alpha\times$momentum per unit length of the string, where α is a constant, find the displacement of the string at any subsequent time t.

- *Solution*

The equation of motion of the string with air resistance $\alpha\rho\partial y/\partial t$ is

$$\rho\frac{\partial^2 y}{\partial t^2} = P\frac{\partial^2 y}{\partial x^2} - \alpha\rho\frac{\partial y}{\partial t} \tag{11.6}$$

where y is the displacement of the string.

To solve this equation we put $y = z\exp(-\alpha t/2)$ and then we get

$$\frac{\partial^2 z}{\partial x^2} = \frac{1}{c^2}\left(\frac{\partial^2 z}{\partial t^2} - \frac{1}{4}\alpha^2 z\right)$$

where $c^2 = P/\rho$. Now separating the x and t variables by setting $z(x,t) = X(x)T(t)$ we see that

$$\frac{1}{X}\frac{d^2 X}{dx^2} = \frac{1}{c^2}\left(\frac{1}{T}\frac{d^2 T}{dt^2} - \frac{1}{4}\alpha^2\right).$$

Both sides must be equal to a constant $-k^2$ and so we get

$$\frac{d^2 X}{dx^2} = -k^2 X, \quad \frac{d^2 T}{dt^2} = -\left(k^2 c^2 - \frac{1}{4}\alpha^2\right)T.$$

Hence, since $y = 0$ when $x = 0$, a possible solution is

$$y = \sin kx\left[A\cos(\sqrt{k^2 c^2 - \alpha^2/4}\,t) + B\sin(\sqrt{k^2 c^2 - \alpha^2/4}\,t)\right]\exp(-\alpha t/2). \tag{11.7}$$

Also $y = 0$ when $x = l$ and so $kl = s\pi$ where s is an integer. Now when $t = 0$ we have $y = a\sin(n\pi x/l)$ and thus we get $A = a$ and $k = n\pi/l$. Further $\partial y/\partial t = 0$ when $t = 0$ which gives $-\alpha A/2 + B\sqrt{k^2c^2 - \alpha^2/4} = 0$. It follows that

$$y(x,t) = a\sin\frac{n\pi x}{l}\left[\cos(\sqrt{k^2c^2 - \alpha^2/4}\,t) + \frac{\alpha}{2\sqrt{k^2c^2 - \alpha^2/4}}\sin(\sqrt{k^2c^2 - \alpha^2/4}\,t)\right]\exp(-\alpha t/2).$$

11.09 A string is fixed at the points $x = 0$ and $x = l$. If the point $x = l/2$ is given a transverse displacement $a\sin\omega t$, show that the mean value of the kinetic energy of the string is

$$\frac{1}{8}\rho\omega ca^2\left(\frac{\omega l}{c} - \sin\frac{\omega l}{c}\right)\Big/\sin^2\frac{\omega l}{2c}$$

where ρ is the mass per unit length of the string.

• *Solution*

If y_1 and y_2 are the displacements of the string for $0 \le x \le l/2$ and $l/2 \le x \le l$ we have

$$y_1(x,t) = a\sin\omega t\frac{\sin(\omega x/c)}{\sin(\omega l/2c)}, \quad y_2(x,t) = a\sin\omega t\frac{\sin[\omega(l-x)/c]}{\sin(\omega l/2c)}.$$

Hence the kinetic energy is given by

$$T = \frac{\rho\omega^2a^2\cos^2\omega t}{2\sin^2(\omega l/2c)}\left[\int_0^{l/2}\sin^2(\omega x/c)\,dx + \int_{l/2}^l\sin^2[\omega(l-x)/c]\,dx\right],$$

that is

$$T = \frac{\rho\omega^2a^2\cos^2\omega t}{4\sin^2(\omega l/2c)}\left(l - \frac{c}{\omega}\sin\frac{\omega l}{c}\right)$$

and since the mean value of $\cos^2\omega t$ is $\frac{1}{2}$ the required result for the mean value of the kinetic energy is obtained.

In the next few problems we shall be considering waves travelling on very long strings which can be regarded as having infinite length.

11.10 A very long straight string at tension $P = \rho c^2$ lies along the x axis . On the negative and positive parts of the axis, the string has uniform line densities ρ and ρ' respectively. A travelling wave $y = A\exp[i\omega(t - x/c)]$ approaches the origin along the section of the string with density ρ and is partly reflected and partly transmitted at the discontinuity in the line density. If B and C are the amplitudes of the reflected and transmitted waves, show that

$$\frac{A}{\sqrt{\rho} + \sqrt{\rho'}} = \frac{B}{\sqrt{\rho} - \sqrt{\rho'}} = \frac{C}{2\sqrt{\rho}}.$$

• *Solution*

Let the displacements of the string for $x < 0$ and $x > 0$ be denoted by y_1 and y_2 respectively.

We may write y_1 as the sum of an incident wave $y_{inc} = A\exp[i\omega(t - x/c)]$ and a reflected wave $y_{refl} = B\exp[i\omega'(t + x/c)]$. Also y_2 is entirely a transmitted wave given by $y_{transm} = C\exp[i\omega''(t - x/c')]$ where $\rho c^2 = \rho'c'^2 = P$.

At $x = 0$ we have $y_1 = y_2$ which, for all time t, gives

$$A \exp i\omega t + B \exp i\omega' t = C \exp i\omega'' t$$

and so $A + B = C$ and $\omega = \omega' = \omega''$.

Also $\partial y_1 / \partial x = \partial y_2 / \partial x$ at $x = 0$ and hence we have $(B - A)/c = -C/c'$.

Now $c = \sqrt{P/\rho}$ and $c' = \sqrt{P/\rho'}$ so that $\sqrt{\rho}(B-A) = -\sqrt{\rho'}C = -\sqrt{\rho'}(A+B)$. Hence $A(\sqrt{\rho} - \sqrt{\rho'}) = B(\sqrt{\rho} + \sqrt{\rho'})$ and $2\sqrt{\rho}A = (\sqrt{\rho} + \sqrt{\rho'})C$ from which the required result follows.

11.11 A bead of mass m is fastened to the point $x = 0$ on a very long stretched string and a travelling wave $A \sin[k(x - ct)]$ approaches the bead. Show that after passing the bead, the energy per unit length is reduced by $1/[1 + (km/2\rho)^2]$, where ρ is the mass per unit length of the string, and that there is a change of phase $\tan^{-1}(km/2\rho)$.

• *Solution*

In this problem the actual displacement associated with the incident travelling wave $A \sin[k(x - ct)]$ is the imaginary part of $A \exp[ik(x - ct)]$.

Let the displacements of the string for $x < 0$ and $x > 0$ be denoted by y_1 and y_2 respectively, and take

$$y_1(x, t) = A \exp[ik(x - ct)] + B \exp[-ik(x + ct)], \quad y_2(x, t) = C \exp[ik(x - ct)]$$

where B and C are complex amplitudes.

The equation of motion of the bead is

$$m \frac{\partial^2 y}{\partial t^2} = P \left(\frac{\partial y_2}{\partial x} - \frac{\partial y_1}{\partial x} \right)$$

at $x = 0$ giving

$$-k^2 c^2 C m = ikP(C - A + B).$$

Also $y_1 = y_2$ at $x = 0$ and so $A + B = C$. Hence $-kc^2 Cm = 2iP(C - A)$ and thus

$$C = \frac{2iPA}{kc^2 m + 2iP} = \frac{2PAe^{i\phi}}{\sqrt{k^2 c^4 m^2 + 4P^2}}$$

where $\tan \phi = kc^2 m/2P = km/2\rho$ since $P = c^2 \rho$, giving the required change of phase.

The energy per unit length of the incident wave is $\frac{1}{2}\rho(kc)^2 A^2$ and that for the transmitted wave is $\frac{1}{2}\rho(kc)^2 |C|^2$. Hence the energy per unit length is diminished in the ratio $|C/A|^2 = 1/[1 + (km/2\rho)^2]$.

11.12 If the displacement of a very long string at the point x and at time t is $y(x, t)$, and at time $t = 0$

$$\frac{y(x, 0)}{A} = \frac{\partial y/\partial t}{B} = \frac{\lambda^2}{x^2 + \lambda^2}$$

where A, B and λ are constants, show that at any subsequent time t

$$y(x, t) = \frac{A\lambda^2(\lambda^2 + x^2 + c^2 t^2)}{(\lambda^2 + x^2 + c^2 t^2)^2 - 4c^2 x^2 t^2} + \frac{\lambda B}{2c} \tan^{-1} \frac{2\lambda ct}{\lambda^2 + x^2 - c^2 t^2}.$$

• *Solution*

The general solution of the wave equation for an infinite string is

$$y(x,t) = f(x - ct) + g(x + ct) \tag{11.8}$$

where f and g are arbitrary functions of $x - ct$ and $x + ct$ respectively.

If $y = u(x)$ and $\partial y/\partial t = v(x)$ at time $t = 0$ then $u(x) = f(x) + g(x)$ and $v(x) = -cf'(x) + cg'(x)$. Hence

$$f(x) = \frac{1}{2}u(x) - \frac{1}{2c}\int^x v(x')\,dx', \quad g(x) = \frac{1}{2}u(x) + \frac{1}{2c}\int^x v(x')\,dx'$$

and so

$$y(x,t) = \frac{1}{2}[u(x - ct) + u(x + ct)] + \frac{1}{2c}\int_{x-ct}^{x+ct} v(x')\,dx'. \tag{11.9}$$

In the present problem we have $u(x) = A\lambda^2/(\lambda^2 + x^2)$ and $v(x) = B\lambda^2/(\lambda^2 + x^2)$. Hence

$$y(x,t) = \frac{A}{2}\left[\frac{\lambda^2}{\lambda^2 + (x - ct)^2} + \frac{\lambda^2}{\lambda^2 + (x + ct)^2}\right] + \frac{B\lambda^2}{2c}\int_{x-ct}^{x+ct}\frac{dx'}{\lambda^2 + x'^2}$$

giving

$$y(x,t) = \frac{A\lambda^2(\lambda^2 + x^2 + c^2t^2)}{(\lambda^2 + x^2 + c^2t^2)^2 - 4x^2c^2t^2} + \frac{\lambda B}{2c}\left(\tan^{-1}\frac{x + ct}{\lambda} - \tan^{-1}\frac{x - ct}{\lambda}\right)$$

which yields the requires solution since

$$\tan^{-1}\frac{x + ct}{\lambda} - \tan^{-1}\frac{x - ct}{\lambda} = \tan^{-1}\frac{(x + ct)/\lambda - (x - ct)/\lambda}{1 + (x^2 - c^2t^2)/\lambda^2} = \tan^{-1}\frac{2\lambda ct}{\lambda^2 + x^2 - c^2t^2}.$$

We shall finish this section with a problem on the small lateral vibrations of a *hanging chain*. Although this is not a vibrating string problem it is quite similar.

11.13 A heavy uniform chain of length l hangs freely from one end and performs small transverse vibrations. If y is the transverse displacement of the chain at a point distant x from the free end, show that the normal vibrations of the chain are given by the expression

$$y = aJ_0\left(2\omega\sqrt{\frac{x}{g}}\right)\cos(\omega t + \epsilon)$$

where $J_0(2\omega\sqrt{l/g}) = 0$.

• *Solution*

The equation of motion of a hanging chain is

$$\rho\frac{\partial^2 y}{\partial t^2} = \frac{\partial}{\partial x}\left(P\frac{\partial y}{\partial x}\right) \tag{11.10}$$

where ρ is the uniform density of the chain and $P = \rho gx$ is the tension in the chain. Hence

$$\frac{\partial^2 y}{\partial t^2} = g\frac{\partial y}{\partial x} + gx\frac{\partial^2 y}{\partial x^2}. \tag{11.11}$$

We now separate the variables by putting $y(x,t) = a\psi(x)\cos(\omega t + \epsilon)$ and then we obtain

$$x\frac{d^2\psi}{dx^2} + \frac{d\psi}{dx} + \frac{\omega^2}{g}\psi = 0.$$

This can be solved by setting $x = gz^2/4\omega^2$ and gives

$$\frac{d^2\psi}{dz^2} + \frac{1}{z}\frac{d\psi}{dz} + \psi = 0$$

which is the zero order Bessel equation having the zero order Bessel function of the first kind solution $\psi = J_0(z)$. Hence

$$\psi(x) = J_0\left(2\omega\sqrt{\frac{x}{g}}\right)$$

which gives the required expression for $y(x,t)$, and since $y = 0$ when $x = l$ we also have $J_0(2\omega\sqrt{l/g}) = 0$.

11-2 Sound waves

The three-dimensional equation for sound waves is

$$\nabla^2\phi = \frac{1}{c^2}\frac{\partial^2\phi}{\partial t^2} \tag{11.12}$$

where ϕ is the velocity potential and c is the velocity of sound waves. However the following problems are one-dimensional in which case the wave equation becomes

$$\frac{\partial^2\phi}{\partial x^2} = \frac{1}{c^2}\frac{\partial^2\phi}{\partial t^2} \tag{11.13}$$

whose solution may be written in the separable form

$$\phi(x,t) = (a\cos kx + b\sin kx)\cos(ckt + \epsilon). \tag{11.14}$$

We begin with some problems on standing waves in tubes or pipes of finite length.

11.14 Find expressions for the velocity potential of stationary sound waves in a cylindrical tube of length l when (i) both ends of the tube are closed, (ii) both ends of the tube are open, (iii) one end of the tube is closed and the other end is open.

• *Solution*

(i) Since the tube is closed at both ends, the velocity u of the air is zero at both ends. Now $u = -\partial\phi/\partial x$ and so we have $\partial\phi/\partial x = 0$ at the ends of the tube, taken to be at $x = 0$ and $x = l$.

Hence, in (11.14), $b = 0$ and $\sin kl = 0$ giving $k = n\pi/l$ ($n = 1, 2, ...$). It follows that the velocity potential takes the form

$$\phi(x,t) = \sum_{n=1}^{\infty} a_n \cos\frac{n\pi x}{l}\cos\left(\frac{n\pi ct}{l} + \epsilon_n\right).$$

(ii) Since the tube is open at both ends, the pressure is constant at both ends. Hence the density of the air is constant at both ends and since the density is given by

$$\rho = \rho_0(1 + s) \tag{11.15}$$

where
$$s = c^{-2} \partial\phi/\partial t \qquad (11.16)$$
is called the *condensation*, we have $\partial\phi/\partial t = 0$ at the ends $x = 0$ and $x = l$.

Hence $a = 0$ and $\sin kl = 0$ giving $k = n\pi/l$ $(n = 1, 2, ...)$. Thus we obtain the expression for the velocity potential
$$\phi(x, t) = \sum_{n=1}^{\infty} b_n \sin \frac{n\pi x}{l} \cos \left(\frac{n\pi ct}{l} + \epsilon_n \right).$$

(iii) Suppose the tube is open at $x = 0$. Then $\partial\phi/\partial t = 0$ at $x = 0$ so that $a = 0$. The tube is closed at $x = l$ and so $\partial\phi/\partial x = 0$ at $x = l$ from which it follows that $\cos kl = 0$ giving $k = (2n + 1)\pi/2l$ $(n = 0, 1, 2, ...)$. Hence the velocity potential takes the form
$$\phi(x, t) = \sum_{n=0}^{\infty} b_n \sin \frac{(2n + 1)\pi x}{2l} \cos \left[\frac{(2n + 1)\pi ct}{2l} + \epsilon_n \right].$$

11.15 A tube of length $2l$ which is closed at both ends, is occupied by equal volumes of two gases having densities ρ_1 and ρ_2, which are separated by a movable disk of negligible mass. Show that for small oscillations of angular frequency ω
$$\rho_1 c_1 \cot \frac{\omega l}{c_1} + \rho_2 c_2 \cot \frac{\omega l}{c_2} = 0$$
where c_1 and c_2 are the wave velocities of the two gases.

• *Solution*

The *displacement* ξ of a stratum of fluid is related to the velocity potential ϕ and the condensation s by
$$\frac{1}{c^2} \frac{\partial\phi}{\partial t} = s = -\frac{\partial\xi}{\partial x}. \qquad (11.17)$$
The displacement satisfies the wave equation for sound waves
$$\frac{\partial^2 \xi}{\partial x^2} = \frac{1}{c^2} \frac{\partial^2 \xi}{\partial t^2} \qquad (11.18)$$
and so, separating the variables, we may write
$$\xi(x, t) = (a \cos \frac{\omega x}{c} + b \sin \frac{\omega x}{c}) \cos(\omega t + \epsilon).$$

Let ξ_1 and ξ_2 be the displacements for $0 \leq x \leq l$ and $l \leq x \leq 2l$. Since the tube is closed at $x = 0$ and $x = 2l$, and $\xi_1 = \xi_2$ at $x = l$ we have
$$\xi_1(x, t) = \frac{b \sin(\omega x/c_1)}{\sin(\omega l/c_1)} \cos(\omega t + \epsilon), \quad \xi_2(x, t) = \frac{b \sin[\omega(2l - x)/c_2]}{\sin(\omega l/c_2)} \cos(\omega t + \epsilon). \qquad (11.19)$$

Now the pressure p is given in terms of the condensation $s = -\partial\xi/\partial x$ by the formula
$$p = p_0 + \rho c^2 s. \qquad (11.20)$$

Since the pressure must be the same on both sides of the light disk we have $\rho_1 c_1^2 \partial\xi_1/\partial x = \rho_2 c_2^2 \partial\xi_2/\partial x$ at $x = l$ and so the required result follows.

11.16 A tube of length $2l$ which is closed at both ends is full of air, which is divided into two parts of equal mass $m/2$ by a thin disk of mass M. Show that for free oscillations of angular frequency ω
$$\frac{\omega l}{c} \tan \frac{\omega l}{c} = \frac{m}{M}.$$

● *Solution*

Since the pressure is given by $p = p_0 - \rho c^2 \partial \xi / \partial x$, the equation of motion of the disk of mass M at $x = l$ is

$$M \frac{\partial^2 \xi}{\partial t^2} = \sigma \rho c^2 \left(\frac{\partial \xi_2}{\partial x} - \frac{\partial \xi_1}{\partial x} \right)$$

where σ is the area of cross section of the tube and ξ_1 and ξ_2 are the displacements for $0 \leq x \leq l$ and $l \leq x \leq 2l$ respectively, given in the previous problem by (11.19) with $c = c_1 = c_2$. Then we have

$$M \omega^2 = 2 \sigma \rho c \omega \cot \frac{\omega l}{c}$$

and since $m/2 = \sigma \rho l$ the required result follows.

11.17 The sounding of the lowest note of an organ pipe closed at one end causes the pressure at the closed end to oscillate about its mean value p_0 with amplitude αp_0. Show that at the open end the air particles oscillate with amplitude $2l\alpha/\pi\gamma$ where l is the length of the organ pipe and γ is the ratio of the specific heats.

● *Solution*

Taking the closed end of the organ pipe to be at $x = 0$ and the open end at $x = l$, we have $\xi = 0$ at $x = 0$ and $\partial \xi / \partial x = 0$ at $x = l$. Hence the displacement takes the form

$$\xi(x,t) = b \sin kx \cos(kct + \epsilon)$$

where $\cos kl = 0$ so that we have $k = (2n+1)\pi/2l$.

The lowest note is given by $n = 0$ and thus in the present problem

$$\xi(x,t) = b \sin \frac{\pi x}{2l} \cos \left(\frac{\pi ct}{2l} + \epsilon \right).$$

Now the pressure $p = p_0 - \rho_0 c^2 \partial \xi / \partial x$ and so at the closed end $x = 0$ we have $p = p_0 - (\pi \rho_0 c^2 b/2l) \cos(\pi ct/2l + \epsilon)$. Hence $\alpha p_0 = \pi \rho_0 c^2 b/2l$ and since $c^2 = \gamma p_0/\rho_0$ it follows that the air particles at the open end $x = l$ oscillate with amplitude $b = 2l\alpha/\pi\gamma$ as required.

11.18 A uniform straight tube of length l is open at one end while the other end is closed by a piston whose displacement is given by $\alpha \sin \omega t$ at time t. Find the kinetic energy of the air within the tube given that c is the velocity of sound in air and M is the mass of the air in the tube.

● *Solution*

We write the displacement in the form

$$\xi(x,t) = \left(a \cos \frac{\omega x}{c} + b \sin \frac{\omega x}{c} \right) \sin \omega t.$$

Let the piston be at the end $x = 0$. Then we have $\xi(0,t) = \alpha \sin \omega t$ and so $a = \alpha$. At the open end $x = l$ we have $\partial \xi / \partial x = 0$ and so $b = a \tan(\omega l/c)$. Hence

$$\xi(x,t) = \alpha \left(\cos \frac{\omega x}{c} + \tan \frac{\omega l}{c} \sin \frac{\omega x}{c} \right) \sin \omega t.$$

Then the kinetic energy of the air in the tube with area of cross section σ is given by

$$T = \frac{1}{2}\rho\sigma\int_0^l \left(\frac{\partial \xi}{\partial t}\right)^2 dx = \frac{1}{2}\frac{M\alpha^2\omega^2}{l}\int_0^l \left[\cos^2\frac{\omega x}{c} + \tan\frac{\omega l}{c}\sin\frac{2\omega x}{c} + \tan^2\frac{\omega l}{c}\sin^2\frac{\omega x}{c}\right]\cos^2\omega t\, dx$$

since $M = \rho\sigma l$, and so carrying out the integrations we get

$$T = \frac{1}{4}M\alpha^2\omega^2\left(\sec^2\frac{\omega l}{c} + \frac{c}{\omega l}\tan\frac{\omega l}{c}\right)\cos^2\omega t.$$

11.19 A tube of unit cross section, which is open at both ends, is divided into two parts of lengths l_1 and l_2 by a thin piston of mass M attached to a spring so that its natural angular frequency of vibration is ω_0. Show that the angular frequency of vibration of the whole system is ω where

$$M(\omega_0^2 - \omega^2) = \rho c\omega\left(\tan\frac{\omega l_1}{c} + \tan\frac{\omega l_2}{c}\right)$$

ρ being the density of air.

- *Solution*

Let ξ_1 and ξ_2 be the displacements of the air on either side of the piston taken to be at $x = 0$. Then we may put

$$\xi_1(x,t) = (a_1\cos\frac{\omega x}{c} + b_1\sin\frac{\omega x}{c})\sin\omega t, \quad \xi_2(x,t) = (a_2\cos\frac{\omega x}{c} + b_2\sin\frac{\omega x}{c})\sin\omega t.$$

Now at $x = -l_1$ and $x = l_2$ we have $\partial\xi_1/\partial x = 0$ and $\partial\xi_2/\partial x = 0$ respectively so that

$$a_1\sin\frac{\omega l_1}{c} + b_1\cos\frac{\omega l_1}{c} = 0, \quad -a_2\sin\frac{\omega l_2}{c} + b_2\cos\frac{\omega l_2}{c} = 0.$$

Also at $x = 0$ we have $\xi_1 = \xi_2$ and so $a_1 = a_2$.
The equation of motion of the piston is

$$M\left(\frac{\partial^2\xi}{\partial x^2} + \omega_0^2\xi\right) = p_1 - p_2$$

where p_1 and p_2 are the pressures on either side of the piston.
Now $p = p_0 + c^2\rho s = p_0 - c^2\rho\partial\xi/\partial x$ and so

$$M\left(\frac{\partial^2\xi}{\partial x^2} + \omega_0^2\xi\right) = \rho c^2\left(\frac{\partial\xi_2}{\partial x} - \frac{\partial\xi_1}{\partial x}\right).$$

Hence we have

$$M(-\omega^2 + \omega_0^2)a_1 = c\rho\omega(b_2 - b_1) = \rho c\omega\left(a_2\tan\frac{\omega l_2}{c} + a_1\tan\frac{\omega l_1}{c}\right)$$

from which the required result follows at once since $a_1 = a_2$.

11.20 A pipe of length l is closed at both ends and the area of its cross section is proportional to the distance x from one end. Show that for small longitudinal oscillations of the gas in the pipe the condensation is proportional to $J_1(kx)\cos kct$ where $J_1(kl) = 0$.

- *Solution*

The wave equation for sound waves in a pipe with cross section $\sigma(x)$ which varies with position x along its length is

$$\frac{\partial}{\partial x}\left[\frac{1}{\sigma}\frac{\partial(\sigma\xi)}{\partial x}\right] = \frac{1}{c^2}\frac{\partial^2\xi}{\partial t^2} \tag{11.21}$$

where ξ is the displacement.

We separate the variables by putting $\xi(x,t) = X(x)\cos kct$ and then setting $\sigma = \alpha x$ we get

$$\frac{d^2 X}{dx^2} + \frac{1}{x}\frac{dX}{dx} + k^2 X = 0$$

whose solution may be written in the form $X(x) = AJ_0(kx)$ where $J_0(v)$ is the Bessel function of zero order of the first kind satisfying the Bessel equation

$$\frac{d^2 X}{dv^2} + \frac{1}{v}\frac{dX}{dv} + X = 0.$$

The condensation is given by $s = -\partial\xi/\partial x = -AkJ_0'(kx)\cos kct$ and since $J_0'(v) = -J_1(v)$ we obtain

$$s(x,t) = AkJ_1(kx)\cos kct.$$

The pipe is closed at both ends and so the condensation must vanish at $x = 0$ and $x = l$. This is satisfied by the above solution at $x = 0$ since $J_1(0) = 0$. The condensation will vanish at $x = l$ if the values of k are chosen so that $J_1(kl) = 0$ as required.

The last problem on sound waves involves a long tube whose length can be regarded as infinite.

11.21 A sound wave travelling along a uniform cylinder in a gas G_1 falls on a freely movable piston of negligible mass on the other side of which is another gas G_2. If the amplitude of the incident wave is A, find the amplitudes of the transmitted and reflected waves, given that c_1 and c_2 are the velocities of sound waves in G_1 and G_2 respectively, and that ρ_1 and ρ_2 are the densities of the two gases.

• *Solution*

We take the incident wave to be $A\exp[i\omega(t - x/c_1)]$, the reflected wave to be $B\exp[i\omega(t + x/c_1)]$ and the transmitted wave to be $C\exp[i\omega(t - x/c_2)]$. Then, if the displacements in G_1 and G_2 are denoted by ξ_1 and ξ_2, we may write

$$\xi_1 = A\exp[i\omega(t - x/c_1)] + B\exp[i\omega(t + x/c_1)], \quad \xi_2 = C\exp[i\omega(t - x/c_2)].$$

Now, taking the piston to be at $x = 0$ we have there $\xi_1 = \xi_2$ and so $A + B = C$.

The pressure $p = p_0 + \rho c^2 s$, where $s = -\partial\xi/\partial x$ is the condensation, is the same on either side of the light piston and so we have $\rho_1 c_1^2 \partial\xi_1/\partial x = \rho_2 c_2^2 \partial\xi_2/\partial x$ at $x = 0$. Hence $\rho_1 c_1(B - A) = -\rho_2 c_2 C$ and so we obtain

$$\frac{B}{A} = \frac{\rho_1 c_1 - \rho_2 c_2}{\rho_1 c_1 + \rho_2 c_2}, \quad \frac{C}{A} = \frac{2\rho_1 c_1}{\rho_1 c_1 + \rho_2 c_2}.$$

11-3 Water waves

We start with some problems on *long waves in shallow water.*
The equation for long waves in shallow water is

$$\frac{\partial^2 \zeta}{\partial x^2} + \frac{\partial^2 \zeta}{\partial y^2} = \frac{1}{c^2}\frac{\partial^2 \zeta}{\partial t^2} \tag{11.22}$$

where $\zeta(x,y,t)$ is the elevation of the surface of the water above the point with coordinates $x, y, z = 0$ at time t and $c^2 = gh$, the depth of the water being denoted by h.

11.22 Show that for long waves in the water in a shallow rectangular tank of constant depth h with sides given by $x = 0, a$ and $y = 0, b$, the elevation $\zeta(x,y,t)$ of the surface at time t is given by

$$\zeta = A\cos\frac{n\pi x}{a}\cos\frac{m\pi y}{b}\cos(\omega t + \epsilon)$$

where n, m are integers satisfying

$$\frac{\omega^2}{c^2} = \pi^2\left(\frac{n^2}{a^2} + \frac{m^2}{b^2}\right)$$

and $c^2 = gh$.

- *Solution*

The equation for long waves in shallow water is given by (11.22).
We can separate the variables by writing $\zeta(x,y,t) = X(x)Y(y)\cos(\omega t + \epsilon)$ and then we get

$$X(x) = a_1\cos k_1 x + b_1\sin k_1 x, \quad Y(y) = a_2\cos k_2 y + b_2\sin k_2 y$$

where $k_1^2 + k_2^2 = \omega^2/c^2$.
If the velocity of the water has components u and v in the x and y directions, we have

$$\frac{\partial u}{\partial t} = -g\frac{\partial \zeta}{\partial x}, \quad \frac{\partial v}{\partial t} = -g\frac{\partial \zeta}{\partial y} \tag{11.23}$$

and so

$$u = -\frac{g}{\omega}\frac{dX}{dx}Y\sin(\omega t + \epsilon), \quad v = -\frac{g}{\omega}X\frac{dY}{dy}\sin(\omega t + \epsilon).$$

Now $u = 0$ for $x = 0, a$ and $v = 0$ for $y = 0, b$. Hence $b_1 = 0, b_2 = 0$ and $k_1 a = n\pi, k_2 b = m\pi$ where n, m are positive integers or zero satisfying $(n\pi/a)^2 + (m\pi/b)^2 = \omega^2/c^2$, and so the required result is obtained.

11.23 A channel of unit width is of depth h, where $h = \alpha x$, α being a constant. Show that for *tidal waves*, or long waves in a shallow channel, the altitude is given by $\zeta = AJ_0(\gamma\sqrt{x})\cos(\omega t + \epsilon)$ where $\gamma^2 = 4\omega^2/\alpha g$. Further show that the wave length of these stationary waves increases with increasing values of x.

- *Solution*

For tidal waves in a channel with variable depth h, the altitude ζ satisfies

$$\frac{\partial}{\partial x}\left(hg\frac{\partial \zeta}{\partial x}\right) = \frac{\partial^2 \zeta}{\partial t^2} \tag{11.24}$$

where in the present problem $h = \alpha x$ so that we have

$$\frac{\partial}{\partial x}\left(x\frac{\partial \zeta}{\partial x}\right) = \frac{1}{\alpha g}\frac{\partial^2 \zeta}{\partial t^2}.$$

Now we separate the variables by putting $\zeta(x,t) = X(x)\cos(\omega t + \epsilon)$ and then we get

$$\frac{d}{dx}\left(x\frac{dX}{dx}\right) + \frac{\omega^2}{\alpha g}X = 0.$$

Next putting $z^2 = \gamma^2 x$ where $\gamma^2 = 4\omega^2/\alpha g$ we obtain

$$\frac{d^2 X}{dz^2} + \frac{1}{z}\frac{dX}{dz} + X = 0$$

which is the zero order Bessel equation. Hence $X = AJ_0(\gamma\sqrt{x})$ as required.

Suppose $x_1 < x_2 < x_3 < x_4 < x_5$ are successive zeros of $J_0(\gamma\sqrt{x})$. Since the distance between successive zeros of $J_0(z)$ approaches π as z becomes large, we can take x_1 to be sufficiently large for

$$\gamma(\sqrt{x_3} - \sqrt{x_1}) \simeq \gamma(\sqrt{x_5} - \sqrt{x_3}) \simeq 2\pi.$$

Then we have

$$\frac{x_5 - x_3}{x_3 - x_1} \simeq \frac{\sqrt{x_5} + \sqrt{x_3}}{\sqrt{x_3} + \sqrt{x_1}} > 1$$

and hence $x_5 - x_3 > x_3 - x_1$ which means that the wave length increases with increasing x.

11.24 An estuary extending from $x = 0$ to $x = l$ has a rectangular cross section of depth αx and breadth βx. The estuary meets the sea at $x = l$, in which there is a tidal oscillation given by $\zeta = \zeta_0\cos(\omega t + \epsilon)$.

Show that in the estuary

$$\zeta = \zeta_0\frac{\sqrt{l}J_1(\gamma\sqrt{x})}{\sqrt{x}J_1(\gamma\sqrt{l})}\cos(\omega t + \epsilon)$$

where $\gamma^2 = 4\omega^2/g\alpha$.

• *Solution*

The equation for the altitude ζ of the free surface of the water for the tidal waves in an estuary of variable depth and breadth is

$$\frac{\partial}{\partial x}\left(\sigma g\frac{\partial \zeta}{\partial x}\right) = b\frac{\partial^2 \zeta}{\partial t^2} \tag{11.25}$$

where σ is the cross section and b is the breadth of the estuary. In the present problem we have $\sigma = \alpha\beta x^2$ and $b = \beta x$ so that the equation becomes

$$\frac{1}{x}\frac{\partial}{\partial x}\left(x^2\frac{\partial \zeta}{\partial x}\right) = \frac{1}{\alpha g}\frac{\partial^2 \zeta}{\partial t^2}.$$

At $x = l$ we have $\zeta = \zeta_0 \cos(\omega t + \epsilon)$ and so we put $\zeta = X(x)\cos(\omega t + \epsilon)$. Then we get

$$\frac{1}{x}\frac{d}{dx}\left(x^2\frac{dX}{dx}\right) + \frac{\omega^2}{\alpha g}X = 0.$$

Now we set $z^2 = \gamma^2 x$ and we obtain

$$\frac{d^2X}{dz^2} + \frac{3}{z}\frac{dX}{dz} + X = 0$$

and if then we put $X(z) = z^{-1}f(z)$ we find that

$$\frac{d^2f}{dz^2} + \frac{1}{z}\frac{df}{dz} + \left(1 - \frac{1}{z^2}\right)f = 0.$$

This is the Bessel equation of the first order and so $f(z) = AJ_1(z)$ where $J_1(z)$ is a first order Bessel function of the first kind. But $\zeta = \zeta_0 \cos(\omega t + \epsilon)$ at $x = l$ and so

$$\zeta(x,t) = \zeta_0 \frac{\sqrt{l}J_1(\gamma\sqrt{x})}{\sqrt{x}J_1(\gamma\sqrt{l})}\cos(\omega t + \epsilon).$$

The remaining problems in this chapter are concerned with *surface waves* in deep water .

In this case the velocity potential satisfies Laplace's equation $\nabla^2\phi = 0$.

At the free surface, situated close to the horizontal plane $z = 0$, we have

$$\frac{\partial^2\phi}{\partial t^2} + g\frac{\partial\phi}{\partial z} = 0 \tag{11.26}$$

where z is the height measured in the upward vertical direction.

The phase velocity c of the surface waves is given by

$$c^2 = \frac{g}{k}\tanh kh \tag{11.27}$$

where the bottom of the water is given by $z = -h$ and $\omega = ck$ is the angular frequency of the oscillations.

11.25 If a canal of rectangular cross section is ended by two vertical walls whose distance apart is $2l$, and if the water is initially at rest and has its plane surface inclined at a small angle α to the length of the canal, show that the altitude ζ of the wave at distance x from the first wall at any time t, is given by

$$\zeta(x,t) = -\frac{8l\alpha}{\pi^2}\sum_{n=0}^{\infty}\frac{1}{(2n+1)^2}\cos\frac{(2n+1)\pi x}{2l}\cos\frac{(2n+1)\pi ct}{2l}$$

where c is the phase velocity of a wave of wave length $4l/(2n+1)$ on an infinitely long canal.

• *Solution*

We separate the variables by writing the altitude in the form

$$\zeta(x,t) = (a\cos kx + b\sin kx)\cos(kct + \epsilon)$$

and the velocity potential as

$$\phi(x, z, t) = f(z)(a \cos kx + b \sin kx) \sin(kct + \epsilon)$$

where

$$\zeta = g^{-1} \partial\phi/\partial t \tag{11.28}$$

at the free surface.

Now $\partial\phi/\partial x = 0$ at $x = 0$ and $x = 2l$ so that we have $b = 0$ and $\sin 2kl = 0$ which gives $k = s\pi/2l$ ($s = 1, 2, ...$). Further $\partial\zeta/\partial t = 0$ at time $t = 0$ and so $\epsilon = 0$. Hence we have

$$\zeta(x, t) = \sum_{s=1}^{\infty} a_s \cos \frac{s\pi x}{2l} \cos \frac{s\pi ct}{2l}.$$

Since the plane surface of the water is inclined at an angle α initially we can put $\zeta = \alpha(x - l)$ when $t = 0$ and so

$$a_s = \frac{\alpha}{l} \int_0^{2a} (x - l) \cos \frac{s\pi x}{2l} \, dx.$$

But

$$\int_0^{2a} \cos \frac{s\pi x}{2l} \, dx = 0, \qquad \int_0^{2a} x \cos \frac{s\pi x}{2l} \, dx = \frac{4l^2}{s^2\pi^2}[(-1)^s - 1]$$

and thus $a_{2n} = 0$ and

$$a_{2n+1} = -\frac{8\alpha l}{\pi^2} \frac{1}{(2n + 1)^2} \qquad (n = 0, 1, 2, ...)$$

which produces the required result.

11.26 Show that for surface waves in a circular tank of radius a and depth h, a possible solution for the velocity potential is given by

$$\phi = D J_n(kr) \cos n\theta \cosh k(z + h) \sin \omega t$$

where k satisfies the equation $J_n'(ka) = 0$ and $\omega^2 = kg \tanh kh$.

- *Solution*

The velocity potential ϕ satisfies Laplace's equation

$$\nabla^2 \phi = 0$$

which in cylindrical polar coordinates r, θ, z becomes

$$\frac{1}{r} \frac{\partial}{\partial r} \left(r \frac{\partial\phi}{\partial r} \right) + \frac{1}{r^2} \frac{\partial^2\phi}{\partial\theta^2} + \frac{\partial^2\phi}{\partial z^2} = 0.$$

We separate the coordinates by putting $\phi(r, \theta, z, t) = R(r)Z(z) \cos n\theta \sin \omega t$ and then we get

$$\frac{1}{R} \left(\frac{d^2R}{dr^2} + \frac{1}{r} \frac{dR}{dr} - \frac{n^2}{r^2} R \right) = -\frac{1}{Z} \frac{d^2Z}{dz^2}.$$

Both sides of this equation must be constant and we put this constant to be $-k^2$. Then $d^2Z/dz^2 = k^2 Z$ whose general solution is $Z = Ae^{kz} + Be^{-kz}$. Now $\partial\phi/\partial z = 0$ at the bottom of the tank $z = -h$ and so $Ae^{-kh} = Be^{kh} = D/2$ giving

$$Z(z) = D \cosh k(z + h).$$

Also we have

$$\frac{d^2R}{dr^2} + \frac{1}{r}\frac{dR}{dr} + \left(k^2 - \frac{n^2}{r^2}\right)R = 0$$

and so we obtain the solution $R(r) = J_n(kr)$ in terms of the nth order Bessel function $J_n(z)$ of the first kind, from which it follows that

$$\phi(r, \theta, z, t) = D J_n(kr) \cosh k(z+h) \cos n\theta \sin \omega t.$$

Since $\partial\phi/\partial r = 0$ at the cylindrical boundary of the tank where $r = a$, the allowed values of k satisfy $J_n'(ka) = 0$.

Also at the free surface of the water in the tank $\partial^2\phi/\partial t^2 + g\partial\phi/\partial z = 0$. Since ζ is small we may set $z = 0$ in this equation and then we get $\omega^2 = gk \tanh kh$ as required.

11.27 If x', y', z' are the coordinates of a particle of fluid relative to its mean position x, y, z in a rectangular tank of depth h with sides $x = 0, a$ and $y = 0, b$, show that for surface waves the path of the particle is the straight line

$$\frac{x'a}{n\pi}\cot\frac{n\pi x}{a} = \frac{y'b}{m\pi}\cot\frac{m\pi y}{b} = -\frac{z'}{k}\coth k(z+h)$$

where

$$k^2 = \pi^2\left(\frac{n^2}{a^2} + \frac{m^2}{b^2}\right)$$

and n, m are integers.

- *Solution*

Since the velocity potential ϕ satisfies Laplace's equation $\nabla^2\phi = 0$ and ϕ satisfies the boundary conditions $\partial\phi/\partial z = 0$ at $z = -h, \partial\phi/\partial x = 0$ at $x = 0, a$ and $\partial\phi/\partial y = 0$ at $y = 0, b$, we may write

$$\phi(x, y, z, t) = A\cos\frac{n\pi x}{a}\cos\frac{m\pi y}{b}\cosh k(z+h)\sin \omega t$$

where $k^2 = \pi^2(n^2/a^2 + m^2/b^2)$. Then we have

$$\frac{dx'}{dt} = -\frac{\partial\phi}{\partial x} = \frac{n\pi}{a}A\sin\frac{n\pi x}{a}\cos\frac{m\pi y}{b}\cosh k(z+h)\sin \omega t,$$

$$\frac{dy'}{dt} = -\frac{\partial\phi}{\partial y} = \frac{m\pi}{b}A\cos\frac{n\pi x}{a}\sin\frac{m\pi y}{b}\cosh k(z+h)\sin \omega t,$$

$$\frac{dz'}{dt} = -\frac{\partial\phi}{\partial z} = -kA\cos\frac{n\pi x}{a}\cos\frac{m\pi y}{b}\sinh k(z+h)\sin \omega t$$

which, on doing the integrations with respect to t, give the required result.

11.28 Waves of small amplitude travel along the surface of the water in a canal of rectangular cross section and depth h bounded by lock gates at a distance l apart. If the waves move along the length of the canal, show that their periods are given by

$$2\sqrt{\frac{\pi l}{ng}\coth\frac{n\pi h}{l}}$$

where n is a positive integer.

- *Solution*

We may write the velocity potential in the form

$$\phi(x, z, t) = Z(z)(A \sin kx + B \cos kx) \cos ckt$$

where x is a coordinate chosen along the length of the canal and z is a coordinate chosen in the upward direction.

Now $\partial\phi/\partial x = 0$ at the two ends of the canal chosen to be at $x = 0, l$. Then we have $A = 0$ and $\sin kl = 0$ which gives $k = n\pi/l$ where n is a positive integer.

The phase velocity c of the surface waves is given by $c^2 = (g/k) \tanh kh$ and since the period of the surface waves is $T = 2\pi/ck = 2l/cn$, the required result follows.

11.29 Simple harmonic surface waves of wave length λ are being propagated in deep water. Show that at a point whose depth below the undisturbed surface is d, the ratio of the pressure when the disturbed surface has altitude ζ to the undisturbed pressure at the same point is given by

$$R = 1 + \frac{\zeta}{d} \exp(-2\pi d/\lambda).$$

- *Solution*

In fluid dynamics, if we put $\mathbf{v} = -\nabla\phi$ and $\omega = \mathbf{0}$ in Euler's equation written in the form (7.5), we obtain $\nabla U = \mathbf{0}$ where $U = \Omega + \frac{1}{2}v^2 + p/\rho - \partial\phi/\partial t$ for an incompressible fluid, from which it follows that

$$\Omega + \tfrac{1}{2}v^2 + \frac{p}{\rho} - \frac{\partial\phi}{\partial t} = F(t)$$

where $F(t)$ is an arbitrary function of the time. This is Bernoulli's equation. Including $F(t)$ in $\partial\phi/\partial t$, setting $\Omega = gz$ and neglecting the square of the velocity v^2, the pressure p is given by

$$\frac{p}{\rho} = \frac{\partial\phi}{\partial t} - gz \tag{11.29}$$

where ρ is the density of the water and z is the vertical coordinate. Hence at depth d where $z = -d$ the undisturbed pressure is given by $p_0 = \rho g d$.

When the water is disturbed so that the altitude is given by $\zeta = a \sin k(x - ct)$ we may express the velocity potential in the form

$$\phi(x, z, t) = aZ(z) \cos k(x - ct)$$

where $d^2Z/dz^2 = k^2Z$ and $dZ/dz = 0$ at the bottom of the canal where $z = -h$. Now $\zeta = g^{-1}\partial\phi/\partial t$ at the free surface $z \simeq 0$ and so

$$Z(z) = \frac{g \cosh k(h + z)}{kc \cosh kh}.$$

Hence, at depth d where $z = -d$ we have

$$\frac{\partial\phi}{\partial t} = ag\frac{\cosh k(h - d)}{\cosh kh} \sin k(x - ct) = g\zeta\frac{\cosh k(h - d)}{\cosh kh} \simeq g\zeta \exp(-kd)$$

since the depth h is very large.

Now $k = 2\pi/\lambda$ where λ is the wave length and so, using Bernoulli's equation (11.29), it follows that

$$R = \frac{p}{p_0} = 1 + \frac{\zeta}{d}\exp(-kd) = 1 + \frac{\zeta}{d}\exp(-2\pi d/\lambda)$$

as required.

11.30 Show that for surface waves $\zeta = a\sin k(x - ct)$ on the water in a canal of depth h, the rate of transmission of energy is given by

$$v = \frac{c}{2}\left(1 + \frac{4\pi h}{\lambda}\operatorname{cosech}\frac{4\pi h}{\lambda}\right)$$

where c is the phase velocity and $\lambda = 2\pi/k$ is the wave length. Further show that v is the same as the group velocity.

• *Solution*

The rate of transmission of energy is found by considering a vertical section of the liquid at right angles to the direction of propagation and determining the rate at which the pressure on one side of this section is doing work on the liquid on the other side. This is

$$\int_{-h}^{0} pu\,dz$$

per unit width of the section, u being the horizontal velocity of the liquid in the direction of propagation.

Neglecting u^2 in Bernoulli's equation we have

$$\frac{p}{\rho} = \frac{\partial\phi}{\partial t} - gz.$$

But, as in the previous problem, we have for the velocity potential

$$\phi = \frac{ag}{kc}\frac{\cosh k(h + z)}{\cosh kh}\cos k(x - ct).$$

Hence

$$\frac{\partial\phi}{\partial t} = ag\frac{\cosh k(h + z)}{\cosh kh}\sin k(x - ct)$$

and

$$u = -\frac{\partial\phi}{\partial x} = \frac{ag}{c}\frac{\cosh k(h + z)}{\cosh kh}\sin k(x - ct).$$

Since the mean value of $\sin k(x - ct)$ is zero and the mean value of $\sin^2 k(x - ct)$ is $\frac{1}{2}$, it follows that the mean rate of working is

$$\int_{-h}^{0} pu\,dz = \frac{1}{2}\frac{\rho a^2 g^2}{c\cosh^2 kh}\int_{-h}^{0}\cosh^2 k(h + z)\,dz.$$

But $2\cosh^2 k(h + z) = 1 + \cosh 2k(h + z)$ and so the mean rate of working is

$$\frac{\rho a^2 g^2}{4c}\operatorname{sech}^2 kh\int_{-h}^{0}[1 + \cosh 2k(h + z)]\,dz = \frac{\rho a^2 gc}{4}(1 + 2kh\operatorname{cosech} 2kh)$$

using

$$c^2 = \frac{g}{k}\tanh kh.$$

But $\frac{1}{2}g\rho a^2$ is the energy of the wave motion per unit length per unit width, to which the potential energy and kinetic energy of the liquid provide equal contributions of $\frac{1}{4}g\rho a^2$, and so the velocity of energy flow is

$$v = \frac{1}{2}c\left(1 + \frac{4\pi h}{\lambda}\operatorname{cosech}\frac{4\pi h}{\lambda}\right).$$

Now the *group velocity* is given by

$$U = \frac{d\omega}{dk} = \frac{d(kc)}{dk} = c + k\frac{dc}{dk}$$

and in the present case we have

$$\frac{dc}{dk} = \frac{c}{2k}\left(-1 + 2kh\operatorname{cosech}2kh\right)$$

from which it follows that

$$U = \frac{1}{2}c(1 + 2kh\operatorname{cosech}2kh).$$

Thus the velocity of energy flow is the same as the group velocity.

12
HEAT CONDUCTION

The first detailed mathematical study of heat conduction was carried out by Fourier and was published in 1822 in his treatise on the *Theorie analytique de la chaleur*. The method he used involved expansions in trigonometric series known as Fourier series, and Fourier transforms discussed in Chapter 9.

The theory of heat conduction is based on *Fourier's Law* $\mathbf{F} = -k\nabla V$ where V is the temperature, k is the *thermal conductivity* and $\mathbf{F}$ is the vector field having the direction of the flow of heat, and magnitude equal to the quantity of heat per unit area per unit time crossing the isothermal surface through the field point.

Then the equation for *heat conduction* is

$$\frac{\partial V}{\partial t} = \kappa \nabla^2 V \tag{12.1}$$

where κ is called the *diffusivity* which is given by $\kappa = k/C\rho$ where ρ is the density and C is the *specific heat* of the body.

The equation (12.1) for the flow of heat through a conducting medium is the same as the equation for the diffusion of a fluid through a second fluid and also for other diffusion phenomena such as the motion of electrons and ions through a gas and the diffusion of neutrons through matter. However we shall not be concerned with these here.

It is important to notice that the equation for heat conduction has a first-order derivative with respect to time so that the time dependence of the solution is quite different from that for the wave equation discussed in the previous chapter. For heat conduction there is an exponential dependence on the time, except in steady state conditions, whereas in the case of wave motion the time dependence is oscillatory.

In the following few problems, dealing with slabs of conducting material, we use Fourier series to obtain the solutions.

12.01 An infinite homogeneous slab has two plane faces at a distance l apart which are maintained at zero temperature. If the slab is at a uniform temperature A initially, find the temperature at any subsequent time t.

• *Solution*

Since this is a one-dimensional problem, the equation for heat conduction (12.1) takes the form

$$\frac{\partial V}{\partial t} = \kappa \frac{\partial^2 V}{\partial x^2}. \tag{12.2}$$

We separate the variables by putting $V(x,t) = T(t)X(x)$ and then we get

$$\frac{1}{T}\frac{dT}{dt} = \frac{\kappa}{X}\frac{d^2 X}{dx^2}.$$

Both sides of this equation must be a constant which we call $-\kappa\omega^2$ and then we see that

$$\frac{dT}{dt} = -\kappa\omega^2 T, \quad \frac{d^2 X}{dx^2} = -\omega^2 x.$$

It follows that we have $V(x,t) = (a\cos\omega x + b\sin\omega x)\exp(-\kappa\omega^2 t)$ as a solution.

Now the temperature $V = 0$ when $x = 0, l$. Hence $a = 0$ and $\sin\omega l = 0$ giving $\omega = n\pi/l$ where n is a positive integer. Thus we have

$$V(x,t) = \sum_{n=1}^{\infty} b_n \sin\frac{n\pi x}{l}\exp\left(-\frac{\kappa n^2\pi^2}{l^2}t\right).$$

Since the temperature $V = A$ at time $t = 0$ it follows that

$$A = \sum_{n=1}^{\infty} b_n \sin\frac{n\pi x}{l}$$

and so

$$b_n = \frac{2}{l}\int_0^l A\sin\frac{n\pi x}{l}\,dx = \frac{2A}{n\pi}[1 - (-1)^n]$$

which yields the result

$$V(x,t) = \frac{4A}{\pi}\sum_{n=0}^{\infty}\frac{1}{2n+1}\sin\left[\frac{(2n+1)\pi x}{l}\right]\exp\left[-\frac{\kappa(2n+1)^2\pi^2}{l^2}t\right].$$

12.02 An infinite slab of homogeneous material bounded by the planes $x = 0$ and $x = l$ is initially at zero temperature. The face $x = 0$ is maintained at temperature A while the face $x = l$ is kept at zero temperature. Derive the temperature distribution within the slab at any time t.

• *Solution*

Let us put

$$V(x,t) = U(x) + W(x,t)$$

where $d^2U/dx^2 = 0$ and $U = A$ at $x = 0$ and $U = 0$ at $x = l$.

Then

$$\frac{\partial W}{\partial t} = \kappa\frac{\partial^2 W}{\partial x^2}$$

and we take $W = 0$ at $x = 0$ and $x = l$ in order to satisfy the boundary conditions, while $W(x,0) = -U(x)$ because the temperature is zero initially.

We see that

$$U(x) = A\left(1 - \frac{x}{l}\right).$$

Now, as in the previous problem, we may put

$$W(x,t) = \sum_{n=1}^{\infty} b_n \sin\frac{n\pi x}{l}\exp\left(-\frac{\kappa n^2\pi^2}{l^2}t\right)$$

where here we have

$$b_n = -\frac{2}{l}\int_0^l U(x)\sin\frac{n\pi x}{l}\,dx = -\frac{2}{l}\int_0^l A\left(1 - \frac{x}{l}\right)\sin\frac{n\pi x}{l}\,dx = -\frac{2A}{n\pi}$$

and hence

$$V(x,t) = A\left(1 - \frac{x}{l}\right) - \frac{2A}{\pi}\sum_{n=1}^{\infty}\frac{1}{n}\sin\frac{n\pi x}{l}\exp\left(-\frac{\kappa n^2\pi^2}{l^2}t\right).$$

12.03 The faces $x = 0, l$ of an infinite slab of homogeneous material are maintained at zero temperature. If the initial temperature distribution in the slab is given by $V(x,0) = Ax$, find the temperature distribution at any subsequent time t.

- *Solution*

Since $V = 0$ at $x = 0$ and $x = l$ we may put

$$V(x,t) = \sum_{n=1}^{\infty} b_n \sin \frac{n\pi x}{l} \exp\left(-\frac{\kappa n^2 \pi^2}{l^2} t\right).$$

Now $V(x,0) = Ax$ and so

$$Ax = \sum_{n=1}^{\infty} b_n \sin \frac{n\pi x}{l}$$

giving

$$b_n = \frac{2A}{l} \int_0^l x \sin \frac{n\pi x}{l}\, dx = -\frac{2Al(-1)^n}{n\pi}.$$

Hence

$$V(x,t) = \frac{2Al}{\pi} \sum_{n=1}^{\infty} \frac{(-1)^{n+1}}{n} \sin \frac{n\pi x}{l} \exp\left(-\frac{\kappa n^2 \pi^2}{l^2} t\right).$$

12.04 The faces $x = 0, l$ of a uniform slab are kept at zero temperature. If the initial temperature distribution is

$$V(x,0) = \alpha x(l - x)$$

find the first two terms of the expansion giving the temperature at time t.

- *Solution*

As before we have

$$V(x,t) = \sum_{s=1}^{\infty} b_s \sin \frac{s\pi x}{l} \exp\left(-\frac{\kappa s^2 \pi^2}{l^2} t\right)$$

where $V(x,0) = \alpha x(l - x)$. Hence

$$\alpha x(l - x) = \sum_{s=1}^{\infty} b_s \sin \frac{s\pi x}{l}$$

and so

$$b_s = \frac{2\alpha}{l} \int_0^l x(l - x) \sin \frac{s\pi x}{l}\, dx = \frac{4\alpha l^2}{s^3 \pi^3}[1 - (-1)^s].$$

It follows that

$$V(x,t) = \frac{8\alpha l^2}{\pi^3} \sum_{n=0}^{\infty} \frac{1}{(2n + 1)^3} \sin \frac{(2n + 1)\pi x}{l} \exp\left[-\frac{\kappa(2n + 1)^2 \pi^2}{l^2} t\right]$$

giving

$$V(x,t) = \frac{8\alpha l^2}{\pi^3}\left[\sin \frac{\pi x}{l} \exp\left(-\frac{\kappa \pi^2}{l^2} t\right) + \frac{1}{27} \sin \frac{3\pi x}{l} \exp\left(-\frac{9\kappa \pi^2}{l^2} t\right) + \ldots\right]$$

which should be compared with the solution obtained in problem 11.01 for a vibrating string fixed at its two ends.

12.05 The temperature throughout a homogeneous sphere of radius a is initially zero. If the surface of the sphere is maintained at a temperature A find the temperature at the centre of the sphere as a function of the time.

• *Solution*

Since the temperature V depends only on the distance r from the centre of the sphere and the time t, the equation of heat conduction takes the form

$$\frac{\partial V}{\partial t} = \kappa \frac{1}{r^2} \frac{\partial}{\partial r} \left(r^2 \frac{\partial V}{\partial r} \right).$$

Putting $V(r,t) = r^{-1} U(r,t)$ and noting that

$$\frac{1}{r^2} \frac{\partial}{\partial r} \left(r^2 \frac{\partial V}{\partial r} \right) = \frac{1}{r^2} \frac{\partial}{\partial r} \left(r \frac{\partial U}{\partial r} - U \right) = \frac{1}{r} \frac{\partial^2 U}{\partial r^2}$$

we obtain

$$\frac{\partial U}{\partial t} = \kappa \frac{\partial^2 U}{\partial r^2}$$

which is the same as the one-dimensional equation of heat conduction.

Since V must be finite at the centre of the sphere, we must have $U = 0$ at $r = 0$. Hence following the analysis given in problems 12.01 and 12.02, we may write

$$U(r,t) = Ar + \sum_{n=1}^{\infty} b_n \sin \frac{n\pi r}{a} \exp \left(-\frac{\kappa n^2 \pi^2}{a^2} t \right)$$

so that $V = A$ when $r = a$ for all t.

Now $V(r, 0) = 0$ for $0 \le r < a$ and so we have

$$Ar = -\sum_{n=1}^{\infty} b_n \sin \frac{n\pi r}{a}$$

which gives

$$b_n = -\frac{2A}{a} \int_0^a r \sin \frac{n\pi r}{a} \, dr = \frac{2A}{n\pi} \left[r \cos \frac{n\pi r}{a} \right]_0^a = \frac{2Aa}{n\pi} (-1)^n.$$

It follows that the temperature distribution inside the sphere is given by

$$V(r,t) = A + \frac{2Aa}{\pi} \frac{1}{r} \sum_{n=1}^{\infty} \frac{(-1)^n}{n} \sin \frac{n\pi r}{a} \exp \left(-\frac{\kappa n^2 \pi^2}{a^2} t \right).$$

Since $r^{-1} \sin(n\pi r/a) \to n\pi/a$ as $r \to 0$ we see that the temperature at the centre of the sphere as a function of the time is

$$V(0,t) = A \left[1 + 2 \sum_{n=1}^{\infty} (-1)^n \exp \left(-\frac{\kappa n^2 \pi^2}{a^2} t \right) \right].$$

The next two problems are concerned with steady state temperature distributions which do not depend on the time.

12.06 A rectangular solid is bounded by the planes $x = 0, x = a, y = 0, y = b$. The boundary plane $x = a$ is kept at temperature $f(y)$ while all the other boundary planes are kept at zero temperature. Find the steady state temperature distribution within the solid.

• *Solution*

When there is a steady state we have $\partial V/\partial t = 0$ and so a steady state temperature distribution satisfies Laplace's equation $\nabla^2 V = 0$, which for a two-dimensional problem becomes

$$\frac{\partial^2 V}{\partial x^2} + \frac{\partial^2 V}{\partial y^2} = 0.$$

We separate the variables by putting $V(x, y) = X(x)Y(y)$ which gives

$$\frac{1}{X}\frac{d^2 X}{dx^2} = -\frac{1}{Y}\frac{d^2 Y}{dy^2}$$

and since each side must be a constant, say ω^2, we obtain

$$\frac{d^2 X}{dx^2} = \omega^2 X, \quad \frac{d^2 Y}{dy^2} = -\omega^2 Y.$$

Hence

$$X = A\cosh\omega x + B\sinh\omega x, \quad Y = A'\cos\omega y + B'\sin\omega y.$$

Now $X = 0$ when $x = 0$ and so $A = 0$. Also $Y = 0$ when $y = 0$ and $y = b$ giving $A' = 0$ and $\omega b = n\pi$.

Also $V(a, y) = f(y)$ and so we may write

$$V(x, y) = \sum_{n=1}^{\infty} b_n \frac{\sinh(n\pi x/b)\sin(n\pi y/b)}{\sinh(n\pi a/b)}$$

where

$$f(y) = \sum_{n=1}^{\infty} b_n \sin\frac{n\pi y}{b}.$$

Hence

$$b_n = \frac{2}{b}\int_0^b f(y)\sin\frac{n\pi y}{b}\,dy.$$

12.07 A semi-circular plate is bounded by the x axis and the upper half circle $x^2 + y^2 = a^2$. If the straight boundary is maintained at zero temperature and the curved boundary at a constant temperature C, find the steady state temperature distribution $V(r, \theta)$ over the plate.

• *Solution*

For a steady state, the temperature distribution satisfies Laplace's equation which, in circular polar coordinates r, θ, takes the form

$$\frac{1}{r}\frac{\partial}{\partial r}\left(r\frac{\partial V}{\partial r}\right) + \frac{1}{r^2}\frac{\partial^2 V}{\partial \theta^2} = 0.$$

Separating the r, θ coordinates then yields the solution

$$V(r, \theta) = (Ar^s + Br^{-s})\sin(s\theta + \epsilon).$$

Since $V(r, \theta)$ vanishes for $\theta = 0$ and $\theta = \pi$ we have $\epsilon = 0$ and $\sin s\pi = 0$ so that s is an integer. Further $V(r, \theta)$ must be finite at $r = 0$ and so $B = 0$. Hence we have

$$V(r, \theta) = \sum_{s=1}^{\infty} A_s r^s \sin s\theta.$$

Now $V(a, \theta) = C$ and so

$$C = \sum_{s=1}^{\infty} A_s a^s \sin s\theta$$

from which it follows that

$$A_s = \frac{2C}{\pi a^s} \int_0^{\pi} \sin s\theta \, d\theta = \frac{2C}{\pi a^s s} [1 - (-1)^s].$$

Hence

$$V(r, \theta) = \frac{4C}{\pi} \sum_{n=1}^{\infty} \frac{1}{2n+1} \left(\frac{r}{a}\right)^{2n+1} \sin(2n+1)\theta.$$

The next set of problems, concerned with semi-infinite and infinite solids, are solved by using Fourier transforms.

12.08 The plane boundary $x = 0$ of a semi-infinite homogeneous solid of diffusivity κ occupying the region $x > 0$ is kept at a constant zero temperature. Initially the temperature throughout the solid is $f(x)$ where $\int_0^{\infty} |f(x)| \, dx < \infty$. Find the temperature distribution at any subsequent time t.

• *Solution*

To solve this problem we shall make a Fourier sine transform. We put

$$\overline{V}(\xi, t) = \sqrt{\frac{2}{\pi}} \int_0^{\infty} \sin \xi x \, V(x, t) \, dx$$

and so

$$\frac{\partial \overline{V}}{\partial t} = \sqrt{\frac{2}{\pi}} \kappa \int_0^{\infty} \sin \xi x \, \frac{\partial^2 V}{\partial x^2} \, dx$$

since $\partial V / \partial t = \kappa \partial^2 V / \partial x^2$.

Now

$$\int_0^{\infty} \sin \xi x \, \frac{\partial^2 V}{\partial x^2} \, dx = \left[\sin \xi x \, \frac{\partial V}{\partial x} - \xi \cos \xi x \, V \right]_0^{\infty} - \xi^2 \int_0^{\infty} \sin \xi x \, V \, dx = -\xi^2 \int_0^{\infty} \sin \xi x \, V \, dx$$

assuming that V and $\partial V / \partial x \to 0$ as $x \to \infty$, and putting $V = 0$ over the boundary surface $x = 0$. Then we obtain

$$\frac{\partial \overline{V}}{\partial t} = -\kappa \xi^2 \overline{V}.$$

This has the solution $\overline{V}(\xi, t) = \overline{V}(\xi, 0) \exp(-\kappa \xi^2 t)$ where $\overline{V}(\xi, 0) = \sqrt{2/\pi} \int_0^{\infty} \sin \xi x \, f(x) \, dx$ and so, making the reciprocal Fourier sine transform, we get

$$V(x, t) = \sqrt{\frac{2}{\pi}} \int_0^{\infty} \sin \xi x \, \overline{V}(\xi, t) \, d\xi = \frac{2}{\pi} \int_0^{\infty} f(x') \, dx' \int_0^{\infty} \sin \xi x \, \sin \xi x' \exp(-\kappa \xi^2 t) \, d\xi.$$

Then

$$V(x, t) = \frac{1}{\pi} \int_0^{\infty} f(x') \, dx' \int_0^{\infty} [\cos \xi(x - x') - \cos \xi(x + x')] \exp(-\kappa \xi^2 t) \, d\xi.$$

Now

$$\int_0^{\infty} \cos b\xi \, \exp(-a\xi^2) \, d\xi = \frac{1}{2} \sqrt{\frac{\pi}{a}} \exp(-b^2/4a) \tag{12.3}$$

and so

$$V(x,t) = \frac{1}{2\sqrt{\pi\kappa t}} \int_0^\infty f(x') \left\{ \exp[-(x-x')^2/4\kappa t] - \exp[-(x+x')^2/4\kappa t] \right\} dx'.$$

12.09 If the boundary $x = 0$ of a homogeneous semi-infinite solid occupying the region $x > 0$ is impervious to heat and the initial temperature throughout the solid is $f(x)$, show that the temperature distribution at time t is

$$V(x,t) = \frac{1}{2\sqrt{\pi\kappa t}} \int_0^\infty f(x') \left\{ \exp\left[-\frac{(x-x')^2}{4\kappa t} \right] + \exp\left[-\frac{(x+x')^2}{4\kappa t} \right] \right\} dx'$$

where $\int_0^\infty |f(x)|\, dx < \infty$.

- *Solution*

Here we use the Fourier cosine transform and put

$$\overline{V}(\xi,t) = \sqrt{\frac{2}{\pi}} \int_0^\infty \cos\xi x\, V(x,t)\, dx.$$

Then we obtain

$$\frac{\partial \overline{V}}{\partial t} = \sqrt{\frac{2}{\pi}}\kappa \int_0^\infty \cos\xi x\, \frac{\partial^2 V}{\partial x^2}\, dx = \sqrt{\frac{2}{\pi}}\kappa \left[\cos\xi x\, \frac{\partial V}{\partial x} + \xi\sin\xi x\, V \right]_0^\infty - \kappa\xi^2 \sqrt{\frac{2}{\pi}} \int_0^\infty \cos\xi x\, V\ dx.$$

Assuming that V and $\partial V/\partial x \to 0$ as $x \to \infty$, and using $\partial V/\partial x = 0$ at $x = 0$, we get $\partial \overline{V}/\partial t = -\kappa\xi^2 \overline{V}$ which has the solution $\overline{V}(\xi,t) = \overline{V}(\xi,0)\exp(-\kappa\xi^2 t)$ where $\overline{V}(\xi,0) = \sqrt{2/\pi} \int_0^\infty \cos\xi x\, f(x)\, dx$.

Now making the reciprocal Fourier cosine transform, we get

$$V(x,t) = \sqrt{\frac{2}{\pi}} \int_0^\infty \cos\xi x\, \overline{V}(\xi,t)\, d\xi = \frac{2}{\pi} \int_0^\infty f(x')\, dx' \int_0^\infty \cos\xi x \cos\xi x'\, \exp(-\kappa\xi^2 t)\, d\xi$$

and then we see that

$$V(x,t) = \frac{1}{\pi} \int_0^\infty f(x')\, dx' \int_0^\infty [\cos\xi(x-x') + \cos\xi(x+x')]\exp(-\kappa\xi^2 t)\, d\xi.$$

Hence using (12.3) we get

$$V(x,t) = \frac{1}{2\sqrt{\pi\kappa t}} \int_0^\infty f(x') \left\{ \exp[-(x-x')^2/4\kappa t] + \exp[-(x+x')^2/4\kappa t] \right\} dx'.$$

12.10 The initial temperature of a homogeneous semi-infinite solid occupying the region $x > 0$ is zero. If the temperature gradient of the plane boundary $x = 0$ of the solid is maintained at the constant value $\partial V/\partial x = A$, find the temperature distribution within the solid at any time t in the form of an integral.

- *Solution*

Making the Fourier cosine transform

$$\overline{V}(\xi,t) = \sqrt{\frac{2}{\pi}} \int_0^\infty \cos\xi x\, V(x,t)\, dx$$

we obtain

$$\frac{\partial \overline{V}}{\partial t} = \sqrt{\frac{2}{\pi}}\kappa \int_0^\infty \cos\xi x \frac{\partial^2 V}{\partial x^2}\, dx = \sqrt{\frac{2}{\pi}}\kappa \left[\cos\xi x \frac{\partial V}{\partial x} + \xi \sin\xi x\, V\right]_0^\infty - \kappa\xi^2 \sqrt{\frac{2}{\pi}}\int_0^\infty \cos\xi x\, V\, dx.$$

Assuming that V and $\partial V/\partial x \to 0$ as $x \to \infty$, and using $\partial V/\partial x = A$ at $x = 0$, we get

$$\frac{\partial \overline{V}}{\partial t} + \kappa\xi^2 \overline{V} = -\kappa A \sqrt{\frac{2}{\pi}}$$

and since $\overline{V}(\xi,0) = 0$ because the initial temperature of the solid is zero, we have

$$\overline{V}(\xi,t) = \sqrt{\frac{2}{\pi}}\frac{A}{\xi^2}\left[\exp(-\kappa\xi^2 t) - 1\right].$$

Hence

$$V(x,t) = \sqrt{\frac{2}{\pi}}\int_0^\infty \cos\xi x\, \overline{V}(\xi,t)\, d\xi = \frac{2A}{\pi}\int_0^\infty \frac{\cos\xi x}{\xi^2}\left[\exp(-\kappa\xi^2 t) - 1\right]\, d\xi.$$

12.11 If the plane $x = 0$ is kept at zero temperature and initially the temperature in the semi-infinite solid occupying the region $x > 0$ is a constant A, show that

$$V(x,t) = \frac{2A}{\sqrt{\pi}}\int_0^{x/2\sqrt{\kappa t}} \exp(-\lambda^2)\, d\lambda.$$

• *Solution*

Because the initial temperature distribution is a constant A it does not vanish in the limit $x \to \infty$. Hence we replace it by $A\exp(-\epsilon x)$ where $\epsilon > 0$. Then we may use the result derived in problem 12.08 which gives

$$V(x,t) = \frac{A}{2\sqrt{\pi\kappa t}}\int_0^\infty \left\{\exp\left[-(x-x')^2/4\kappa t\right] - \exp\left[-(x+x')^2/4\kappa t\right]\right\}\exp(-\epsilon x')\, dx'.$$

We can now let $\epsilon \to 0$, and then making the substitutions $\lambda = (x' \pm x)/2\sqrt{\kappa t}$ we get

$$V(x,t) = \frac{A}{\sqrt{\pi}}\int_{-x/2\sqrt{\kappa t}}^{x/2\sqrt{\kappa t}} \exp(-\lambda^2)\, d\lambda$$

and the required result follows.

12.12 If the temperature in an infinite solid is initially $f(x)$ show that at time t the temperature distribution is given by

$$V(x,t) = \frac{1}{2\sqrt{\pi\kappa t}}\int_{-\infty}^\infty f(x')\exp\left[-\frac{(x-x')^2}{4\kappa t}\right]\, dx'.$$

Hence show that if

$$f(x) = \begin{cases} A & (x > 0) \\ B & (x < 0) \end{cases}$$

then

$$V(x,t) = \frac{A+B}{2} + \frac{A-B}{\sqrt{\pi}}\int_0^{x/2\sqrt{\kappa t}} \exp(-\lambda^2)\, d\lambda.$$

• *Solution*

Here we make an exponential Fourier transform by putting

$$\overline{V}(\xi,t) = \frac{1}{\sqrt{2\pi}} \int_{-\infty}^{\infty} \exp(i\xi x)\, V(x,t)\, dx.$$

Then we have

$$\frac{\partial \overline{V}}{\partial t} = \frac{\kappa}{\sqrt{2\pi}} \int_{-\infty}^{\infty} \exp(i\xi x) \frac{\partial^2 V}{\partial x^2}\, dx = \frac{\kappa}{\sqrt{2\pi}} \left[\exp(i\xi x)\left(\frac{\partial V}{\partial x} - i\xi V\right)\right]_{-\infty}^{\infty} - \frac{\kappa \xi^2}{\sqrt{2\pi}} \int_{-\infty}^{\infty} \exp(i\xi x)\, V\, dx.$$

Assuming that V and $\partial V/\partial x \to 0$ as $x \to \pm\infty$ we get $\partial \overline{V}/\partial t = -\kappa\xi^2\overline{V}$ and so $\overline{V}(\xi,t) = \overline{V}(\xi,0)\exp(-\kappa\xi^2 t)$. But

$$\overline{V}(\xi,0) = \frac{1}{\sqrt{2\pi}} \int_{-\infty}^{\infty} \exp(i\xi x')\, f(x')\, dx'$$

and thus using the reciprocal exponential Fourier transform we get

$$V(x,t) = \frac{1}{\sqrt{2\pi}} \int_{-\infty}^{\infty} \exp(-i\xi x)\, \overline{V}(\xi,t)\, d\xi = \frac{1}{2\pi} \int_{-\infty}^{\infty} f(x')\, dx' \int_{-\infty}^{\infty} \exp(-\kappa\xi^2 t)\, \exp[-i\xi(x-x')]\, d\xi$$

which gives

$$V(x,t) = \frac{1}{\pi} \int_{-\infty}^{\infty} f(x')\, dx' \int_{0}^{\infty} \exp(-\kappa\xi^2 t)\, \cos[\xi(x-x')]\, d\xi .$$

Now using the integral (12.3) given in problem 12.08 we obtain

$$V(x,t) = \frac{1}{2\sqrt{\pi\kappa t}} \int_{-\infty}^{\infty} f(x')\, \exp[-(x-x')^2/4\kappa t]\, dx'.$$

Hence we have

$$V(x,t) = \frac{B}{2\sqrt{\pi\kappa t}} \int_{-\infty}^{0} \exp[-(x-x')^2/4\kappa t]\, dx' + \frac{A}{2\sqrt{\pi\kappa t}} \int_{0}^{\infty} \exp[-(x-x')^2/4\kappa t]\, dx'$$

which may be rewritten as

$$V(x,t) = \frac{B}{\sqrt{\pi}} \int_{-\infty}^{-x/2\sqrt{\kappa t}} \exp(-\lambda^2)\, d\lambda + \frac{A}{\sqrt{\pi}} \int_{-x/2\sqrt{\kappa t}}^{\infty} \exp(-\lambda^2)\, d\lambda.$$

Now $\int_0^\infty \exp(-\lambda^2)\, d\lambda = \sqrt{\pi}/2$ and so the required result follows.

The last two problems involve cooling through *radiation* as well as conduction of heat.

12.13 A thin rod of uniform cross section σ, having perimeter p and surface emissivity e is situated in air at zero temperature. If the ends of the rod at $x = -l$ and $x = l$ are impermeable to heat, and the mid-point of the rod is maintained at the constant temperature A, show that when the steady state is reached the temperature distribution is given by

$$V(x) = A \cosh\left[\sqrt{\frac{ep}{k\sigma}}(l-|x|)\right] \Big/ \cosh\left[\sqrt{\frac{ep}{k\sigma}}l\right]$$

where k is the thermal conductivity.

● *Solution*

The equation for heat conduction with radiation from the surface of a rod takes the form

$$\frac{\partial V}{\partial t} = \kappa \frac{\partial^2 V}{\partial x^2} - hV \tag{12.4}$$

where $\kappa = k/C\rho$ is the diffusivity and $h = ep/C\rho\sigma$, the density of the rod being ρ and its specific heat being C.

When the steady state is reached we have $\partial V/\partial t = 0$ and then

$$\frac{\partial^2 V}{\partial x^2} = \frac{h}{\kappa}V = \frac{ep}{k\sigma}V$$

which has the solution

$$V(x) = a\cosh\left(-\sqrt{\frac{ep}{k\sigma}}|x| + \epsilon\right)$$

using the symmetry about the centre of the rod at $x = 0$.

Since the ends of the rod at $x = l$ and $x = -l$ are impermeable to heat, the temperature at the ends must satisfy $dV/dx = 0$. Hence $\sinh(-\sqrt{ep/k\sigma}\,l + \epsilon) = 0$ and so $\epsilon = \sqrt{ep/k\sigma}\,l$. Also the middle point of the rod at $x = 0$ is maintained at the temperature A so that $a = A/\cosh(\sqrt{ep/k\sigma}\,l)$. The required result now follows.

12.14 A thin rod of uniform cross section and length l is situated in air at zero temperature. If the initial temperature of the rod is given by

$$V(x,0) = \alpha x(l - x)$$

and both ends of the rod are maintained at zero temperature, find the temperature distribution at time t if cooling occurs through radiation from the sides.

● *Solution*

The equation for heat conduction for the rod with cooling through radiation from the sides is given in the previous problem by (12.4).

To solve this equation we put $V(x,t) = \exp(-ht)U(x,t)$ where h is defined in the previous problem. Then we see that U satisfies

$$\frac{\partial U}{\partial t} = \kappa \frac{\partial^2 U}{\partial x^2}$$

which has the same form as the heat conduction equation without radiation.

Since both ends of the rod are maintained at zero temperature we can put

$$U(x,t) = \sum_{s=1}^{\infty} b_s \sin\frac{s\pi x}{l} \exp\left(-\frac{\kappa s^2 \pi^2}{l^2}t\right).$$

From the temperature distribution at time $t = 0$ we have

$$\alpha x(l - x) = \sum_{s=1}^{\infty} b_s \sin\frac{s\pi x}{l}$$

and so

$$b_s = \frac{2\alpha}{l}\int_0^l x(l - x)\sin\frac{s\pi x}{l}\,dx = \frac{2\alpha}{l}\left\{\frac{2l^3}{s^3\pi^3}[1 - (-1)^s]\right\}.$$

Thus we get

$$V(x,t) = \frac{8\alpha l^2}{\pi^3}\exp(-ht)\sum_{n=0}^{\infty}\frac{1}{(2n+1)^3}\sin\frac{(2n+1)\pi x}{l}\exp\left[-\frac{\kappa(2n+1)^2\pi^2}{l^2}t\right].$$

which is just $\exp(-ht)$ times the result obtained in problem 12.04.

13
TENSOR ANALYSIS

The theory of *tensor analysis* was developed by Gregorio Ricci-Curbastro (1853-1925) in his *absolute differential calculus* of 1844, Tullio Levi-Civita (1873-1941), and Luigi Bianchi (1856-1928) in his *differential geometry*. The word tensor came from its use in elasticity but we shall not discuss this here since our interest in tensor analysis is mainly concerned with its use by Einstein in the *theory of relativity* which is the subject of the next chapter.

13-1 Cartesian tensors

We begin by considering *Cartesian tensors* for an n-dimensional *Euclidean space E_n* whose points are given by the ordered set of n coordinates $(x_1, x_2, ..., x_n)$.

We introduce an *orthogonal transformation* given by

$$x'_i = a_{ij}x_j, \quad x_j = x'_i a_{ij} \tag{13.1}$$

where the a_{ij} are constants and satisfy $a_{ki}a_{kj} = \delta_{ij}$ and $a_{ik}a_{jk} = \delta_{ij}$, using the convention that summations are implied by repeated dummy suffixes.

Then any set of n^r quantities

$$A'_{i_1 i_2 ... i_r} = a_{i_1 j_1} a_{i_2 j_2} ... a_{i_r j_r} A_{j_1 j_2 ... j_r} \tag{13.2}$$

are the components of a *tensor* of *rank r*.

Vectors are tensors of the first rank.

A *tensor density* or *pseudo-tensor* $A_{i_1 ... i_r}$ transforms according to

$$A'_{i_1 i_2 ... i_r} = |A| \, a_{i_1 j_1} a_{i_2 j_2} ... a_{i_r j_r} A_{j_1 j_2 ... j_r} \tag{13.3}$$

where $|A| = \det A = \pm 1$ is the determinant of the transformation matrix $A = (a_{ij})$

The *Levi-Civita tensor density* has special importance. It is skew -symmetric with respect to every pair of indices and is denoted by $e_{i_1 i_2 ... i_r}$. All of its components vanish except those for which all the indices are different. If any pair of indices are interchanged $e_{i_1 i_2 ... i_r}$ changes sign. We choose $e_{12...r} = 1$.

In three-dimensional Euclidean space E_3, the Levi-Civita tensor density is e_{ijk}. It is 0 if any two of ijk are equal, it is 1 if ijk are all different and in the order 12312 (even permutation), and it is -1 if ijk are all different and in the order 21321 (odd permutation).

We start with an elementary problem to illustrate these definitions.

13.01 If A_i is a first rank and B_{ij} is a second rank Cartesian tensor, establish the transformation properties of

$$(i) \ C_{ijk} = \frac{\partial B_{ij}}{\partial x_k}, \quad (ii) \ D_i = e_{ijk}\frac{\partial A_k}{\partial x_j}.$$

• *Solution*

(i) Since $x_j = x'_i a_{ij}$ by (13.1) it follows that

$$\frac{\partial x_j}{\partial x'_i} = a_{ij}.$$

Hence

$$C'_{ijk} = \frac{\partial B'_{ij}}{\partial x'_k} = \frac{\partial (a_{i\alpha} a_{j\beta} B_{\alpha\beta})}{\partial x_\gamma} \frac{\partial x_\gamma}{\partial x'_k} = a_{k\gamma} a_{i\alpha} a_{j\beta} \frac{\partial B_{\alpha\beta}}{\partial x_\gamma} = a_{i\alpha} a_{j\beta} a_{k\gamma} C_{\alpha\beta\gamma}$$

and so, by (13.2), we see that $C_{\alpha\beta\gamma}$ transforms like a tensor of rank 3.

(ii) We have

$$D'_i = e'_{ijk} \frac{\partial A'_k}{\partial x'_j} = e'_{ijk} \frac{\partial (a_{kk_1} A_{k_1})}{\partial x_{j_1}} \frac{\partial x_{j_1}}{\partial x'_j} = e'_{ijk} a_{jj_1} a_{kk_1} \frac{\partial A_{k_1}}{\partial x_{j_1}} = |A| \, a_{i\alpha} a_{j\beta} a_{k\gamma} e_{\alpha\beta\gamma} a_{jj_1} a_{kk_1} \frac{\partial A_{k_1}}{\partial x_{j_1}}$$

since e_{ijk} is a tensor density which transforms according to (13.3), from which it follows that

$$D'_i = |A| \, a_{i\alpha} e_{\alpha\beta\gamma} \delta_{\beta j_1} \delta_{\gamma k_1} \frac{\partial A_{k_1}}{\partial x_{j_1}} = |A| \, a_{i\alpha} e_{\alpha\beta\gamma} \frac{\partial A_\gamma}{\partial x_\beta} = |A| \, a_{i\alpha} D_\alpha$$

and so D_i transforms like a vector density.

13.02 If T_{ij} is a Cartesian tensor of rank 2 show that the eigenvalues λ_p defined by

$$\det(T_{ij} - \lambda_p \delta_{ij}) = 0$$

are invariants under orthogonal transformation.

• *Solution*

The transformed secular equation is

$$\det(T'_{ij} - \lambda'_p \delta'_{ij}) = 0.$$

Now

$$\det(T'_{ij} - \lambda'_p \delta'_{ij}) = \det[a_{i\alpha} a_{j\beta}(T_{\alpha\beta} - \lambda'_p \delta_{\alpha\beta})] = \det[A(T - \lambda'_p I)\tilde{A}] = \det A \det \tilde{A} \det(T - \lambda'_p I)$$

and since $\det A \det \tilde{A} = 1$ it follows that $\det(T'_{ij} - \lambda'_p \delta'_{ij}) = \det(T_{\alpha\beta} - \lambda'_p \delta_{\alpha\beta})$ and so

$$\det(T_{ij} - \lambda'_p \delta_{ij}) = 0.$$

Thus the eigenvalues λ_p are invariants.

13.03 In the 3-dimensional Euclidean space E_3 show that

(i) $e_{ikl} e_{imn} = \delta_{km} \delta_{nl} - \delta_{kn} \delta_{lm}$,   (ii) $e_{ikl} e_{ikm} = 2\delta_{lm}$,   (iii) $e_{ikl} e_{ikl} = 6$.   (13.4)

• *Solution*

(i) We see that $e_{ikl}e_{imn} = e_{1kl}e_{1mn} + e_{2kl}e_{2mn} + e_{3kl}e_{3mn}$ which is 0 if $k = l$ or if $m = n$, is 1 if $k = m, l = n$ and $k \neq l$, and is -1 if $k = n, l = m$ and $k \neq l$. Hence

$$e_{ikl}e_{imn} = \delta_{km}\delta_{nl} - \delta_{kn}\delta_{lm}$$

which is skew-symmetric with respect to k, l and with respect to m, n.

(ii) From part (i) we see that

$$e_{ikl}e_{ikm} = \delta_{kk}\delta_{lm} - \delta_{km}\delta_{lk} = 3\delta_{lm} - \delta_{lm} = 2\delta_{lm}$$

noting that $\delta_{kk} = \delta_{11} + \delta_{22} + \delta_{33} = 3$.

(iii) From part (ii) we get

$$e_{ikl}e_{ikl} = 2\delta_{ll} = 6.$$

13.04 In E_3 show that

(i) $\nabla^2 V = \text{div grad } V = \dfrac{\partial^2 V}{\partial x_i \partial x_i}$,   (ii) curl curl $\mathbf{A}$ = grad div $\mathbf{A} - \nabla^2 \mathbf{A}$.

- *Solution*

(i) We have

$$\text{div } A = \frac{\partial A_i}{\partial x_i}, \quad (\text{grad } V)_i = \frac{\partial V}{\partial x_i}$$

and so

$$\text{div grad } V = \frac{\partial}{\partial x_i}\left(\frac{\partial V}{\partial x_i}\right) = \frac{\partial^2 V}{\partial x_i \partial x_i} = \nabla^2 V.$$

(ii) Let $\mathbf{B} = \text{curl curl } \mathbf{A} = \text{curl } \mathbf{C}$ where $\mathbf{C} = \text{curl } \mathbf{A}$.
Now, from the definition of curl $\mathbf{A}$ we have

$$C_1 = \frac{\partial A_3}{\partial x_2} - \frac{\partial A_2}{\partial x_3}, \quad C_2 = \frac{\partial A_1}{\partial x_3} - \frac{\partial A_3}{\partial x_1}, \quad C_3 = \frac{\partial A_2}{\partial x_1} - \frac{\partial A_1}{\partial x_2}$$

and these can be written as

$$C_i = e_{ijk} A_{k,j}$$

where

$$A_{i,j} = \frac{\partial A_i}{\partial x_j}.$$

It follows that

$$B_i = e_{ijk} C_{k,j} = e_{ijk} e_{klm} A_{m,lj}$$

where $A_{m,lj} = \partial^2 A_m / \partial x_l \partial x_j$. Thus, from part (i) of (13.4) given in problem 13.03 we have

$$B_i = e_{kij} e_{klm} A_{m,lj} = (\delta_{il}\delta_{jm} - \delta_{im}\delta_{jl}) A_{m,lj} = A_{j,ij} - A_{i,jj}.$$

Now $A_{i,jj} = \partial^2 A_i / \partial x_j \partial x_j = \nabla^2 A_i$ and $A_{j,ij} = \partial A_{j,j} / \partial x_i = \partial \text{ div } \mathbf{A} / \partial x_i$. Hence $B_i = \partial \text{ div } \mathbf{A} / \partial x_i - \nabla^2 A_i$ and so

$$\mathbf{B} = \text{grad div } \mathbf{A} - \nabla^2 \mathbf{A}$$

which yields the required result.

13.05 In E_3 show that the vector triple product is given by

$$\mathbf{A} \times (\mathbf{B} \times \mathbf{C}) = \mathbf{A.C}\ \mathbf{B} - \mathbf{A.B}\ \mathbf{C}$$

- *Solution*

Let $\mathbf{B} \times \mathbf{C} = \mathbf{D}$ and then $\mathbf{A} \times (\mathbf{B} \times \mathbf{C}) = \mathbf{A} \times \mathbf{D}$.

Now $(\mathbf{A} \times \mathbf{D})_1 = A_2 D_3 - A_3 D_2 = e_{1jk} A_j D_k$ and so we can write in general $(\mathbf{A} \times \mathbf{D})_i = e_{ijk} A_j D_k$ where $D_k = e_{klm} B_l C_m$. Hence we obtain

$$(\mathbf{A} \times \mathbf{D})_i = e_{ijk} A_j e_{klm} B_l C_m = (\delta_{il}\delta_{jm} - \delta_{im}\delta_{jl}) A_j B_l C_m$$

using part (i) of (13.4) given in problem 13.03. Thus we see that

$$(\mathbf{A} \times \mathbf{D})_i = A_j B_i C_j - A_j B_j C_i = B_i \mathbf{A}.\mathbf{C} - C_i \mathbf{A}.\mathbf{B}$$

and the required result follows at once.

13-2 Contravariant and covariant tensors

We now consider an n-dimensional *Riemannian space* R_n in which any point P is specified by n coordinates $(x^1, x^2, ..., x^n)$ and the distance ds between two neighbouring points is given by

$$ds^2 = g_{ij} dx^i dx^j \qquad (13.5)$$

where the g_{ij} are arbitrary functions of the coordinates x^i. The expression $g_{ij} dx^i dx^j$ is the *metric* of the space.

If it is possible to obtain a transformation to a new coordinate system $y^i(x^1, x^2, ..., x^n)$ $(i = 1, 2, ..., n)$ such that $ds^2 = dy^i dy^i$ then the space is *Euclidean*.

Suppose the coordinates of the point P referred to another frame of reference are $x'^i = x'^i(x^1, x^2, ..., x^n)$. Then

$$dx'^i = \frac{\partial x'^i}{\partial x^j} dx^j.$$

Any set of quantities A^i which transform in the same way as the components dx^i of an infinitesimal displacement vector, that is according to

$$A'^i = \frac{\partial x'^i}{\partial x^j} A^j \qquad (13.6)$$

are the components of a *contravariant vector*.

A scalar quantity V is invariant under a change of coordinates so that

$$V'(x'^1, x'^2, ..., x'^n) = V(x^1, x^2, ..., x^n).$$

Then

$$\frac{\partial V'}{\partial x'^i} = \frac{\partial x^j}{\partial x'^i} \frac{\partial V}{\partial x^j}$$

and any set of quantities B_i which transform in the same way as $\partial V/\partial x^i$, that is according to

$$B'_i = \frac{\partial x^j}{\partial x'^i} B_j \qquad (13.7)$$

are the components of a *covariant vector*.

Any set of n^2 quantities A^{ij} transforming according to

$$A'^{ij} = \frac{\partial x'^i}{\partial x^k} \frac{\partial x'^j}{\partial x^l} A^{kl} \qquad (13.8)$$

is a *contravariant tensor* of rank 2.

Any set of n^2 quantities A_{ij} transforming according to

$$A'_{ij} = \frac{\partial x^k}{\partial x'^i} \frac{\partial x^l}{\partial x'^j} A_{kl} \tag{13.9}$$

is a *covariant tensor* of rank 2.

Any set of n^2 quantities A_j^i transforming according to

$$A'^i_j = \frac{\partial x'^i}{\partial x^k} \frac{\partial x^l}{\partial x'^j} A_l^k \tag{13.10}$$

is a *mixed tensor* of rank 2.

The *fundamental mixed tensor*

$$\delta_j^i = \begin{cases} 1 & j = i \\ 0 & j \neq i \end{cases} \tag{13.11}$$

has the same components referred to all frames, that is $\delta'^i_j = \delta^i_j$.

The *general transformation law* of tensors is

$$A'^{i_1 \ldots i_r}_{j_1 \ldots j_s} = \frac{\partial x'^{i_1}}{\partial x^{\alpha_1}} \cdots \frac{\partial x'^{i_r}}{\partial x^{\alpha_r}} \frac{\partial x^{\beta_1}}{\partial x'^{j_1}} \cdots \frac{\partial x^{\beta_s}}{\partial x'^{j_s}} A^{\alpha_1 \ldots \alpha_r}_{\beta_1 \ldots \beta_s} \tag{13.12}$$

where $A^{i_1 \ldots i_r}_{j_1 \ldots j_s}$ is a mixed tensor of rank $r + s$.

An important theorem is the *Quotient theorem* which states the following.

Suppose that $A^{i_1 \ldots i_r}_{j_1 \ldots j_s}$ is a given matrix and $B^{j_l \ldots j_s \alpha_1 \ldots}_{i_k \ldots i_r \beta_1 \ldots}$ is an arbitrary tensor. If the inner product

$$C^{i_1 \ldots i_{k-1} \alpha_1 \ldots}_{j_1 \ldots j_{l-1} \beta_1 \ldots} = A^{i_1 \ldots i_{k-1} i_k \ldots i_r}_{j_1 \ldots j_{l-1} j_l \ldots j_s} B^{j_l \ldots j_s \alpha_1 \ldots}_{i_k \ldots i_r \beta_1 \ldots}$$

is a tensor, then $A^{i_1 \ldots i_r}_{j_1 \ldots j_s}$ is a tensor also.

13.06 If A_i is a covariant vector, prove that

$$B_{ij} = A_{i,j} - A_{j,i}$$

where $A_{i,j} = \partial A_i / \partial x^j$ and $A_{j,i} = \partial A_j / \partial x^i$, transforms like a covariant tensor of rank 2.

- *Solution*

We have

$$A'_i = \frac{\partial x^j}{\partial x'^i} A_j$$

and so

$$B'_{ij} = \frac{\partial A'_i}{\partial x'^j} - \frac{\partial A'_j}{\partial x'^i} = \frac{\partial}{\partial x'^j} \left(\frac{\partial x^k}{\partial x'^i} A_k \right) - \frac{\partial}{\partial x'^i} \left(\frac{\partial x^l}{\partial x'^j} A_l \right).$$

This yields

$$B'_{ij} = \frac{\partial^2 x^k}{\partial x'^j \partial x'^i} A_k - \frac{\partial^2 x^l}{\partial x'^i \partial x'^j} A_l + \frac{\partial x^k}{\partial x'^i} \frac{\partial A_k}{\partial x'^j} - \frac{\partial x^l}{\partial x'^j} \frac{\partial A_l}{\partial x'^i}$$

giving

$$B'_{ij} = \frac{\partial x^k}{\partial x'^i} \frac{\partial x^l}{\partial x'^j} \left(\frac{\partial A_k}{\partial x^l} - \frac{\partial A_l}{\partial x^k} \right) = \frac{\partial x^k}{\partial x'^i} \frac{\partial x^l}{\partial x'^j} B_{kl}$$

which proves the result since (13.9) is satisfied.

13.07 The metric for a spherical surface of unit radius is

$$ds^2 = d\theta^2 + \sin^2\theta \, d\phi^2$$

where θ, ϕ are spherical polar coordinates. Show that the only non-zero Chistoffel symbols are

$$\left\{ \begin{matrix} 1 \\ 2\ 2 \end{matrix} \right\} = -\sin\theta\cos\theta, \quad \left\{ \begin{matrix} 2 \\ 1\ 2 \end{matrix} \right\} = \left\{ \begin{matrix} 2 \\ 2\ 1 \end{matrix} \right\} = \cot\theta. \tag{13.13}$$

Further show that the only non-vanishing components of the covariant curvature tensor B_{ijkl} are

$$B_{1212} = -B_{1221} = B_{2121} = -B_{2112} = \sin^2\theta$$

and that the components of the metrical Ricci tensor are

$$R_{12} = R_{21} = 0, \quad R_{11} = -1, \quad R_{22} = -\sin^2\theta.$$

Also show that the curvature scalar $R = -2$.

- *Solution*

Set $x^1 = \theta$ and $x^2 = \phi$. Then $g_{11} = 1$, $g_{22} = \sin^2\theta$, $g_{12} = g_{21} = 0$ and so $g^{11} = 1$, $g^{22} = 1/g_{22} = 1/\sin^2\theta$.

Now the *Christoffel symbol of the second kind* is defined by

$$\left\{ \begin{matrix} s \\ i\ j \end{matrix} \right\} = \frac{1}{2} g^{sk} \left(\frac{\partial g_{jk}}{\partial x^i} + \frac{\partial g_{ki}}{\partial x^j} - \frac{\partial g_{ij}}{\partial x^k} \right). \tag{13.14}$$

where g_{ij} is the *fundamental covariant tensor*, g^{ij} is the *fundamental contravariant tensor* satisfying

$$g^{ij} g_{jk} = \delta^i_k \tag{13.15}$$

and $\det g_{ij} = |g_{ij}| \neq 0$. The metric tensors g_{ij} and g^{ij} are symmetric.

Hence

$$\left\{ \begin{matrix} 1 \\ 2\ 2 \end{matrix} \right\} = \frac{1}{2} g^{11} \left(-\frac{\partial g_{22}}{\partial x^1} \right) = -\frac{1}{2} \frac{\partial}{\partial\theta} (\sin^2\theta) = -\sin\theta\cos\theta,$$

$$\left\{ \begin{matrix} 2 \\ 1\ 2 \end{matrix} \right\} = \left\{ \begin{matrix} 2 \\ 2\ 1 \end{matrix} \right\} = \frac{1}{2} g^{22} \frac{\partial g_{22}}{\partial x^1} = \frac{1}{2\sin^2\theta} \frac{\partial}{\partial\theta} (\sin^2\theta) = \frac{\cos\theta}{\sin\theta} = \cot\theta$$

while the other three-index symbols all vanish as can be readily verified.

Now the *curvature tensor* is

$$B^i_{jkl} = \left\{ \begin{matrix} i \\ m\ k \end{matrix} \right\} \left\{ \begin{matrix} m \\ j\ l \end{matrix} \right\} - \left\{ \begin{matrix} i \\ m\ l \end{matrix} \right\} \left\{ \begin{matrix} m \\ j\ k \end{matrix} \right\} + \frac{\partial}{\partial x^k} \left\{ \begin{matrix} i \\ j\ l \end{matrix} \right\} - \frac{\partial}{\partial x^l} \left\{ \begin{matrix} i \\ j\ k \end{matrix} \right\} \tag{13.16}$$

and the *covariant curvature tensor* is given by

$$B_{ijkl} = g_{in} B^n_{jkl}. \tag{13.17}$$

Hence

$$B_{1212} = g_{11} B^1_{212} = g_{11} \left[\left\{ \begin{matrix} 1 \\ m\ 1 \end{matrix} \right\} \left\{ \begin{matrix} m \\ 2\ 2 \end{matrix} \right\} - \left\{ \begin{matrix} 1 \\ m\ 2 \end{matrix} \right\} \left\{ \begin{matrix} m \\ 2\ 1 \end{matrix} \right\} + \frac{\partial}{\partial x^1} \left\{ \begin{matrix} 1 \\ 2\ 2 \end{matrix} \right\} - \frac{\partial}{\partial x^2} \left\{ \begin{matrix} 1 \\ 2\ 1 \end{matrix} \right\} \right]$$

which gives

$$B_{1212} = -\left\{\begin{matrix} 1 \\ 2\ 2 \end{matrix}\right\}\left\{\begin{matrix} 2 \\ 2\ 1 \end{matrix}\right\} + \frac{\partial}{\partial x^1}\left\{\begin{matrix} 1 \\ 2\ 2 \end{matrix}\right\} = \cos^2\theta - \frac{\partial}{\partial\theta}(\sin\theta\cos\theta) = \sin^2\theta.$$

Further $B_{1212} = -B_{1221} = B_{2121} = -B_{2112}$.

The *metrical Ricci tensor* is given by

$$R_{jk} = B^i_{jki} = \left\{\begin{matrix} i \\ m\ k \end{matrix}\right\}\left\{\begin{matrix} m \\ j\ i \end{matrix}\right\} - \left\{\begin{matrix} i \\ m\ i \end{matrix}\right\}\left\{\begin{matrix} m \\ j\ k \end{matrix}\right\} + \frac{\partial}{\partial x^k}\left\{\begin{matrix} i \\ j\ i \end{matrix}\right\} - \frac{\partial}{\partial x^i}\left\{\begin{matrix} i \\ j\ k \end{matrix}\right\} \qquad (13.18)$$

and so we get $R_{12} = R_{21} = 0$,

$$R_{11} = \left\{\begin{matrix} i \\ m\ 1 \end{matrix}\right\}\left\{\begin{matrix} m \\ 1\ i \end{matrix}\right\} - \left\{\begin{matrix} i \\ m\ i \end{matrix}\right\}\left\{\begin{matrix} m \\ 1\ 1 \end{matrix}\right\} + \frac{\partial}{\partial x^1}\left\{\begin{matrix} i \\ 1\ i \end{matrix}\right\} - \frac{\partial}{\partial x^i}\left\{\begin{matrix} i \\ 1\ 1 \end{matrix}\right\},$$

$$R_{22} = \left\{\begin{matrix} i \\ m\ 2 \end{matrix}\right\}\left\{\begin{matrix} m \\ 2\ i \end{matrix}\right\} - \left\{\begin{matrix} i \\ m\ i \end{matrix}\right\}\left\{\begin{matrix} m \\ 2\ 2 \end{matrix}\right\} + \frac{\partial}{\partial x^2}\left\{\begin{matrix} i \\ 2\ i \end{matrix}\right\} - \frac{\partial}{\partial x^i}\left\{\begin{matrix} i \\ 2\ 2 \end{matrix}\right\},$$

so that

$$R_{11} = \cot^2\theta + \frac{\partial}{\partial\theta}\cot\theta = -1, \qquad R_{22} = -\cos^2\theta + \frac{\partial}{\partial\theta}(\sin\theta\cos\theta) = -\sin^2\theta.$$

The *curvature scalar* is defined according to

$$R = R^j_j = g^{ji}R_{ij} \qquad (13.19)$$

and so we have in the present problem

$$R = R^1_1 + R^2_2 = g^{1i}R_{i1} + g^{2i}R_{i2} = g^{11}R_{11} + g^{22}R_{22} = R_{11} + \frac{R_{22}}{g_{22}} = -2.$$

13.08 The points of a spherical surface of unit radius are given by the spherical polar coordinates θ, ϕ. A vector **A** is equal to a unit vector with components $A^\theta = 1$ and $A^\phi = 0$ at $\theta = \theta_0$ and $\phi = 0$. Find **A** after it has been transported once around the circle $\theta = \theta_0$. What is the magnitude of **A**?

• *Solution*

A *parallel displacement of a contravariant vector* A^i to a neighbouring point without changing its magnitude or direction produces a vector $A^i + \delta A^i$ where

$$\delta A^i = -\Gamma^i_{jk}A^j dx^k, \qquad (13.20)$$

the set of quantities Γ^i_{jk} being an *affinity* which specify an *affine connection* .

We put $x^1 = \theta$ and $x^2 = \phi$. Then $A^1 = 1$ and $A^2 = 0$ at $\theta = \theta_0, \phi = 0$.

A parallel displacement along the circle $\theta = \theta_0$ has $d\theta = 0$ and so $dx^1 = 0, dx^2 = d\phi$. For a metric affinity we have $\Gamma^i_{jk} = \{_{j\ k}^{\ i}\}$ and then $\delta A^i = -\{_{j\ k}^{\ i}\}A^j dx^k$ from which it follows, using (13.13), that

$$\delta A^1 = -\left\{\begin{matrix} 1 \\ j\ 2 \end{matrix}\right\}A^j d\phi = -\left\{\begin{matrix} 1 \\ 2\ 2 \end{matrix}\right\}A^2 d\phi = \sin\theta_0\cos\theta_0 A^2 d\phi,$$

$$\delta A^2 = -\left\{\begin{matrix} 2 \\ j\ 2 \end{matrix}\right\}A^j d\phi = -\left\{\begin{matrix} 2 \\ 1\ 2 \end{matrix}\right\}A^1 d\phi = -\cot\theta_0 A^1 d\phi.$$

Thus we get

$$\frac{\delta A^1}{d\phi} = \sin\theta_0 \cos\theta_0 A^2, \quad \frac{\delta A^2}{d\phi} = -\frac{\cos\theta_0}{\sin\theta_0} A^1$$

and so

$$\frac{\delta^2 A^1}{d\phi^2} = -\cos^2\theta_0 A^1.$$

Hence

$$A^1 = \cos(\phi\cos\theta_0)$$

since $A^1 = 1$ when $\phi = 0$, and

$$A^2 = -\frac{\sin(\phi\cos\theta_0)}{\sin\theta_0}$$

since $A^2 = 0$ when $\phi = 0$.

After a parallel displacement once round the circle we get

$$A^1 = \cos(2\pi\cos\theta_0), \quad A^2 = -\frac{\sin(2\pi\cos\theta_0)}{\sin\theta_0}.$$

The magnitude A of the vector is given by

$$A = (g_{ij}A^iA^j)^{1/2} = \left[(A^1)^2 + \sin^2\theta_0\,(A^2)^2\right]^{1/2} = \left[\cos^2(\phi\cos\theta_0) + \sin^2(\phi\cos\theta_0)\right]^{1/2} = 1$$

and so the magnitude does not change as the consequence of the parallel displacement although the components of the vector change.

13.09 A 3-dimensional Euclidean space E_3 has a coordinate system x_i $(i = 1, 2, 3)$ specified by the *natural basis* vectors

$$\mathbf{e}_i = \frac{\partial\mathbf{r}}{\partial x^i}$$

where $\mathbf{r}$ is the position vector of the point with coordinates (x^1, x^2, x^3).

(i) Show that the vectors $\mathbf{e}_i$ transform according to the transformation law for the components of a covariant vector.

(ii) Show that the metric tensor is given by

$$g_{ij} = \mathbf{e}_i.\mathbf{e}_j$$

Further prove that

$$\frac{\partial\mathbf{e}_i}{\partial x^j} = \frac{\partial\mathbf{e}_j}{\partial x^i} = \left\{\begin{matrix} l \\ i\,j \end{matrix}\right\}\mathbf{e}_l$$

where

$$\left\{\begin{matrix} l \\ i\,j \end{matrix}\right\} = g^{lk}[i\,j, k]$$

and

$$[i\,j, k] = \frac{1}{2}\left(\frac{\partial g_{jk}}{\partial x^i} + \frac{\partial g_{ki}}{\partial x^j} - \frac{\partial g_{ij}}{\partial x^k}\right)$$

are the Christoffel symbols of the second and first kinds respectively and $g_{ml}g^{lk} = \delta_m^k$.

(iii) In addition, show that the *covariant derivative* $g_{ik;j}$ of the metric tensor is zero.

- *Solution*

(i) We choose a rectangular Cartesian frame of reference and write

$$\mathbf{r} = \mathbf{b}_j y^j$$

where the $\mathbf{b}_j$ ($j = 1, 2, 3$) form a set of orthonormal basis vectors (usually denoted by $\mathbf{i}, \mathbf{j}, \mathbf{k}$) which are independent of position. Then we obtain

$$d\mathbf{r} = \mathbf{b}_j dy^j = \mathbf{b}_j \frac{\partial y^j}{\partial x^i} dx^i.$$

Now

$$d\mathbf{r} = \frac{\partial \mathbf{r}}{\partial x^i} dx^i = \mathbf{e}_i dx^i$$

and so, since the dx^i are arbitrary, we obtain

$$\mathbf{e}_i = \frac{\partial y^j}{\partial x^i} \mathbf{b}_j$$

which is the transformation law (13.7) for the components of a covariant vector where $(\mathbf{b}_1, \mathbf{b}_2, \mathbf{b}_3)$ are the components referred to the frame with rectangular Cartesian coordinates y^j and $(\mathbf{e}_1, \mathbf{e}_2, \mathbf{e}_3)$ are the components referred to the frame with curvilinear coordinates x^i.

(ii) We have

$$ds^2 = d\mathbf{r}.d\mathbf{r} = \frac{\partial \mathbf{r}}{\partial x^i} dx^i . \frac{\partial \mathbf{r}}{\partial x^j} dx^j = g_{ij} dx^i dx^j$$

where the metric tensor is

$$g_{ij} = \frac{\partial \mathbf{r}}{\partial x^i} . \frac{\partial \mathbf{r}}{\partial x^j} = \mathbf{e}_i.\mathbf{e}_j$$

and we see that

$$\frac{\partial \mathbf{e}_i}{\partial x^j} = \frac{\partial^2 \mathbf{r}}{\partial x^j \partial x^i} = \frac{\partial \mathbf{e}_j}{\partial x^i}.$$

Now

$$\frac{\partial g_{ij}}{\partial x^k} = \frac{\partial \mathbf{e}_i}{\partial x^k}.\mathbf{e}_j + \mathbf{e}_i.\frac{\partial \mathbf{e}_j}{\partial x^k}$$

and so the *Christoffel symbol of the first kind*

$$[ij, k] = \frac{1}{2} \left(\frac{\partial g_{jk}}{\partial x^i} + \frac{\partial g_{ki}}{\partial x^j} - \frac{\partial g_{ij}}{\partial x^k} \right) \tag{13.21}$$

becomes

$$[ij, k] = \frac{1}{2} \left(\frac{\partial \mathbf{e}_i}{\partial x^j} + \frac{\partial \mathbf{e}_j}{\partial x^i} \right).\mathbf{e}_k + \frac{1}{2} \left(\frac{\partial \mathbf{e}_k}{\partial x^j} - \frac{\partial \mathbf{e}_j}{\partial x^k} \right).\mathbf{e}_i + \frac{1}{2} \left(\frac{\partial \mathbf{e}_k}{\partial x^i} - \frac{\partial \mathbf{e}_i}{\partial x^k} \right).\mathbf{e}_j = \frac{\partial \mathbf{e}_i}{\partial x^j}.\mathbf{e}_k$$

Now the *Christoffel symbol of the second kind* is given by

$$\begin{Bmatrix} l \\ i \ j \end{Bmatrix} = g^{lk} [ij, k] \tag{13.22}$$

and so

$$\begin{Bmatrix} l \\ i \ j \end{Bmatrix} = g^{lk} \frac{\partial \mathbf{e}_i}{\partial x^j}.\mathbf{e}_k$$

But $g_{ml}g^{lk} = \delta_m^k$ and so

$$g_{ml}\begin{Bmatrix} l \\ i\ j \end{Bmatrix} = \frac{\partial \mathbf{e}_i}{\partial x^j}.\mathbf{e}_m,$$

that is

$$\mathbf{e}_m.\mathbf{e}_l\begin{Bmatrix} l \\ i\ j \end{Bmatrix} = \frac{\partial \mathbf{e}_i}{\partial x^j}.\mathbf{e}_m$$

for $m = 1, 2, 3$ and hence

$$\mathbf{e}_l\begin{Bmatrix} l \\ i\ j \end{Bmatrix} = \frac{\partial \mathbf{e}_i}{\partial x^j}.$$

(iii) We have

$$g_{ik,j} = \frac{\partial}{\partial x^j}(\mathbf{e}_i.\mathbf{e}_k) = \frac{\partial \mathbf{e}_i}{\partial x^j}.\mathbf{e}_k + \mathbf{e}_i.\frac{\partial \mathbf{e}_k}{\partial x^j} = \mathbf{e}_k.\mathbf{e}_l\begin{Bmatrix} l \\ i\ j \end{Bmatrix} + \mathbf{e}_i.\mathbf{e}_l\begin{Bmatrix} l \\ k\ j \end{Bmatrix}$$

which gives

$$g_{ik;j} = g_{ik,j} - \begin{Bmatrix} l \\ i\ j \end{Bmatrix}g_{lk} - \begin{Bmatrix} l \\ k\ j \end{Bmatrix}g_{il} = 0, \tag{13.23}$$

that is the covariant derivative $g_{ik;j}$ of the metric tensor g_{ik} is zero, which is known as *Ricci's theorem*.

13.10 Show that

$$A_{i;jk} - A_{i;kj} = B_{ijk}^l A_l + (\Gamma_{kj}^l - \Gamma_{jk}^l)A_{i;l} \tag{13.24}$$

Hence prove that B_{ijk}^l is a tensor and that covariant differentiations are commutative in a space for which $B_{ijk}^l = 0$ and the affinity Γ_{kj}^l is symmetric.

• *Solution*

The covariant derivative of the covariant vector A_i is given by

$$A_{i;j} = A_{i,j} - \Gamma_{ij}^l A_l \tag{13.25}$$

and the covariant derivative of $A_{i;j}$ is

$$A_{i;jk} = \frac{\partial}{\partial x_k}A_{i;j} - \Gamma_{ik}^l A_{l;j} - \Gamma_{jk}^l A_{i;l} \tag{13.26}$$

remembering that the sign before an affinity is $-$ for a covariant index in the definition of a covariant derivative.
Hence

$$A_{i;jk} = A_{i,jk} - \frac{\partial \Gamma_{ij}^l}{\partial x^k}A_l - \Gamma_{ij}^l A_{l,k} - \Gamma_{ik}^l A_{l;j} - \Gamma_{jk}^l A_{i;l}.$$

Similarly

$$A_{i;kj} = A_{i,kj} - \frac{\partial \Gamma_{ik}^l}{\partial x^j}A_l - \Gamma_{ik}^l A_{l,j} - \Gamma_{ij}^l A_{l;k} - \Gamma_{kj}^l A_{i;l}$$

and so

$$A_{i;jk} - A_{i;kj} = \left(\frac{\partial \Gamma_{ik}^l}{\partial x^j} - \frac{\partial \Gamma_{ij}^l}{\partial x^k}\right)A_l + \Gamma_{ij}^l(A_{l;k} - A_{l,k}) - \Gamma_{ik}^l(A_{l;j} - A_{l,j}) + (\Gamma_{kj}^l - \Gamma_{jk}^l)A_{i;l}.$$

It follows, using (13.25), that

$$A_{i;jk} - A_{i;kj} = \left(\frac{\partial\Gamma_{ik}^l}{\partial x^j} - \frac{\partial\Gamma_{ij}^l}{\partial x^k}\right)A_l + \Gamma_{ik}^l\Gamma_{lj}^m A_m - \Gamma_{ij}^l\Gamma_{lk}^m A_m + (\Gamma_{kj}^l - \Gamma_{jk}^l)A_{i;l}$$

which may be rewritten as

$$A_{i;jk} - A_{i;kj} = \left(\frac{\partial\Gamma_{ik}^l}{\partial x^j} - \frac{\partial\Gamma_{ij}^l}{\partial x^k} + \Gamma_{ik}^m\Gamma_{mj}^l - \Gamma_{ij}^m\Gamma_{mk}^l\right)A_l + (\Gamma_{kj}^l - \Gamma_{jk}^l)A_{i;l}$$

and this gives the required result (13.24) since

$$B_{ijk}^l = \Gamma_{ik}^m\Gamma_{mj}^l - \Gamma_{ij}^m\Gamma_{mk}^l + \frac{\partial\Gamma_{ik}^l}{\partial x^j} - \frac{\partial\Gamma_{ij}^l}{\partial x^k}$$

is the *curvature tensor or Riemann tensor*.

Although an affinity is not a tensor, the difference of two affinities is a tensor and so the second term on the right of (13.24) is a tensor of rank 3. The term on the left is a tensor of rank 3. Since A_l is an arbitrary vector, it follows by the *quotient theorem* that B_{ijk}^l is a tensor of rank 4.

If the affinity is symmetric so that $\Gamma_{kj}^l = \Gamma_{jk}^l$, and if $B_{ijk}^l = 0$, we get $A_{i;jk} = A_{i;kj}$ and thus covariant derivatives are commutative.

13.11 If g_{ij} is a metric tensor satisfying $(g_{ij}A^j)_{;k} = g_{ij}A^j_{\ ;k}$ show that $g_{ij;k} = 0$. Further show that $g^{ij}_{\ ;k} = 0$.

● *Solution*

We have

$$(g_{ij}A^j)_{;k} = g_{ij;k}A^j + g_{ij}A^j_{\ ,k} = g_{ij}A^j_{\ ;k}$$

and so

$$g_{ij;k}A^j = 0.$$

Since A^j is arbitrary it follows that $g_{ij;k} = 0$ which is Ricci's theorem demonstrated in part (iii) of problem 13.09.

Now the metric tensor g_{ij} satisfies $g_{lm}g^{mj} = \delta_l^j$ and so

$$g_{lm;k}g^{mj} + g_{lm}g^{mj}_{\ ;k} = 0.$$

Hence using $g^{il}g_{lm} = \delta_m^i$ we obtain

$$g^{ij}_{\ ;k} = -g^{il}g_{lm;k}g^{mj}$$

and since $g_{lm;k} = 0$ it follows that $g^{ij}_{\ ;k} = 0$.

13.12 By differentiating $g^{ij}g_{jk} = \delta_k^i$ where g_{ij} is a symmetric metric tensor, show that

$$\frac{\partial g^{im}}{\partial x^l} = -g^{mk}g^{ij}\frac{\partial g_{jk}}{\partial x^l}$$

and hence prove that

$$\frac{\partial g^{im}}{\partial x^l} + g^{ij}\begin{Bmatrix}m\\j\ l\end{Bmatrix} + g^{mj}\begin{Bmatrix}i\\j\ l\end{Bmatrix} = 0.$$

• *Solution*

We have

$$\frac{\partial}{\partial x^l}(g^{ij}g_{jk}) = 0$$

and so

$$\frac{\partial g^{ij}}{\partial x^l}g_{jk} + g^{ij}\frac{\partial g_{jk}}{\partial x^l} = 0.$$

The first result now follows since $g_{jk}g^{km} = \delta_j^m$.

Further we have, from the definition of the Christoffel symbol of the second kind given by (13.14),

$$g^{ij}\begin{Bmatrix} m \\ j\ l \end{Bmatrix} + g^{mj}\begin{Bmatrix} i \\ j\ l \end{Bmatrix} = \frac{1}{2}g^{ij}g^{mk}\left(\frac{\partial g_{lk}}{\partial x^j} + \frac{\partial g_{kj}}{\partial x^l} - \frac{\partial g_{jl}}{\partial x^k}\right) + \frac{1}{2}g^{mj}g^{ik}\left(\frac{\partial g_{lk}}{\partial x^j} + \frac{\partial g_{kj}}{\partial x^l} - \frac{\partial g_{jl}}{\partial x^k}\right)$$

and so, since the first and last terms in the brackets cancel out, we get

$$g^{ij}\begin{Bmatrix} m \\ j\ l \end{Bmatrix} + g^{mj}\begin{Bmatrix} i \\ j\ l \end{Bmatrix} = g^{ij}g^{mk}\frac{\partial g_{jk}}{\partial x^l}.$$

The second result now follows from the first result.

13.13 In a particular coordinate system the affinity is given by

$$\Gamma_{jk}^i = \delta_j^i \frac{\partial \phi}{\partial x^k} + \delta_k^i \frac{\partial \psi}{\partial x^j}$$

where ϕ and ψ are functions of position. Prove that the curvature tensor B_{jkl}^i is a function of ψ only.

If $\psi = -\ln a_m x^m$ show that $B_{jkl}^i = 0$.

• *Solution*

The curvature tensor is given in terms of the affinity by

$$B_{jkl}^i = \Gamma_{mk}^i \Gamma_{jl}^m - \Gamma_{ml}^i \Gamma_{jk}^m + \frac{\partial \Gamma_{jl}^i}{\partial x^k} - \frac{\partial \Gamma_{jk}^i}{\partial x^l}$$

and so

$$\begin{aligned}
B_{jkl}^i = {} & \left(\delta_m^i \partial\phi/\partial x^k + \delta_k^i \partial\psi/\partial x^m\right)\left(\delta_j^m \partial\phi/\partial x^l + \delta_l^m \partial\psi/\partial x^j\right) \\
& - \left(\delta_m^i \partial\phi/\partial x^l + \delta_l^i \partial\psi/\partial x^m\right)\left(\delta_j^m \partial\phi/\partial x^k + \delta_k^m \partial\psi/\partial x^j\right) \\
& + \partial/\partial x^k \left(\delta_j^i \partial\phi/\partial x^l + \delta_l^i \partial\psi/\partial x^j\right) - \partial/\partial x^l \left(\delta_j^i \partial\phi/\partial x^k + \delta_k^i \partial\psi/\partial x^j\right)
\end{aligned}$$

giving

$$B_{jkl}^i = \delta_k^i \frac{\partial\psi}{\partial x^l}\frac{\partial\psi}{\partial x^j} - \delta_l^i \frac{\partial\psi}{\partial x^k}\frac{\partial\psi}{\partial x^j} + \delta_l^i \frac{\partial^2\psi}{\partial x^k \partial x^j} - \delta_k^i \frac{\partial^2\psi}{\partial x^l \partial x^j}$$

which is a function of ψ only.

Now suppose that $\psi = -\ln a_m x^m$. Then

$$B_{jkl}^i = \frac{\delta_k^i a_l a_j}{(a_m x^m)^2} - \frac{\delta_l^i a_k a_j}{(a_m x^m)^2} + \frac{\delta_l^i a_j a_k}{(a_m x^m)^2} - \frac{\delta_k^i a_j a_l}{(a_m x^m)^2} = 0.$$

13.14 If a 2-dimensional Riemannian space has the metric

$$ds^2 = g_{11}(dx^1)^2 + g_{22}(dx^2)^2$$

show that

$$B_{1212} = -\frac{1}{2}\sqrt{g}\left[\frac{\partial}{\partial x^1}\left(\frac{1}{\sqrt{g}}\frac{\partial g_{22}}{\partial x^1}\right) + \frac{\partial}{\partial x^2}\left(\frac{1}{\sqrt{g}}\frac{\partial g_{11}}{\partial x^2}\right)\right]$$

where $g = |g_{ik}|$ is the determinant of (g_{ik}).

- *Solution*

We have, using the analysis given in problem 13.07, that

$$B_{1212} = g_{11}B^1_{211} = g_{11}\left[\begin{Bmatrix}1\\m\,1\end{Bmatrix}\begin{Bmatrix}m\\2\,2\end{Bmatrix} - \begin{Bmatrix}1\\m\,2\end{Bmatrix}\begin{Bmatrix}m\\2\,1\end{Bmatrix} + \frac{\partial}{\partial x^1}\begin{Bmatrix}1\\2\,2\end{Bmatrix} - \frac{\partial}{\partial x^2}\begin{Bmatrix}1\\2\,1\end{Bmatrix}\right].$$

Now

$$\begin{Bmatrix}1\\2\,2\end{Bmatrix} = \frac{1}{2}g^{11}\left(-\frac{\partial g_{22}}{\partial x^1}\right), \quad \begin{Bmatrix}1\\2\,1\end{Bmatrix} = \frac{1}{2}g^{11}\left(\frac{\partial g_{11}}{\partial x^2}\right)$$

and so

$$\frac{\partial}{\partial x^1}\begin{Bmatrix}1\\2\,2\end{Bmatrix} - \frac{\partial}{\partial x^2}\begin{Bmatrix}1\\2\,1\end{Bmatrix} = -\frac{1}{2}\left[\frac{\partial}{\partial x^1}\left(g^{11}\frac{\partial g_{22}}{\partial x^1}\right) + \frac{\partial}{\partial x^2}\left(g^{11}\frac{\partial g_{11}}{\partial x^2}\right)\right].$$

Also

$$\begin{Bmatrix}1\\m\,1\end{Bmatrix}\begin{Bmatrix}m\\2\,2\end{Bmatrix} = \frac{1}{2}g^{11}\frac{\partial g_{11}}{\partial x^1}\frac{1}{2}g^{11}\left(-\frac{\partial g_{22}}{\partial x^1}\right) + \frac{1}{2}g^{11}\frac{\partial g_{11}}{\partial x^2}\frac{1}{2}g^{22}\left(\frac{\partial g_{22}}{\partial x^2}\right),$$

$$\begin{Bmatrix}1\\m\,2\end{Bmatrix}\begin{Bmatrix}m\\2\,1\end{Bmatrix} = \frac{1}{2}g^{11}\frac{\partial g_{11}}{\partial x^2}\frac{1}{2}g^{11}\frac{\partial g_{11}}{\partial x^2} - \frac{1}{2}g^{11}\frac{\partial g_{22}}{\partial x^1}\frac{1}{2}g^{22}\frac{\partial g_{22}}{\partial x^1}.$$

Since $g_{12} = 0$ we have that $g = g_{11}g_{22}$. Also $g^{11} = g_{11}^{-1}$ and $g^{22} = g_{22}^{-1}$ so that $g = (g^{11}g^{22})^{-1}$. Then we get after some analysis

$$B_{1212} = -\frac{1}{2}\left[\frac{\partial^2 g_{22}}{(\partial x^1)^2} + \frac{\partial^2 g_{11}}{(\partial x^2)^2} - \frac{1}{2}\left(\frac{1}{g_{11}}\frac{\partial g_{11}}{\partial x^1} + \frac{1}{g_{22}}\frac{\partial g_{22}}{\partial x^1}\right)\frac{\partial g_{22}}{\partial x^1} - \frac{1}{2}\left(\frac{1}{g_{11}}\frac{\partial g_{11}}{\partial x^2} + \frac{1}{g_{22}}\frac{\partial g_{22}}{\partial x^2}\right)\frac{\partial g_{11}}{\partial x^2}\right]$$

so that

$$B_{1212} = -\frac{1}{2}\left[\frac{\partial^2 g_{22}}{(\partial x^1)^2} + \frac{\partial^2 g_{11}}{(\partial x^2)^2} - \frac{1}{2g}\frac{\partial g}{\partial x^1}\frac{\partial g_{22}}{\partial x^1} - \frac{1}{2g}\frac{\partial g}{\partial x^2}\frac{\partial g_{11}}{\partial x^2}\right]$$

which can be rewritten in the required form.

13.15 If Γ^i_{jk} is a symmetric affinity, show that

$$\widehat{\Gamma}^i_{jk} = \Gamma^i_{jk} + \delta^i_j A_k + \delta^i_k A_j$$

is a symmetric affinity.

If B^i_{jkl}, $\widehat{B}^i_{jkl}$ are the curvature tensors with affinities Γ^i_{jk}, $\widehat{\Gamma}^i_{jk}$ respectively, show that

$$\widehat{B}^i_{jkl} = B^i_{jkl} + \delta^i_k A_{jl} - \delta^i_l A_{jk} + \delta^i_j(A_{kl} - A_{lk})$$

where $A_{ij} = A_i A_j - A_{i;j}$.

If A_i is the gradient of a scalar, show that we have

$$\widehat{B}^i_{jil} - \widehat{B}^i_{lij} = B^i_{jil} - B^i_{lij}.$$

• *Solution*

We see that $\delta^i_j A_k + \delta^i_k A_j$ is a tensor of rank 3 with two covariant indices and one contravariant index. If such a tensor is added to an affinity we obtain an affinity. Hence $\widehat{\Gamma}^i_{jk}$ is an affinity. It is clearly symmetric since $\widehat{\Gamma}^i_{jk} = \widehat{\Gamma}^i_{kj}$.

Now

$$B^i_{jkl} = \Gamma^i_{mk}\Gamma^m_{jl} - \Gamma^i_{ml}\Gamma^m_{jk} + \frac{\partial \Gamma^i_{jl}}{\partial x^k} - \frac{\partial \Gamma^i_{jk}}{\partial x^l}$$

from which it follows after some analysis that

$$\widehat{B}^i_{jkl} = B^i_{jkl} + \delta^i_k\left(A_jA_l + \Gamma^m_{jl}A_m - \frac{\partial A_j}{\partial x^l}\right) - \delta^i_l\left(A_jA_k + \Gamma^m_{jk}A_m - \frac{\partial A_j}{\partial x^k}\right) + \delta^i_j\left(\frac{\partial A_l}{\partial x^k} - \frac{\partial A_k}{\partial x^l}\right).$$

But

$$A_{j;k} = \frac{\partial A_j}{\partial x^k} - \Gamma^m_{jk}A_m$$

and so

$$\widehat{B}^i_{jkl} = B^i_{jkl} + \delta^i_l(A_{j;k} - A_jA_k) - \delta^i_k(A_{j;l} - A_jA_l) + \delta^i_j\left(\frac{\partial A_l}{\partial x^k} - \frac{\partial A_k}{\partial x^l}\right),$$

that is

$$\widehat{B}^i_{jkl} = B^i_{jkl} + \delta^i_k A_{jl} - \delta^i_l A_{jk} + \delta^i_j(A_{l;k} + \Gamma^m_{lk}A_m - A_{k;l} - \Gamma^m_{kl}A_m)$$

which gives the first result.

Hence

$$\widehat{B}^i_{jil} - \widehat{B}^i_{lij} = B^i_{jil} - B^i_{lij} + (n+1)(A_{jl} - A_{lj})$$

since $\delta^i_i = n$, the dimension of the Riemannian space, and

$$A_{jl} - A_{lj} = A_{l;j} - A_{j;l} = \frac{\partial A_l}{\partial x^j} - \frac{\partial A_j}{\partial x^l}.$$

If $A_j = \partial V/\partial x^j$ where V is a scalar, we see that

$$A_{jl} - A_{lj} = \frac{\partial^2 V}{\partial x^j \partial x^l} - \frac{\partial^2 V}{\partial x^l \partial x^j} = 0$$

and so

$$\widehat{B}^i_{jil} - \widehat{B}^i_{lij} = B^i_{jil} - B^i_{lij}$$

as required.

13.16 If X^i ($i = 1, 2, 3, 4$) are rectangular Cartesian coordinates in a 4-dimensional Euclidean space , show that

$$X^1 = R\cos\theta, \quad X^2 = R\sin\theta\cos\phi, \quad X^3 = R\sin\theta\sin\phi\cos\psi, \quad X^4 = R\sin\theta\sin\phi\sin\psi$$

are the parametric equations of a hypersphere of radius R.

Choosing (θ, ϕ, ψ) as coordinates on the hypersphere, show that the metric for this 3-dimensional space is

$$ds^2 = R^2[d\theta^2 + \sin^2\theta(d\phi^2 + \sin^2\phi\, d\psi^2)].$$

Prove that in this 3-space

$$B_{1212} = R^2\sin^2\theta, \quad B_{2323} = R^2\sin^4\theta\sin^2\phi, \quad B_{3131} = R^2\sin^2\theta\sin^2\phi.$$

All the other components of B_{ijkl} are zero.

Hence demonstrate that

$$B_{ijkl} = K(g_{ik}g_{jl} - g_{il}g_{jk})$$

where $K = R^{-2}$.

- *Solution*

We see that

$$(X^1)^2+(X^2)^2+(X^3)^2+(X^4)^2 = R^2[\cos^2\theta+\sin^2\theta\cos^2\phi+\sin^2\theta\sin^2\phi(\cos^2\psi+\sin^2\psi)] = R^2$$

and so we have parametric equations of a hypersphere of radius R.

On the hypersphere

$$ds^2 = R^2\{[d(\cos\theta)]^2 + [d(\sin\theta\cos\phi)]^2 + [d(\sin\theta\sin\phi\cos\psi)]^2 + [d(\sin\theta\sin\phi\sin\psi)]^2\}$$

which yields
$$ds^2 = R^2(d\theta^2 + \sin^2\theta\, d\phi^2 + \sin^2\theta\sin^2\phi\, d\psi^2).$$

Thus, taking $x^1 = \theta, x^2 = \phi, x^3 = \psi$, we have $g_{11} = R^2, g_{22} = R^2\sin^2\theta$, $g_{33} = R^2\sin^2\theta\sin^2\phi$ and $g_{ij} = 0$ for $i \neq j$.

Then, using (13.14), we get

$$\left\{{1 \atop 2\,2}\right\} = -\sin\theta\cos\theta, \quad \left\{{2 \atop 1\,2}\right\} = \left\{{2 \atop 2\,1}\right\} = \cot\theta,$$

$$\left\{{1 \atop 3\,3}\right\} = -\sin\theta\cos\theta\sin^2\phi, \quad \left\{{3 \atop 1\,3}\right\} = \left\{{3 \atop 3\,1}\right\} = \cot\theta,$$

$$\left\{{2 \atop 3\,3}\right\} = -\sin\phi\cos\phi, \quad \left\{{3 \atop 3\,2}\right\} = \left\{{3 \atop 2\,3}\right\} = \cot\phi,$$

while the rest of the Christoffel symbols of the second kind are all zero.

Hence, using (13.17), we find that

$$B_{1212} = R^2\left[\cos^2\theta - \frac{\partial}{\partial\theta}(\sin\theta\cos\theta)\right] = R^2\sin^2\theta,$$

$$B_{2323} = R^2\sin^2\theta\left[-\cos^2\theta\sin^2\phi + \cos^2\phi - \frac{\partial}{\partial\phi}(\sin\phi\cos\phi)\right] = R^2\sin^4\theta\sin^2\phi,$$

$$B_{3131} = R^2\sin^2\theta\sin^2\phi\left[-\cot^2\theta - \frac{\partial}{\partial\theta}\cot\theta\right] = R^2\sin^2\theta\sin^2\phi.$$

Now $g_{ik}g_{jl} - g_{il}g_{jk} \neq 0$ if $i = k$ and $j = l$ but $i \neq j$, or $i = l$ and $j = k$ but $i \neq j$. For all other cases $g_{ik}g_{jl} - g_{il}g_{jk} = 0$. We see that

$$g_{11}g_{22} - g_{12}g_{21} = g_{11}g_{22} = R^4\sin^2\theta = R^2B_{1212},$$

$$g_{22}g_{33} - g_{23}g_{32} = g_{22}g_{33} = R^4\sin^4\theta\sin^2\phi = R^2B_{2323},$$

$$g_{33}g_{11} - g_{31}g_{13} = g_{33}g_{11} = R^4\sin^2\theta\sin^2\phi = R^2B_{3131}.$$

We can obtain the other non-vanishing components of the curvature tensor by using the symmetry relations: $B_{ijkl} = -B_{jikl}$ and $B_{ijkl} = -B_{ijlk}$.

Hence we get

$$B_{ijkl} = \frac{1}{R^2}(g_{ik}g_{jl} - g_{il}g_{jk}).$$

13.17 In a Riemannian 4-space whose metric is

$$ds^2 = -[e^{2\lambda}(dx^1)^2 + (dx^2)^2 + (dx^3)^2] + e^{2\mu}(dx^4)^2,$$

where $\lambda = \lambda(x^1)$ and $\mu = \mu(x^1)$, prove that the curvature tensor

$$B^i_{jkl} = 0$$

if and only if $\mu'' - \lambda'\mu' + \mu'^2 = 0$ where the primes denote differentiations with respect to x^1.

If $\mu = -\lambda$ show that for a flat space

$$\mu = \frac{1}{2}\ln(a + bx^1)$$

where a, b are constants.

- *Solution*

We have here $g_{11} = -e^{2\lambda}, g_{22} = g_{33} = -1, g_{44} = e^{2\mu}$ and $g_{ij} = 0$ if $i \neq j$ where $\lambda = \lambda(x^1)$ and $\mu = \mu(x^1)$. Then we find that

$$\left\{\begin{matrix}1\\1\ 1\end{matrix}\right\} = \lambda', \quad \left\{\begin{matrix}1\\4\ 4\end{matrix}\right\} = e^{2(\mu-\lambda)}\mu', \quad \left\{\begin{matrix}4\\4\ 1\end{matrix}\right\} = \left\{\begin{matrix}4\\1\ 4\end{matrix}\right\} = \mu'$$

and all the other 3-index symbols are zero.

The indices of B^i_{jkl} must be 1 or 4. We obtain

$$B^1_{414} = e^{2(\mu-\lambda)}(\mu'^2 - \lambda'\mu' + \mu''), \quad B^4_{114} = \mu'^2 - \lambda'\mu' + \mu''$$

and since B^i_{jkl} is skew-symmetric with respect to k, l we have $B^1_{441} = -B^1_{414}$ and $B^4_{141} = -B^4_{114}$. All the other components of the curvature tensor vanish.

We now see that all the components of the curvature tensor vanish if and only if

$$\mu'' - \lambda'\mu' + \mu'^2 = 0.$$

Finally consider the special case when $\lambda = -\mu$. Then for a flat R_4 space, for which all the components of the curvature tensor are zero, μ necessarily satisfies $\mu'' + 2\mu'^2 = 0$. Now $\mu'' = d(\frac{1}{2}\mu'^2)/d\mu$ and so we have

$$\frac{d}{d\mu}\left(\frac{1}{2}e^{4\mu}\mu'^2\right) = 0.$$

Integrating gives $e^{4\mu}\mu'^2 = b^2/4$ where b is a constant, which yields

$$e^{2\mu}\frac{d\mu}{dx^1} = \frac{b}{2}$$

and this gives

$$e^{2\mu} = a + bx^1$$

where a is another constant, and so the required result is obtained.

14

THEORY OF RELATIVITY

The *theory of relativity* was developed by Albert Einstein (1879-1955) in a series of papers published from 1905 to 1917, and followed the earlier work of Hendrik Antoon Lorentz (1853-1928) published in 1904, and George Francis Fitzgerald (1851-1901). It is divided into two parts known as the *Special theory of relativity* and the *General theory of relativity*.

14-1 Special theory of relativity

The special theory is based on the *Lorentz transformation* between two inertial frames of reference S and S' moving relative to each other, and uses a light pulse starting from an initial common origin to derive the transformation.

The Lorentz transformation between two rectangular Cartesian frames of reference S and S' with parallel axes having relative velocity u takes the form

$$x = \gamma(x' + ut'), \quad y = y', \quad z = z', \quad t = \gamma\left(t' + \frac{ux'}{c^2}\right); \tag{14.1}$$

$$x' = \gamma(x - ut), \quad y' = y, \quad z' = z, \quad t' = \gamma\left(t - \frac{ux}{c^2}\right) \tag{14.2}$$

where c is the velocity of light *in vacuo* and

$$\gamma = \frac{1}{\sqrt{1 - u^2/c^2}} \tag{14.3}$$

is the *Lorentz factor*.

Here x, y, z, t and x', y', z', t' are the spatial coordinates and the time referred to S and S' respectively, and the origin O' of S' is moving with speed u along the x-axis referred to the origin O of S. At time $t = t' = 0$ the origins O and O' coincide.

Our first problem involves light pulses.

14.01 Two light pulses, separated by a distance d, are moving along the x-axis of a frame of reference S. Show that referred to a frame S', moving with speed u in the direction of the x-axis relative to S, the distance between the pulses is

$$d\sqrt{\frac{c+u}{c-u}}$$

• *Solution*

Let x_1', x_2' be the x' coordinates referred to S' at time t' of the light pulses. Referred to S the light pulses are associated with times t_1, t_2 where, by the Lorentz transformation (14.1),

$$t_1 = \gamma\left(t' + \frac{u}{c^2}x_1'\right), \quad t_2 = \gamma\left(t' + \frac{u}{c^2}x_2'\right).$$

Since the light pulses are separated by a distance d at a given time t we may set

$$x_1 = t_1 c + a, \quad x_2 = t_2 c + d + a$$

for their x coordinates referred to S.

Now, by the Lorentz transformation (14.1),

$$x_1 = \gamma(x_1' + ut'), \quad x_2 = \gamma(x_2' + ut')$$

and so we have

$$\frac{\gamma}{c}(x_1' + ut') = \gamma(t' + \frac{u}{c^2}x_1') + \frac{a}{c}, \quad \frac{\gamma}{c}(x_2' + ut') = \gamma(t' + \frac{u}{c^2}x_2') + \frac{d+a}{c}.$$

Hence

$$\frac{\gamma}{c}d' = \frac{\gamma u}{c^2}d' + \frac{d}{c}$$

where $d' = x_2' - x_1'$ is the distance between the light pulses referred to S'. It follows that

$$d' = \frac{d}{\gamma(1 - u/c)} = d\sqrt{\frac{1 + u/c}{1 - u/c}} \ .$$

using (14.3), which yields the required result.

14.02 If v is the relativistic sum of collinear velocities v' and u, show that

$$V = V' + U$$

where

$$V = c\tanh^{-1}\frac{v}{c}, \quad V' = c\tanh^{-1}\frac{v'}{c}, \quad U = c\tanh^{-1}\frac{u}{c}$$

• *Solution*

The *Einstein formulae* for the *addition of velocities* are

$$v_x = \frac{v_x' + u}{1 + uv_x'/c^2}, \quad v_y = \frac{\sqrt{1 - u^2/c^2}\,v_y'}{1 + uv_x'/c^2}, \quad v_z = \frac{\sqrt{1 - u^2/c^2}\,v_z'}{1 + uv_x'/c^2}; \quad (14.4)$$

$$v_x' = \frac{v_x - u}{1 - uv_x/c^2}, \quad v_y' = \frac{\sqrt{1 - u^2/c^2}\,v_y}{1 - uv_x/c^2}, \quad v_z' = \frac{\sqrt{1 - u^2/c^2}\,v_z}{1 - uv_x/c^2}. \quad (14.5)$$

Hence the relativistic sum for the collinear velocities v' and u is given by

$$v = \frac{v' + u}{1 + v'u/c^2}.$$

Now we have

$$v = c\tanh\frac{V}{c}, \quad v' = c\tanh\frac{V'}{c}, \quad u = c\tanh\frac{U}{c}$$

and so

$$\tanh(V/c) = \frac{\tanh(V'/c) + \tanh(U/c)}{1 + \tanh(V'/c)\tanh(U/c)} = \tanh[(V' + U)/c].$$

Hence $V = V' + U$ as required.

14.03 Show that

$$\frac{\gamma(v)}{\gamma(v')} = \gamma(u)\left(1 + \frac{\mathbf{v'}.\mathbf{u}}{c^2}\right),$$

$$\frac{\gamma(v')}{\gamma(v)} = \gamma(u)\left(1 - \frac{\mathbf{v}.\mathbf{u}}{c^2}\right),$$

where

$$\gamma(u) = \left(1 - \frac{u^2}{c^2}\right)^{-\frac{1}{2}}.$$

• *Solution*

From the Einstein formulae (14.4) and (14.5) for the addition of velocities given in the previous problem it follows that

$$1 - v^2/c^2 = \frac{(1 - u^2/c^2)(1 - v'^2/c^2)}{(1 + v'_x u/c^2)^2} \tag{14.6}$$

and so $\gamma(v) = \gamma(u)\gamma(v')(1 + v'_x u/c^2)$. Hence

$$\frac{\gamma(v)}{\gamma(v')} = \gamma(u)(1 + v'_x u/c^2).$$

Since the relative velocity of the two frames of reference is parallel to the x-axis we have $\mathbf{v'}.\mathbf{u} = v'_x u$ and so the first result follows.

The second result is given by interchanging primed and unprimed quantities and replacing $\mathbf{u}$ by $-\mathbf{u}$.

14.04 Show that the acceleration $d\mathbf{v}/dt$ relative to S of a point which is instantaneously at rest in S', is given in terms of the acceleration $d\mathbf{v'}/dt'$ relative to S' by

$$\frac{dv_x}{dt} = \left(1 - \frac{u^2}{c^2}\right)^{\frac{3}{2}} \frac{dv'_x}{dt'},$$

$$\frac{dv_y}{dt} = \left(1 - \frac{u^2}{c^2}\right) \frac{dv'_y}{dt'},$$

$$\frac{dv_z}{dt} = \left(1 - \frac{u^2}{c^2}\right) \frac{dv'_z}{dt'}.$$

• *Solution*

From the Einstein formulae for the addition of velocities (14.4) and (14.5) stated in problem 14.02 we obtain

$$dv_x = \left(1 - \frac{u^2}{c^2}\right) dv'_x, \quad dv_y = \sqrt{1 - \frac{u^2}{c^2}}\, dv'_y, \quad dv_z = \sqrt{1 - \frac{u^2}{c^2}}\, dv'_z$$

setting $v'_x = v'_y = v'_z = 0$ since the point is instantaneously at rest in S'.

Now $dt = dt'/\sqrt{1 - u^2/c^2}$ using (14.1) and so we get the required acceleration formulae.

14.05 A particle moves from rest at the origin in the frame S along the x-axis so that it has constant acceleration α relative to its instantaneous rest frame. If v is the speed of the particle at time t relative to S, show that

$$\alpha = \left(1 - \frac{v^2}{c^2}\right)^{-\frac{3}{2}} \frac{dv}{dt} = \frac{d[v\gamma(v)]}{dt}$$

so that $\alpha t = v\gamma(v)$ and

$$\frac{v}{c} = \frac{\alpha t}{(\alpha^2 t^2 + c^2)^{\frac{1}{2}}}.$$

Hence show that

$$\alpha x^2 + 2c^2 x - \alpha c^2 t^2 = 0$$

which is a rectangular hyperbola in an (x, ct) diagram.

Prove that light signals emitted after time $t = c/\alpha$ at the origin will never reach the particle.

- *Solution*

From the previous problem we have $dv_x/dt = (1 - u^2/c^2)^{3/2} dv'_x/dt'$. In the present problem we have $v_x = v = u$, $v_y = v_z = 0$ and $dv'_x/dt' = \alpha$. Hence $dv/dt = (1 - v^2/c^2)^{3/2}\alpha$ so that

$$\alpha = \left(1 - \frac{v^2}{c^2}\right)^{-3/2} \frac{dv}{dt} = \frac{d}{dt}\left(\frac{v}{\sqrt{1 - v^2/c^2}}\right) = \frac{d}{dt}[v\gamma(v)].$$

Now $v = 0$ when $t = 0$ and so $\alpha t = v\gamma(v)$. Hence $\alpha^2 t^2 = v^2/(1 - v^2/c^2)$ and thus $v/c = \alpha t/(\alpha^2 t^2 + c^2)^{1/2}$ as required. Since $v = dx/dt$ we get

$$\frac{x}{c} = \frac{1}{\alpha}(\alpha^2 t^2 + c^2)^{1/2} - \frac{c}{\alpha}$$

and so

$$(\alpha x + c^2)^2 = c^2(\alpha^2 t^2 + c^2)$$

which yields $\alpha x^2 + 2c^2 x - \alpha c^2 t^2 = 0$ as required. This is the equation of a rectangular hyperbola with asymptotes $x \pm ct = 0$ in an (x, ct) diagram.

Suppose a light signal is emitted at time t_0. The x-coordinate of the light signal at time t will be $c(t - t_0)$ and at this time it will have reached the particle if

$$\alpha c^2 (t - t_0)^2 + 2c^3(t - t_0) - \alpha c^2 t^2 = 0,$$

that is

$$t = \frac{t_0(2c - \alpha t_0)}{2(c - \alpha t_0)}.$$

Since $t \to \infty$ as $t_0 \to c/\alpha$ from below, it follows that t_0 must not exceed c/α if the light pulse is to reach the particle.

This can also be seen as follows. At time t the particle has travelled a distance $x = -c^2/\alpha + c\sqrt{c^2/\alpha^2 + t^2}$ whereas the light pulse has travelled a distance $x_0 = (t - t_0)c < (t - c/\alpha)c < x$ if it is emitted at any time $t_0 > c/\alpha$.

14.06 Derive the *relativistic aberration formulae* for light signals

$$(i) \ \cos\theta' = \frac{\cos\theta - u/c}{1 - (u/c)\cos\theta}, \quad (ii) \ \sin\theta' = \frac{\sqrt{1 - u^2/c^2}\,\sin\theta}{1 - (u/c)\cos\theta}, \quad (iii) \ \frac{\tan\frac{1}{2}\theta'}{\tan\frac{1}{2}\theta} = \sqrt{\frac{c + u}{c - u}}.$$

$$(14.7)$$

• *Solution*

Consider a particle whose motion is confined to the common plane $O(x,y)$ and $O'(x',y')$ of the frames of reference S and S' respectively. If its direction of motion makes an angle θ with Ox and θ' with $O'x'$, then

$$v_x = v\cos\theta, \quad v_y = v\sin\theta; \quad v_x' = v'\cos\theta', \quad v_y' = v'\sin\theta'$$

where v and v' are the speeds of the particle relative to S and S' respectively. Hence, using the addition formulae for velocities (14.4) and (14.5) given in problem 14.02, we obtain

$$\tan\theta = \frac{\sqrt{1-u^2/c^2}\,v'\sin\theta'}{v'\cos\theta'+u}, \quad \tan\theta' = \frac{\sqrt{1-u^2/c^2}\,v\sin\theta}{v\cos\theta-u}.$$

In the case of a light signal we have $v = v' = c$ and then we get

$$\tan\theta = \frac{\sqrt{1-u^2/c^2}\,\sin\theta'}{\cos\theta'+u/c}, \quad \tan\theta' = \frac{\sqrt{1-u^2/c^2}\,\sin\theta}{\cos\theta-u/c}.$$

Now

$$\left(1-u^2/c^2\right)\sin^2\theta + (\cos\theta - u/c)^2 = [1-(u/c)\cos\theta]^2$$

and so we get the aberration formulae (i) and (ii) of (14.7) for $\cos\theta'$ and $\sin\theta'$.

From these we obtain

$$\tan\tfrac{1}{2}\theta' = \frac{\sin\frac{1}{2}\theta'}{\cos\frac{1}{2}\theta'} = \sqrt{\frac{1-\cos\theta'}{1+\cos\theta'}} = \sqrt{\frac{(1+u/c)(1-\cos\theta)}{(1-u/c)(1+\cos\theta)}} = \sqrt{\frac{1+u/c}{1-u/c}}\tan\tfrac{1}{2}\theta$$

and so aberration formula (iii) of (14.7) follows.

14.07 Show that the ratio of the solid angles subtended in S and S' by a narrow cone of light rays converging on the coincident origins of these frames at angles θ, θ' with the x-axis is given by

$$\frac{d\Omega}{d\Omega'} = \left(\frac{d\theta}{d\theta'}\right)^2 = \frac{[1+(u/c)\cos\theta]^2}{1-u^2/c^2}.$$

• *Solution*

We have

$$\frac{d\Omega}{d\Omega'} = \frac{\sin\theta\,d\theta}{\sin\theta'\,d\theta'} = \frac{d(\cos\theta)}{d(\cos\theta')}$$

and using the aberration formula (i) of (14.7) for $\cos\theta'$ obtained in the previous problem but with u replaced by $-u$ we obtain

$$d(\cos\theta') = \frac{(1-u^2/c^2)d(\cos\theta)}{[1+(u/c)\cos\theta]^2}$$

which gives

$$\frac{d\Omega}{d\Omega'} = \frac{[1+(u/c)\cos\theta]^2}{1-u^2/c^2} = \frac{\sin^2\theta}{\sin^2\theta'}$$

using the aberration formula (ii) of (14.7) for $\sin\theta'$ obtained in the previous problem with u replaced by $-u$. Hence

$$\frac{d\theta}{d\theta'} = \frac{\sin\theta}{\sin\theta'}$$

and the required results follow.

14.08 A uniform source of light is fixed in a frame S'. Show that the light in S is contained in a narrow forward cone for large u and that half the photons are emitted into a cone having semi-angle

$$\alpha = \cos^{-1}\frac{u}{c}.$$

• *Solution*

We have seen in problem 14.06 that

$$\tan\frac{1}{2}\theta = \sqrt{\frac{c-u}{c+u}}\,\tan\frac{1}{2}\theta'.$$

For a given value of $\theta' < \pi$ we have that $\tan\frac{1}{2}\theta'$ is a certain positive number. As $u \to c$ we find that

$$\theta = 2\tan^{-1}\left[\frac{\sqrt{c^2-u^2}}{c+u}\,\tan\frac{1}{2}\theta'\right] \to \sqrt{1-\frac{u^2}{c^2}}\,\tan\frac{1}{2}\theta'$$

which can be made as small as we wish by letting u approach sufficiently close to the speed of light c. Hence if the light is emitted uniformly in all directions in S', the light is concentrated in a narrow forward cone in S as $u \to c$.

In S' half the photons are uniformly distributed within the region $0 \le \theta' \le \pi/2$ and $0 \le \phi' \le 2\pi$.

From the aberration formula

$$\cos\theta = \frac{\cos\theta' + u/c}{1 + (u/c)\cos\theta'}$$

which corresponds to formula (i) of (14.7), we see that $\theta = 0$ when $\theta' = 0$ and $\cos\theta = u/c$ when $\theta' = \pi/2$. It follows that half the photons are emitted into a cone having semi-angle $\alpha = \cos^{-1}(u/c)$.

In the next problem we require a formula for the *relativistic Doppler effect*.

To obtain such a formula we consider a light source P having coordinates x, y, z, t referred to a frame S and suppose that P is moving with speed u parallel to the Ox axis such that $x = ut$, $y = $ constant, $z = 0$.

If T is the time when an observer A at the origin O receives a light signal emitted by P at time t, we have $OP = r = c(T - t)$ so that $r^2 = x^2 + y^2 + z^2 = c^2(T - t)^2$. Hence $x u\, dt = c^2(T - t)(dT - dt)$ giving

$$dT = \left(1 + \frac{xu}{rc}\right)dt.$$

If dt' is the time interval corresponding to dt for an observer B at rest with respect to P, then

$$dt' = \sqrt{1 - \frac{u^2}{c^2}}\,dt$$

from the Lorentz transformation formula (14.1) for t, and so

$$\frac{dT}{dt'} = \frac{1 + xu/rc}{\sqrt{1 - u^2/c^2}}.$$

Now let us choose dt' to be the period of the light ray emitted by P as measured by B so that dT is the period as measured by A. Suppose ν and ν' are the frequencies of the light waves as observed by A and B respectively. Then $\nu' = 1/dt'$ and $\nu = 1/dT$, and so we get the relativistic Doppler effect formula

$$\nu' = \nu \frac{1 + u_r/c}{\sqrt{1 - u^2/c^2}} \tag{14.8}$$

where $u_r = xu/r$ is the radial velocity of the light source at the time when the light is emitted.

14.09 A plane mirror moves in the direction of its normal with uniform speed u referred to a frame S. A light ray having frequency ν_1 is incident on the mirror at an angle α_1 to the normal and is reflected with frequency ν_2 at an angle α_2 to the normal. Show that

$$\frac{\sin \alpha_1}{\cos \alpha_1 + u/c} = \frac{\sin \alpha_2}{\cos \alpha_2 - u/c}$$

and

$$\frac{\nu_1}{\nu_2} = \frac{c - u \cos \alpha_2}{c + u \cos \alpha_1}.$$

- *Solution*

Referred to the frame S' moving with the mirror, the frequency ν' of the light ray is unchanged and the angles of incidence and reflection α' are equal.

Using the aberration formulae (i) and (ii) of (14.7) given in problem 14.06 for a light ray moving away from the origin O' we have

$$\tan \alpha' = \frac{\sqrt{1 - u^2/c^2} \sin \alpha_2}{\cos \alpha_2 - u/c}$$

while for a ray moving towards O' we have

$$\tan \alpha' = \frac{\sqrt{1 - u^2/c^2} \sin \alpha_1}{\cos \alpha_1 + u/c}$$

and hence the first result follows.

Now using the relativistic Doppler effect formula (14.8) derived above we have for the incident ray

$$\nu' = \nu_1 \frac{1 + (u/c) \cos \alpha_1}{\sqrt{1 - u^2/c^2}}$$

while for the reflected ray

$$\nu' = \nu_2 \frac{1 - (u/c) \cos \alpha_2}{\sqrt{1 - u^2/c^2}}$$

and the second result follows.

14.10 Show that the *four-acceleration* is given by

$$\left(\frac{dV_i}{d\tau}\right) = \left(\gamma^2 \mathbf{a} + \frac{\gamma^4}{c^2} \mathbf{v}\mathbf{v}.\mathbf{a}, \, i \frac{\gamma^4}{c} \mathbf{v}.\mathbf{a}\right)$$

where τ is the *proper time* and $\mathbf{a} = d\mathbf{v}/dt = d^2\mathbf{r}/dt^2$ is the three-dimensional acceleration.

- *Solution*

We consider the motion of a particle in the 4-dimensional Euclidean space -time continuum E_4 known as *Minkowski space-time* or *four-space*. The coordinates of the particle are given by the four-vector $(x_i) = (\mathbf{r}, ict)$ where $\mathbf{r}$ is the three-dimensional position vector of the particle and t is the time referred to an inertial frame of reference S.

The proper time τ is the time measured by a clock following the motion of a particle and by the Lorentz transformation (14.1) we have $d\tau = (1 - v^2/c^2)^{1/2}dt$ where v is the speed of the particle. Then $V_i = dx_i/d\tau$ are also the components of a four-vector which is called the *four-velocity* and is given by

$$(V_i) = (\gamma \mathbf{v}, i\gamma c) \qquad (14.9)$$

where $\mathbf{v} = d\mathbf{r}/dt$ is the three-dimensional velocity.

Then the four-acceleration is given by

$$\left(\frac{dV_i}{d\tau}\right) = \left(\gamma \frac{d(\gamma \mathbf{v})}{dt}, i\gamma c \frac{d\gamma}{dt}\right) = \left(\gamma^2 \mathbf{a} + \gamma \mathbf{v}\frac{d\gamma}{dt}, i\gamma c \frac{d\gamma}{dt}\right).$$

Now $d\gamma/dt = (\gamma^3/c^2)\mathbf{v}.d\mathbf{v}/dt = (\gamma^3/c^2)\mathbf{v}.\mathbf{a}$ and so the required result follows.

14.11 Prove that

$$A_i A_i = \gamma^4[\mathbf{a}^2 + \frac{\gamma^2}{c^2}(\mathbf{a}.\ \mathbf{v})^2]$$

where $A_i = dV_i/d\tau$ is the four-acceleration and $\gamma = 1/\sqrt{1 - v^2/c^2}$.

For motion in a straight line show that

$$A_i = \gamma^6 \left(\frac{dv}{dt}\right)^2.$$

- *Solution*

From the preceding problem we see that

$$A_i A_i = \left(\gamma^2 \mathbf{a} + \frac{\gamma^4}{c^2}\mathbf{v}\ \mathbf{v}.\mathbf{a}\right)^2 - \frac{\gamma^8}{c^2}(\mathbf{v}.\mathbf{a})^2 = \gamma^4\mathbf{a}^2 + \frac{\gamma^6}{c^2}(\mathbf{v}.\mathbf{a})^2(2 + \gamma^2 v^2/c^2 - \gamma^2)$$

from which the first result follows at once.

For motion in a straight line $\mathbf{v}.\mathbf{a} = v dv/dt$ so that we get

$$A_i A_i = \gamma^4 \left(\frac{dv}{dt}\right)^2 \left(1 + \frac{\gamma^2 v^2}{c^2}\right) = \gamma^6 \left(\frac{dv}{dt}\right)^2.$$

14.12 If a particle moves according to the equation of motion

$$\frac{dA_i}{d\tau} = \frac{\alpha^2}{c^2}V_i$$

where α^2 is a scalar, V_i is the four-velocity and $A_i = dV_i/d\tau$ is the four-acceleration, show that $V_i A_i = 0$ and $A_i A_i = \alpha^2$.

- *Solution*

We see that $V_i A_i = V_i dV_i/d\tau = \frac{1}{2}d(V_i V_i)/d\tau$. Now $V_i V_i = \gamma^2(v^2 - c^2) = -c^2$ from (14.9) and so $V_i A_i = 0$.

Also we have $A_i A_i + V_i dA_i/d\tau = d(V_i A_i)/d\tau = 0$ and so, using the given equation of motion, we obtain

$$A_i A_i = -V_i \frac{dA_i}{d\tau} = -\frac{\alpha^2}{c^2} V_i V_i = \alpha^2$$

since $V_i V_i = -c^2$.

14.13 A rocket propels itself in a straight line starting with velocity v_0 and ending with velocity v_1. If the gas ejected by the rocket has constant backward velocity $w(\ll c)$ relative to the instantaneous rest-frame of the rocket, show that the ratio of the initial to the final rest masses of the rocket is

$$\left[\frac{(c + v_1)(c - v_0)}{(c - v_1)(c + v_0)} \right]^{c/2w}.$$

- *Solution*

If M is the rest mass of the rocket then, relative to the instantaneous rest frame of the rocket, we have by *conservation of linear momentum*

$$w dM + M dv' = 0$$

since $w/c \ll 1$ where dv' is the change in the velocity of the rocket referred to its instantaneous rest frame.

Now from the first of the relativistic addition formulae (14.5) for velocities we have

$$dv' = \frac{dv}{1 - v^2/c^2}$$

where v is the speed of the rocket and so

$$w \frac{dM}{M} = -\frac{dv}{1 - v^2/c^2} = -\frac{c}{2}\left(\frac{1}{c - v} + \frac{1}{c + v} \right) dv.$$

Hence

$$w \ln \frac{M_1}{M_0} = \frac{c}{2}\left[\ln \frac{c - v_1}{c - v_0} - \ln \frac{c + v_1}{c + v_0} \right]$$

where M_0 and M_1 are the initial and final rest masses of the rocket. The required result follows readily.

14.14 A particle of mass M at rest disintegrates into two particles having rest masses M_1, M_2. Show that the energies E_1, E_2 of the two particles are given by

$$E_1 = \frac{c^2}{2M}\left(M^2 + M_1^2 - M_2^2 \right), \quad E_2 = \frac{c^2}{2M}\left(M^2 + M_2^2 - M_1^2 \right).$$

- *Solution*

The relation between the relativistic energy E and momentum $\mathbf{p}$ of a particle of mass M is

$$E = c(M^2 c^2 + \mathbf{p}^2)^{1/2}. \tag{14.10}$$

Hence we have

$$E_1^2 = c^2(M_1^2 c^2 + \mathbf{p}_1^2), \quad E_2^2 = c^2(M_2^2 c^2 + \mathbf{p}_2^2)$$

where by conservation of momentum $\mathbf{p}_1 + \mathbf{p}_2 = 0$ and so $\mathbf{p}_1^2 = \mathbf{p}_2^2$ giving $E_1^2 - E_2^2 = c^4(M_1^2 - M_2^2)$.

Now by conservation of energy

$$E_1 + E_2 = Mc^2$$

and hence we get

$$E_1 - E_2 = \frac{c^2(M_1^2 - M_2^2)}{M}$$

from which the required expressions for E_1 and E_2 follow at once.

14.15 A photon is scattered through an angle θ as the result of a collision with an electron which is initially at rest. Determine the shift in the wave length of the photon given that the rest mass of the electron is m_0.

• *Solution*

This problem requires the use of the quantum mechanical formulae for the energy E and momentum p of a quantum of light, or *photon*, of frequency ν. They are $E = h\nu$ and $p = h\nu/c$ where h is *Planck's constant*.

By conservation of energy we have, using (14.10),

$$h(\nu - \nu') = c\sqrt{m_0^2 c^2 + p^2} - m_0 c^2$$

where ν and ν' are the frequencies of the photon before and after the collision, and p is the momentum of the scattered electron.

By conservation of momentum we have, resolving in the direction of motion of the incident photon

$$\frac{h\nu}{c} = \frac{h\nu'}{c}\cos\theta + p\cos\phi$$

and perpendicular to this direction

$$\frac{h\nu'}{c}\sin\theta = p\sin\phi$$

where ϕ is the angle which the direction of motion of the scattered electron makes with the direction of incidence of the photon.

Hence

$$p^2 = \frac{h^2}{c^2}[(\nu - \nu'\cos\theta)^2 + (\nu'\sin\theta)^2]$$

and so, using the conservation of energy formula, we get

$$[h(\nu - \nu') + m_0 c^2]^2 = m_0^2 c^4 + h^2[(\nu - \nu'\cos\theta)^2 + (\nu'\sin\theta)^2].$$

Simplification of this leads to

$$m_0 c^2(\nu - \nu') = h\nu\nu'(1 - \cos\theta)$$

and since $\lambda = c/\nu$ and $\lambda' = c/\nu'$ we obtain

$$\lambda' - \lambda = \frac{h}{m_0 c}(1 - \cos\theta) = 2\lambda_0 \sin^2\frac{\theta}{2}$$

where $\lambda_0 = h/m_0 c$ is the *Compton wave length*.

14.16 A particle having rest mass m_0 and charge q moves under the action of a uniform electric field F in the Ox direction. If the particle is projected from the origin O with velocity v_0 in the Oy direction, perpendicular to Ox, show that the path of the particle is

$$x = \frac{m_0\gamma_0 c^2}{qF}\left(\cosh\frac{qFy}{m_0\gamma_0 v_0 c} - 1\right)$$

where $\gamma_0 = 1/\sqrt{1 - v_0^2/c^2}$.

- *Solution*

The equations of motion of the particle are

$$\frac{dp_x}{dt} = qF, \quad \frac{dp_y}{dt} = 0.$$

Hence $p_x = qFt$ since $v_x = 0$ when $t = 0$. Also, using the relativistic formula for the momentum $\mathbf{p} = \gamma m_0 \mathbf{v}$, we see that $p_y = \gamma_0 m_0 v_0$ since $v_y = v_0$ when $t = 0$.

The energy of the particle is given by $E = c(m_0^2 c^2 + p^2)^{1/2} = \gamma m_0 c^2$ where p is its momentum. Hence at time $t = 0$ the energy of the particle is $E_0 = c[m_0^2 c^2 + (\gamma_0 m_0 v_0)^2]^{1/2} = \gamma_0 m_0 c^2$ and so we have

$$E = [E_0^2 + (cqFt)^2]^{1/2}.$$

Now $p_x = \gamma m_0 dx/dt$ and this gives

$$\frac{dx}{dt} = \frac{qFt}{\gamma m_0} = \frac{c^2 qFt}{E} = \frac{c^2 qFt}{[E_0^2 + (cqFt)^2]^{1/2}}$$

from which it follows that

$$x = \frac{[E_0^2 + (cqFt)^2]^{1/2} - E_0}{qF}$$

since $x = 0$ when $t = 0$.

Also

$$\frac{dy}{dt} = \frac{p_y}{\gamma m_0} = \frac{m_0\gamma_0 v_0 c^2}{E} = \frac{m_0\gamma_0 v_0 c^2}{[E_0^2 + (cqFt)^2]^{1/2}}$$

and so we get

$$y = \frac{m_0\gamma_0 v_0 c}{qF}\sinh^{-1}\frac{cqFt}{E_0}$$

since $y = 0$ when $t = 0$.

Consequently

$$cqFt = E_0\sinh\frac{qFy}{m_0\gamma_0 v_0 c}$$

and hence

$$\left[E_0^2 + (cqFt)^2\right]^{1/2} = E_0\cosh\frac{qFy}{m_0\gamma_0 v_0 c}$$

using $\cosh^2\theta = \sinh^2\theta + 1$, from which follows the required equation for the path of the charged particle.

14.17 If F_{ij} is the electromagnetic field tensor and V_i is the four-velocity, show that for $i = 1, 2, 3$

$$F_i = \frac{q}{c}F_{ij}V_j$$

leads to the Lorentz force

$$\mathbf{f} = q\mathbf{E} + \frac{q}{c}\mathbf{v} \times \mathbf{H},$$

and for $i = 4$ we obtain $\mathbf{f} \cdot \mathbf{v} = q\mathbf{E} \cdot \mathbf{v}$.

- *Solution*

The *electromagnetic field tensor* is skew-symmetric. Using Cartesian tensors it satisfies $F_{ji} = -F_{ij}$ and its components are given by

$$(F_{23}, F_{31}, F_{12}) = \mathbf{H}, \quad (F_{41}, F_{42}, F_{43}) = i\mathbf{E}$$

where $\mathbf{E}$ and $\mathbf{H}$ are the electric and magnetic field vectors respectively.

Since the four-velocity is given by $(V_i) = (\gamma\mathbf{v}, i\gamma c)$ we have, using the convention that a repeated dummy suffix implies a summation,

$$F_1 = \frac{q}{c}F_{1j}V_j = \frac{\gamma q}{c}(H_z v_y - H_y v_z + cE_x) = \gamma q[E_x + \frac{1}{c}(\mathbf{v} \times \mathbf{H})_x] = \gamma f_x$$

where f_x is the x-component of the Lorentz force. Together with the analogous results for $i = 2, 3$ we see that

$$F_i = \frac{q}{c}F_{ij}V_j = \gamma f_i \quad (i = 1, 2, 3).$$

Consequently it gives the spatial part of the *four-force*

$$(F_i) = (\gamma\mathbf{f}, \frac{i\gamma}{c}\mathbf{f} \cdot \mathbf{v}) \tag{14.11}$$

where $\mathbf{f}$ is the Lorentz force.

For $i = 4$ we find

$$F_4 = \frac{q}{c}F_{4j}V_j = i\frac{q}{c}\gamma\mathbf{E} \cdot \mathbf{v}$$

and this is in accordance with the 4th component of the four-force if $q\mathbf{E} \cdot \mathbf{v} = \mathbf{f} \cdot \mathbf{v}$ where $\mathbf{f}$ is the Lorentz force.

Thus the Lorentz force is established.

14.18 Prove that

$$F_{ij}F_{ij} = 2(H^2 - E^2)$$

and hence show that $H^2 - E^2$ is an invariant with respect to Lorentz transformations.

- *Solution*

We have $F_{ij}F_{ij} = \text{trace } F\tilde{F}$ where $\tilde{F}$ is the transpose of the electromagnetic field tensor

$$F = \begin{pmatrix} 0 & H_z & -H_y & -iE_x \\ -H_z & 0 & H_x & -iE_y \\ H_y & -H_x & 0 & -iE_z \\ iE_x & iE_y & iE_z & 0 \end{pmatrix} \tag{14.12}$$

and so the $(1,1),(2,2),(3,3),(4,4)$ diagonal elements of $F\tilde{F}$ are respectively

$$H_z^2 + H_y^2 - E_x^2, \quad H_z^2 + H_x^2 - E_y^2, \quad H_y^2 + H_x^2 - E_z^2, \quad -E_x^2 - E_y^2 - E_z^2$$

and their sum is

$$2(H_x^2 + H_y^2 + H_z^2 - E_x^2 - E_y^2 - E_z^2) = 2(H^2 - E^2).$$

Since this is the trace of $F\tilde{F}$ we obtain $F_{ij}F_{ij} = 2(H^2 - E^2)$ as required.

Now $F_{ij}F_{ij}$ is the product of two Cartesian tensors of rank 2 which has been contracted twice reducing the rank by 4. Hence the rank of $F_{ij}F_{ij}$ is 0 and so it is a scalar invariant.

14.19 Prove that

$$e_{ijkl}F_{ij}F_{kl} = -8i\mathbf{E}.\mathbf{H}$$

and hence show that $\mathbf{E}.\mathbf{H}$ is an invariant scalar density with respect to Lorentz transformations.

- *Solution*

We have that e_{ijkl} has $4! = 24$ non-vanishing components. Since $F_{ji} = -F_{ij}$ and $e_{ijkl} = -e_{jikl}$, there are 8 components which are equal to $e_{1234}F_{12}F_{34} = -iH_zE_z$, there are another 8 components equal to $e_{1342}F_{13}F_{42} = -iH_yE_y$ and a further 8 components equal to $e_{1423}F_{14}F_{23} = -iH_xE_x$, using (14.12). Altogether we have

$$e_{ijkl}F_{ij}F_{kl} = -8i(H_zE_z + H_yE_y + H_xE_x) = -8i\mathbf{E}.\mathbf{H}$$

Since $e_{ijkl}F_{ij}F_{kl}$ is a tensor density of rank $4 + 2 + 2 = 8$ which has been contracted 4 times reducing the rank by 8, its rank is 0 and so it is an invariant scalar density, that is a pseudoscalar.

14.20 Show that the energy-momentum tensor

$$S_{ij} = \frac{1}{4\pi}F_{ik}F_{jk} - \frac{1}{16\pi}\delta_{ij}F_{kl}F_{kl},$$

where F_{ij} is the electromagnetic field tensor, is given by

$$S_{ij} = -\frac{1}{4\pi}(E_iE_j + H_iH_j) + \frac{1}{8\pi}\delta_{ij}(E^2 + H^2)$$

for $i,j = 1,2,3$; $S_{i4} = S_{4i}$ for $i = 1,2,3$ are the components of the vector $i\mathbf{S}/c$ where

$$\mathbf{S} = \frac{c}{4\pi}\mathbf{E} \times \mathbf{H}$$

is the Poynting vector; and $S_{44} = -U$ where

$$U = \frac{1}{8\pi}(E^2 + H^2)$$

is the energy density of the electromagnetic field.

- *Solution*

First we evaluate S_{ij} for $i, j = 1, 2, 3$ and $i = j$.

We have shown in the previous problem that $F_{kl}F_{kl} = 2(H^2 - E^2)$.

Now using (14.12) we have $F_{ik}F_{ik} = H^2 - H_i^2 - E_i^2$ $(i = 1, 2, 3)$ and so, no summation over i being implied, we see that

$$S_{ii} = \frac{1}{4\pi}(H^2 - H_i^2 - E_i^2) - \frac{1}{8\pi}(H^2 - E^2) = -\frac{1}{4\pi}(E_i^2 + H_i^2) + \frac{1}{8\pi}(E^2 + H^2) \quad (i = 1, 2, 3).$$

Next we evaluate S_{ij} for $i, j = 1, 2, 3$ and $i \neq j$.

We have $F_{ik}F_{jk} = -(E_i E_j + H_i H_j)$ and so we see that

$$S_{ij} = -\frac{1}{4\pi}(E_i E_j + H_i H_j) + \frac{1}{8\pi}\delta_{ij}(E^2 + H^2) \quad (i, j = 1, 2, 3 \quad i \neq j).$$

Thus the formula for the energy momentum tensor is established.

Further $S_{i4} = S_{4i} = F_{ik}F_{4k}/4\pi$ $(i = 1, 2, 3)$ so that

$$S_{14} = S_{41} = i(E_y H_z - E_z H_y)/4\pi,$$
$$S_{24} = S_{42} = i(E_z H_x - E_x H_z)/4\pi,$$
$$S_{34} = S_{43} = i(E_x H_y - E_y H_x)/4\pi,$$

and these are the x, y, z components of $i\mathbf{E} \times \mathbf{H}/4\pi$, that is $i\mathbf{S}/c$ where $\mathbf{S} = (c/4\pi)\mathbf{E} \times \mathbf{H}$ is the Poynting vector.

Also

$$S_{44} = \frac{1}{4\pi}F_{4k}F_{4k} - \frac{1}{16\pi}F_{kl}F_{kl} = -\frac{1}{4\pi}E^2 - \frac{1}{8\pi}(H^2 - E^2) = -\frac{1}{8\pi}(E^2 + H^2)$$

and so $S_{44} = -U$ where U is the energy density $(E^2 + H^2)/8\pi$.

14.21 (i) The four-force due to the electromagnetic field acting on the charge per unit proper volume is given by the force density

$$D_i = \frac{\rho}{c}F_{ij}V_j$$

where $\rho = n_0 q$ is the proper charge density corresponding to n_0 particles of charge q per unit proper volume, and V_i is the four-velocity. Show that

$$D_i = \frac{1}{c}F_{ij}J_j = \frac{1}{4\pi}F_{ij}F_{jk,k}$$

where J_i is the four-current density.

(ii) Show that

$$F_{ik,j}F_{jk} - \frac{1}{2}\delta_{ij}F_{kl}F_{kl,j} = \frac{1}{2}\left(F_{ik,j} + F_{ji,k} + F_{kj,i}\right)F_{jk} = 0.$$

(iii) Hence show that the divergence of the energy momentum tensor S_{ij} is given by $S_{ij,j} = -D_i$.

- *Solution*

(i) The *four-current* is given by

$$(J_i) = (\rho V_i) \tag{14.13}$$

where the four-velocity $(V_i) = (\gamma\mathbf{v}, i\gamma c)$ and ρ is the density of the charge as measured by an observer in its rest frame. Then we have

$$D_i = \frac{\rho}{c}F_{ij}V_j = \frac{1}{c}F_{ij}J_j.$$

Now the Maxwell electromagnetic equations

$$\operatorname{curl}\mathbf{H} - \frac{1}{c}\frac{\partial\mathbf{E}}{\partial t} = \frac{4\pi}{c}\mathbf{j}. \quad \operatorname{div}\mathbf{E} = 4\pi\rho$$

can be expressed in the relativistically invariant form

$$F_{ij,j} = \frac{4\pi}{c}J_i \tag{14.14}$$

where $F_{ij,j} = \partial F_{ij}/\partial x_j$, and so, as required, we find that

$$D_i = \frac{1}{4\pi}F_{ij}F_{jk,k}.$$

(ii) Since F_{ij} is skew-symmetric, we see that

$$F_{ik,j}F_{jk} = F_{ij,k}F_{kj} = F_{ji,k}F_{jk}$$

and hence

$$F_{ik,j}F_{jk} = \frac{1}{2}(F_{ik,j} + F_{ji,k})F_{jk}.$$

Also

$$\delta_{ij}F_{kl}F_{kl,j} = F_{kl}F_{kl,i} = -F_{jk}F_{kj,i}$$

and so

$$F_{ik,j}F_{jk} - \frac{1}{2}\delta_{ij}F_{kl}F_{kl,j} = \frac{1}{2}\left(F_{ik,j} + F_{ji,k} + F_{kj,i}\right)F_{jk}.$$

Now the Maxwell electromagnetic field equations

$$\operatorname{div}\mathbf{H} = 0, \quad \operatorname{curl}\mathbf{E} + \frac{1}{c}\frac{\partial\mathbf{H}}{\partial t} = \mathbf{0}$$

can be written in the relativistically invariant form

$$F_{ik,j} + F_{ji,k} + F_{kj,i} = 0 \tag{14.15}$$

and so the required result is found.

(iii) From the previous problem we have

$$S_{ij} = \frac{1}{4\pi}F_{ik}F_{jk} - \frac{1}{16\pi}\delta_{ij}F_{kl}F_{kl}$$

and so

$$S_{ij,j} = \frac{1}{4\pi}F_{ik,j}F_{jk} + \frac{1}{4\pi}F_{ik}F_{jk,j} - \frac{1}{8\pi}\delta_{ij}F_{kl}F_{kl,j}.$$

Now using the results obtained in (ii) we get

$$S_{ij,j} = \frac{1}{4\pi}F_{ik}F_{jk,j} = -\frac{1}{4\pi}F_{ij}F_{jk,k} = -D_i$$

as required.

14-2 General theory of relativity

The *general theory of relativity* originated with the work of Einstein and was published in 1916. The theory is based on the general tensor analysis using contravariant and covariant tensors developed by Ricci and Levi-Civita which we looked at in chapter 13.

We now analyze some typical problems involving various metrics having the general form $ds^2 = g_{ij}dx^i dx^j$ with $i, j = 1, 2, 3, 4$ characterizing the *space-time continuum*.

14.22 If space-time has the metric

$$ds^2 = -e^{2\lambda}[(dx^1)^2 + (dx^2)^2] - (x^2)^2 e^{-2\mu}(dx^3)^2 + e^{2\mu}(dx^4)^2$$

where λ and μ are functions of x^1 and x^2 only, show that the field equation $R_{44} = 0$ holding in free space yields

$$\frac{\partial^2\mu}{(\partial x^1)^2} + \frac{\partial^2\mu}{(\partial x^2)^2} + \frac{1}{x^2}\frac{\partial\mu}{\partial x^2} = 0.$$

• *Solution*

Here we have $g_{11} = g_{22} = -e^{2\lambda}$, $g_{33} = -(x^2)^2 e^{-2\mu}$, $g_{44} = e^{2\mu}$ and $g_{ij} = 0$ for $i \neq j$.
Now

$$R_{44} = \begin{Bmatrix} i \\ m\ 4 \end{Bmatrix}\begin{Bmatrix} m \\ 4\ i \end{Bmatrix} - \begin{Bmatrix} i \\ m\ i \end{Bmatrix}\begin{Bmatrix} m \\ 4\ 4 \end{Bmatrix} + \frac{\partial}{\partial x^4}\begin{Bmatrix} i \\ 4\ i \end{Bmatrix} - \frac{\partial}{\partial x^i}\begin{Bmatrix} i \\ 4\ 4 \end{Bmatrix}$$

where

$$\begin{Bmatrix} s \\ i\ j \end{Bmatrix} = \frac{1}{2}g^{sk}\left(\frac{\partial g_{jk}}{\partial x^i} + \frac{\partial g_{ki}}{\partial x^j} - \frac{\partial g_{ij}}{\partial x^k}\right).$$

We obtain

$$\begin{Bmatrix} 1 \\ 4\ 4 \end{Bmatrix} = \frac{1}{2}e^{-2\lambda}\frac{\partial e^{2\mu}}{\partial x^1}, \quad \begin{Bmatrix} 2 \\ 4\ 4 \end{Bmatrix} = \frac{1}{2}e^{-2\lambda}\frac{\partial e^{2\mu}}{\partial x^2}, \quad \begin{Bmatrix} 3 \\ 4\ 4 \end{Bmatrix} = \begin{Bmatrix} 4 \\ 4\ 4 \end{Bmatrix} = 0$$

and so

$$\frac{\partial}{\partial x^i}\begin{Bmatrix} i \\ 4\ 4 \end{Bmatrix} = \frac{1}{2}\left[\frac{\partial}{\partial x^1}\left(e^{-2\lambda}\frac{\partial e^{2\mu}}{\partial x^1}\right) + \frac{\partial}{\partial x^2}\left(e^{-2\lambda}\frac{\partial e^{2\mu}}{\partial x^2}\right)\right],$$

that is

$$\frac{\partial}{\partial x^i}\begin{Bmatrix} i \\ 4\ 4 \end{Bmatrix} = \left[\frac{\partial^2\mu}{(\partial x^1)^2} + \frac{\partial^2\mu}{(\partial x^2)^2} + 2\left(\frac{\partial\mu}{\partial x^1}\right)^2 + 2\left(\frac{\partial\mu}{\partial x^2}\right)^2 - 2\frac{\partial\lambda}{\partial x^1}\frac{\partial\mu}{\partial x^1} - 2\frac{\partial\lambda}{\partial x^2}\frac{\partial\mu}{\partial x^2}\right]e^{2\mu-2\lambda}.$$

Also, since the g_{ij} do not depend on x^4, we have

$$\frac{\partial}{\partial x^4}\begin{Bmatrix} i \\ 4\ i \end{Bmatrix} = 0.$$

Now

$$\begin{Bmatrix} 1 \\ 1\ 1 \end{Bmatrix} = \frac{1}{2g_{11}}\frac{\partial g_{11}}{\partial x^1} = \frac{\partial\lambda}{\partial x^1}, \quad \begin{Bmatrix} 2 \\ 2\ 1 \end{Bmatrix} = \begin{Bmatrix} 2 \\ 1\ 2 \end{Bmatrix} = \frac{1}{2g_{22}}\frac{\partial g_{22}}{\partial x^1} = \frac{\partial\lambda}{\partial x^1},$$

$$\begin{Bmatrix} 3 \\ 3\ 1 \end{Bmatrix} = \begin{Bmatrix} 3 \\ 1\ 3 \end{Bmatrix} = \frac{1}{2g_{33}}\frac{\partial g_{33}}{\partial x^1} = -\frac{\partial\mu}{\partial x^1}, \quad \begin{Bmatrix} 4 \\ 4\ 1 \end{Bmatrix} = \begin{Bmatrix} 4 \\ 1\ 4 \end{Bmatrix} = \frac{1}{2g_{44}}\frac{\partial g_{44}}{\partial x^1} = \frac{\partial\mu}{\partial x^1},$$

$$\left\{\begin{matrix}1\\1\ 2\end{matrix}\right\}=\left\{\begin{matrix}1\\2\ 1\end{matrix}\right\}=\frac{1}{2g_{11}}\frac{\partial g_{11}}{\partial x^2}=\frac{\partial\lambda}{\partial x^2},\quad\left\{\begin{matrix}2\\2\ 2\end{matrix}\right\}=\frac{1}{2g_{22}}\frac{\partial g_{22}}{\partial x^2}=\frac{\partial\lambda}{\partial x^2},$$

$$\left\{\begin{matrix}3\\3\ 2\end{matrix}\right\}=\left\{\begin{matrix}3\\2\ 3\end{matrix}\right\}=\frac{1}{2g_{33}}\frac{\partial g_{33}}{\partial x^2}=-\frac{\partial\mu}{\partial x^2}+\frac{1}{x^2},\quad\left\{\begin{matrix}4\\4\ 2\end{matrix}\right\}=\left\{\begin{matrix}4\\2\ 4\end{matrix}\right\}=\frac{1}{2g_{44}}\frac{\partial g_{44}}{\partial x^2}=\frac{\partial\mu}{\partial x^2},$$

and all other Christoffel symbols are zero.

Thus we get

$$\left\{\begin{matrix}i\\m\ 4\end{matrix}\right\}\left\{\begin{matrix}m\\4\ i\end{matrix}\right\}=2\left[\left(\frac{\partial\mu}{\partial x^1}\right)^2+\left(\frac{\partial\mu}{\partial x^2}\right)^2\right]e^{2\mu-2\lambda},$$

$$\left\{\begin{matrix}i\\m\ i\end{matrix}\right\}\left\{\begin{matrix}m\\4\ 4\end{matrix}\right\}=\left[2\frac{\partial\lambda}{\partial x_1}\frac{\partial\mu}{\partial x_1}+\left(2\frac{\partial\lambda}{\partial x^2}+\frac{1}{x^2}\right)\frac{\partial\mu}{\partial x^2}\right]e^{2\mu-2\lambda}.$$

Hence

$$R_{44}=-\left[\frac{\partial^2\mu}{(\partial x^1)^2}+\frac{\partial^2\mu}{(\partial x^2)^2}+\frac{1}{x^2}\frac{\partial\mu}{\partial x^2}\right]e^{2\mu-2\lambda}$$

and it follows that the field equation $R_{44}=0$ holding in free space gives the required equation for μ.

14.23 Find the geodesics in space-time for the metric

$$ds^2=e^{2\lambda x}[-(dx^2+dy^2+dz^2)+c^2dt^2]$$

where λ is a constant.

If

$$v^2=\left(\frac{dx}{dt}\right)^2+\left(\frac{dy}{dt}\right)^2+\left(\frac{dz}{dt}\right)^2$$

and if $v=v_0$ when $x=0$, show that

$$c^2-v^2=(c^2-v_0^2)e^{2\lambda x}.$$

• *Solution*

In this problem we have $g_{11}=g_{22}=g_{33}=-e^{2\lambda x^1}$, $g_{44}=e^{2\lambda x^1}$ and $g_{ij}=0$ for $i\neq j$ where $x^1=x,x^2=y,x^3=z,x^4=ct$.

The equations for geodesics are

$$\frac{d^2x^l}{ds^2}+\left\{\begin{matrix}l\\i\ j\end{matrix}\right\}\frac{dx^i}{ds}\frac{dx^j}{ds}=0 \qquad (14.16)$$

where ds is an element of arc along the geodesic.

We find that

$$\left\{\begin{matrix}1\\1\ 1\end{matrix}\right\}=\frac{1}{2g_{11}}\frac{\partial g_{11}}{\partial x^1}=\lambda,\quad\left\{\begin{matrix}1\\2\ 2\end{matrix}\right\}=\left\{\begin{matrix}1\\3\ 3\end{matrix}\right\}=-\lambda,\quad\left\{\begin{matrix}1\\4\ 4\end{matrix}\right\}=\lambda,$$

$$\left\{\begin{matrix}2\\2\ 1\end{matrix}\right\}=\left\{\begin{matrix}2\\1\ 2\end{matrix}\right\}=\frac{1}{2g_{22}}\frac{\partial g_{22}}{\partial x^1}=\lambda,\quad\left\{\begin{matrix}3\\3\ 1\end{matrix}\right\}=\left\{\begin{matrix}3\\1\ 3\end{matrix}\right\}=\lambda,\quad\left\{\begin{matrix}4\\4\ 1\end{matrix}\right\}=\left\{\begin{matrix}4\\1\ 4\end{matrix}\right\}=\lambda,$$

and the other Christoffel symbols are zero.

Hence we get the following equations for the geodesics

$$\frac{d^2x}{ds^2} + \lambda\left[\left(\frac{dx}{ds}\right)^2 - \left(\frac{dy}{ds}\right)^2 - \left(\frac{dz}{ds}\right)^2 + \left(\frac{dx^4}{ds}\right)^2\right] = 0,$$

$$\frac{d^2y}{ds^2} + 2\lambda\frac{dx}{ds}\frac{dy}{ds} = 0, \quad \frac{d^2z}{ds^2} + 2\lambda\frac{dx}{ds}\frac{dz}{ds} = 0, \quad \frac{d^2x^4}{ds^2} + 2\lambda\frac{dx}{ds}\frac{dx^4}{ds} = 0.$$

The last of these equations can be written

$$\frac{d^2t}{ds^2} = -2\lambda\frac{dx}{ds}\frac{dt}{ds}$$

which, on carrying out an integration with respect to s, yields

$$\frac{dt}{ds} = Ae^{-2\lambda x}$$

where A is a constant.

Now $ds^2 = (c^2dt^2 - dx^2 - dy^2 - dz^2)e^{2\lambda x}$ and so we also have

$$\left(\frac{ds}{dt}\right)^2 = (c^2 - v^2)e^{2\lambda x}$$

since $v^2 = (dx/dt)^2 + (dy/dt)^2 + (dz/dt)^2$. Hence

$$c^2 - v^2 = (c^2 - v_0^2)e^{2\lambda x}$$

using $v = v_0$ when $x = 0$.

14.24 For the space-time having the metric

$$ds^2 = e^{2\lambda}[(dx^1)^2 + (dx^2)^2 + (dx^3)^2 + (dx^4)^2]$$

where λ is a function of x^1, x^2, x^3, x^4 show that the scalar curvature $R = 0$ if and only if

$$\lambda_{ii} + \lambda_i\lambda_i = 0$$

where $\lambda_i = \partial\lambda/\partial x^i$ and $\lambda_{ii} = \partial^2\lambda/(\partial x^i)^2$.

If λ is a function of $r = [(x^1)^2 + (x^2)^2 + (x^3)^2]^{\frac{1}{2}}$ only, prove that

$$\lambda'' + \frac{2}{r}\lambda' + (\lambda')^2 = 0$$

where the primes denote differentiations with respect to r.

• *Solution*

We have $g_{11} = g_{22} = g_{33} = g_{44} = e^{2\lambda}$ and $g_{ij} = 0$ if $i \neq j$.

The scalar curvature is given by $R = R_k^k = g^{kj}R_{jk} = g^{kj}B_{jki}^i = e^{-2\lambda}B_{jji}^i$ and so using the formula (13.16) for the curvature tensor we have

$$R = e^{-2\lambda}\left[\begin{Bmatrix} i \\ m\ j \end{Bmatrix}\begin{Bmatrix} m \\ j\ i \end{Bmatrix} - \begin{Bmatrix} i \\ m\ i \end{Bmatrix}\begin{Bmatrix} m \\ j\ j \end{Bmatrix} + \frac{\partial}{\partial x^j}\begin{Bmatrix} i \\ j\ i \end{Bmatrix} - \frac{\partial}{\partial x^i}\begin{Bmatrix} i \\ j\ j \end{Bmatrix}\right]$$

where

$$\left\{ {s \atop i\ j} \right\} = \frac{1}{2} g^{sk} \left(\frac{\partial g_{jk}}{\partial x^i} + \frac{\partial g_{ki}}{\partial x^j} - \frac{\partial g_{ij}}{\partial x^k} \right).$$

Now, without the summations being implied by the repetition of the indices, and putting $\lambda_i = \partial \lambda / \partial x^i$, we obtain

$$\left\{ {i \atop i\ i} \right\} = \frac{1}{2g_{ii}} \frac{\partial g_{ii}}{\partial x^i} = \lambda_i, \quad \left\{ {i \atop j\ i} \right\} = \left\{ {i \atop i\ j} \right\} = \frac{1}{2g_{ii}} \frac{\partial g_{ii}}{\partial x^j} = \lambda_j,$$

$$\left\{ {i \atop j\ j} \right\} = \frac{1}{2g_{ii}} \left(-\frac{\partial g_{jj}}{\partial x^i} \right) = -\lambda_i \quad (j \neq i),$$

and $\left\{ {k \atop i\ j} \right\} = 0$ if i, j, k are all different.

Then, with the inclusion of the summations implied by the repetition of the indices, we have

$$\frac{\partial}{\partial x^j} \left\{ {i \atop j\ i} \right\} = 4\lambda_{jj}, \quad \frac{\partial}{\partial x^i} \left\{ {i \atop j\ j} \right\} = -2\lambda_{ii}, \quad \left\{ {i \atop m\ i} \right\} \left\{ {m \atop j\ j} \right\} = -8\lambda_m \lambda_m$$

and

$$\left\{ {i \atop m\ j} \right\} \left\{ {m \atop j\ i} \right\} = \sum_m \left\{ {m \atop m\ j} \right\} \left\{ {m \atop j\ m} \right\} + \sum_{j \neq i} \left\{ {i \atop j\ j} \right\} \left\{ {j \atop j\ i} \right\} + \sum_{i \neq m} \left\{ {i \atop m\ i} \right\} \left\{ {m \atop i\ i} \right\}$$

giving

$$\left\{ {i \atop m\ j} \right\} \left\{ {m \atop j\ i} \right\} = 4\lambda_j \lambda_j - 3\lambda_i \lambda_i - 3\lambda_i \lambda_i = -2\lambda_i \lambda_i.$$

Hence

$$R = 6e^{-2\lambda} (\lambda_i \lambda_i + \lambda_{ii})$$

and so $R = 0$ if and only if $\lambda_i \lambda_i + \lambda_{ii} = 0$ which is the result required.

If $\lambda = \lambda(r)$ we have $\partial r / \partial x^i = r^{-1} x^i$ $(i = 1, 2, 3)$ and so $\lambda_i = r^{-1} x^i \lambda'$ for $i = 1, 2, 3$ and $\lambda_i = 0$ for $i = 4$. Hence

$$\lambda_i \lambda_i = \frac{(x^1)^2 + (x^2)^2 + (x^3)^2}{r^2} (\lambda')^2 = (\lambda')^2$$

$$\lambda_{ii} = \frac{\partial}{\partial x^i} \left(\frac{x^i}{r} \lambda' \right) = \frac{3}{r} \lambda' - \frac{(x^1)^2 + (x^2)^2 + (x^3)^2}{r^3} \lambda' + \frac{(x^1)^2 + (x^2)^2 + (x^3)^2}{r^2} \lambda'' = \frac{2}{r} \lambda' + \lambda''.$$

Hence the scalar curvature R vanishes if and only if $\lambda_i \lambda_i + \lambda_{ii} = \lambda'' + 2r^{-1} \lambda' + (\lambda')^2 = 0$ as required.

14.25 Show that the scalar curvature is zero for the space-time metric

$$ds^2 = e^{2\lambda} (dr^2 + r^2 d\theta^2 + r^2 \sin^2\theta \, d\phi^2 - c^2 dt^2)$$

where

$$\lambda = \ln \left(1 + \frac{m}{r} \right).$$

• *Solution*

If we set $x^1 = x = r\sin\theta\cos\phi, x^2 = y = r\sin\theta\sin\phi, x^3 = z = r\cos\theta, x^4 = ict$ we obtain

$$ds^2 = e^{2\lambda}[(dx^1)^2 + (dx^2)^2 + (dx^3)^2 + (dx^4)^2]$$

This is the same as the previous problem and so the scalar curvature vanishes if and only if

$$\lambda'' + 2r^{-1}\lambda' + (\lambda')^2 = 0.$$

Now $\lambda = \ln(1 + mr^{-1})$ and so $\lambda' = -mr^{-2}(1 + mr^{-1})^{-1}$ and $\lambda'' = 2mr^{-3}(1 + mr^{-1})^{-1} - m^2r^{-4}(1 + mr^{-1})^{-2}$ from which it follows that $\lambda'' + 2r^{-1}\lambda' + (\lambda')^2 = 0$. Hence the scalar curvature R is zero.

The last three problems involve the calculus of variations and geodesics in space-time. Thus they could be placed in the final chapter on *variational principles* but it is convenient to include them in this chapter amongst our problems on the theory of relativity.

The metric for an n-dimensional *Riemannian space* whose points are specified by n coordinates $x^1, x^2, ..., x^n$ is given by

$$ds^2 = g_{ij}dx^i dx^j$$

where ds is the distance between two neighbouring points in the space and the g_{ij} are arbitrary functions of the coordinates x^i.

If $n = 4$ and the coordinates x^1, x^2, x^3, x^4 specify position in space and time we have a *space-time continuum*.

In the following problem the coordinates are chosen to be the spherical polar coordinates r, θ, ϕ and the time t so that we have $x^1 = r, x^2 = \theta, x^3 = \phi, x^4 = t$ with $g_{11} = e^{2\lambda}, g_{22} = e^{2\lambda}r^2, g_{33} = e^{2\lambda}r^2\sin^2\theta, g_{44} = -c^2e^{2\lambda}$ and $g_{ij} = 0$ for $i \neq j$.

14.26 Show that the geodesics of the space-time continuum having the metric

$$ds^2 = e^{2\lambda}(dr^2 + r^2d\theta^2 + r^2\sin^2\theta\,d\phi^2 - c^2dt^2),$$

where $\lambda = \ln(1 + m/r)$, satisfy the equations

$$r^2\sin^2\theta\frac{d\phi}{ds} = ae^{-2\lambda}, \quad \frac{dt}{ds} = be^{-2\lambda}$$

where a, b are constants. If $\phi = 0, d\phi/ds = 0$ initially, show that $\phi = 0$ at all points along the geodesic and that

$$r^2\frac{d\theta}{ds} = he^{-2\lambda}$$

where h is a constant.

- *Solution*

The geodesics can be determined by solving the variational problem $\delta \int L\,ds = 0$ where the Lagrangian function L is given by

$$L = e^{2\lambda}(r'^2 + r^2\theta'^2 + r^2\sin^2\theta\,\phi'^2 - c^2t'^2)$$

and the primes indicate differentiations with respect to the arc distance s in space-time.

The Lagrange equations (8.2) in classical dynamics can be generalized to hold for a general Lagrangian function $L(x; y_1, ..., y_n; y_1', ..., y_n')$ where $y_r' = dy/dx$. They then become the *Euler-Lagrange equations*

$$\frac{d}{dx}\left(\frac{\partial L}{\partial y_r'}\right) - \frac{\partial L}{\partial y_r} = 0 \qquad (r = 1, ..., n)$$

and arise from the variational problem for the integral $I = \int L\, dx$.

The Euler-Lagrange's equations for the present variational problem are

$$\frac{d}{ds}\left(\frac{\partial L}{\partial t'}\right) - \frac{\partial L}{\partial t} = 0, \quad \frac{d}{ds}\left(\frac{\partial L}{\partial r'}\right) - \frac{\partial L}{\partial r} = 0, \quad \frac{d}{ds}\left(\frac{\partial L}{\partial \theta'}\right) - \frac{\partial L}{\partial \theta} = 0, \quad \frac{d}{ds}\left(\frac{\partial L}{\partial \phi'}\right) - \frac{\partial L}{\partial \phi} = 0$$

where the arc distance s takes the place of x here.

Also we have $L = 1$ and so

$$e^{2\lambda}(r'^2 + r^2\theta'^2 + r^2\sin^2\theta\,\phi'^2 - c^2t'^2) = 1.$$

From the equation for t we obtain $d(-2c^2t'e^{2\lambda})/ds = 0$ and so $t'e^{2\lambda}$ =constant along a geodesic, that is

$$\frac{dt}{ds} = be^{-2\lambda}$$

where b is a constant.

From the equation for ϕ we obtain $d(2\phi'r^2\sin^2\theta\,e^{2\lambda})/ds = 0$ and so $\phi'r^2\sin^2\theta\,e^{2\lambda}$ =constant, that is

$$r^2\sin^2\theta\frac{d\phi}{ds} = ae^{-2\lambda}$$

where a is a constant.

Now if $d\phi/ds = 0$ initially it follows that $a = 0$ and so $d\phi/ds = 0$ always. Hence ϕ =constant and if $\phi = 0$ initially also, we see that $\phi = 0$ at all points along the geodesic.

The equation for θ gives

$$\frac{d}{ds}(2\theta'r^2e^{2\lambda}) = 2\sin\theta\cos\theta\,r^2\phi'^2e^{2\lambda}$$

and since $\phi' = 0$ always we obtain $d(\theta'r^2e^{2\lambda})/ds = 0$ so that $\theta'r^2e^{2\lambda}$ =constant along a geodesic which gives

$$r^2\frac{d\theta}{ds} = he^{-2\lambda}$$

where h is a constant.

14.27 The metric for a space-time continuum is

$$ds^2 = kc^2dt^2 - k^{-1}dr^2 - r^2d\theta^2 - r^2\sin^2\theta\,d\phi^2$$

where $k = 1 - r^2/R^2$ and R is a constant.

Obtain the differential equations satisfied by the null-geodesics and show that along the null-geodesics in the plane $\theta = \pi/2$

$$a\frac{dr}{d\phi} = r(r^2 - a^2)^{\frac{1}{2}}$$

where a is a constant. Deduce that the paths of the light rays are straight lines in the plane $\theta = \pi/2$ taking r, ϕ as plane polar coordinates.

- *Solution*

The null-geodesics are solutions of the variational problem $\delta \int L\,dq = 0$ where the Lagrangian function is given by

$$L = kc^2 t'^2 - k^{-1}r'^2 - r^2\theta'^2 - r^2\sin^2\theta\,\phi'^2$$

where the primes denote differentiations with respect to the parameter q specifying position in the space-time continuum.

The Euler-Lagrange equation for t is

$$\frac{d}{dq}\left(\frac{\partial L}{\partial t'}\right) - \frac{\partial L}{\partial t} = 0$$

which gives $d(2kc^2 t')/dq = 0$ and so $kt' =$ constant.

The Euler-Lagrange equation for ϕ is

$$\frac{d}{dq}\left(\frac{\partial L}{\partial \phi'}\right) - \frac{\partial L}{\partial \phi} = 0$$

which gives $d(-2r^2\sin^2\theta\,\phi')/dq = 0$ and so $r^2\sin^2\theta\,\phi' =$ constant.

For a *null-geodesic* $ds = 0$ and so we see that

$$kc^2\left(\frac{dt}{d\phi}\right)^2 - k^{-1}\left(\frac{dr}{d\phi}\right)^2 - r^2\left(\frac{d\theta}{d\phi}\right)^2 - r^2\sin^2\theta = 0$$

In the plane $\theta = \pi/2$ we have $d\theta/d\phi = 0$ and $dt/d\phi = t'/\phi' = br^2/k$ where b is a constant. Hence

$$\left(\frac{dr}{d\phi}\right)^2 = k\left[kc^2\left(\frac{t'}{\phi'}\right)^2 - r^2\right] = r^2(c^2b^2r^2 - k) = r^2\left(\frac{r^2}{a^2} - 1\right)$$

where $a = R/\sqrt{c^2b^2R^2 + 1}$ is a constant, and so

$$a\frac{dr}{d\phi} = r(r^2 - a^2)^{1/2}.$$

To solve this equation we put $r = u^{-1}$. Then $dr/d\phi = -u^{-2}du/d\phi$ and so

$$a^2\left(\frac{du}{d\phi}\right)^2 = 1 - a^2u^2$$

from which it follows that $au = \cos(\phi+\epsilon)$ where ϵ is an arbitrary constant. Choosing the polar axis so that $\epsilon = 0$ gives $a = r\cos\phi$ which is the equation of a straight line whose perpendicular distance from the origin is a.

14.28 The metric for a space-time continuum is

$$ds^2 = c^2 dt^2 - (1 - r^2/a^2)^{-1}dr^2 - r^2 d\theta^2 - r^2\sin^2\theta\,d\phi^2.$$

Obtain the equations for null geodesics and show that in the plane $\theta = \pi/2$, the null geodesics satisfy

$$\left(\frac{dr}{d\phi}\right)^2 = r^2(1 - r^2/a^2)(r^2/b^2 - 1)$$

where b is a constant.

Deduce that the paths of light rays in the plane $\theta = \pi/2$ are the ellipses $x^2/a^2 + y^2/b^2 = 1$ where x, y are rectangular Cartesian coordinates.

Further show that the time taken by a photon to make a complete circuit of the ellipse is $2\pi a/c$.

• *Solution*

To determine the equations for a null geodesic we introduce a parameter q and solve the variational problem $\delta \int L \, dq = 0$ where

$$L = c^2 t'^2 - (1 - r^2/a^2)^{-1} r'^2 - r^2 \theta'^2 - r^2 \sin^2\theta \, \phi'^2$$

is the Lagrangian function and the primes denote differentiations with respect to q.

The Euler-Lagrange equation for t is

$$\frac{d}{dq}\left(\frac{\partial L}{\partial t'}\right) - \frac{\partial L}{\partial t} = 0$$

which gives $d(2c^2 t')/dq = 0$ and so $t' = $ constant.

The Euler-Lagrange equation for ϕ is

$$\frac{d}{dq}\left(\frac{\partial L}{\partial \phi'}\right) - \frac{\partial L}{\partial \phi} = 0$$

which gives $d(-2r^2 \sin^2\theta \, \phi')/dq = 0$ and so $r^2 \sin^2\theta \, \phi' = $ constant.

Now for a null geodesic $ds = 0$ and so we have

$$c^2 \left(\frac{dt}{d\phi}\right)^2 - \left(1 - \frac{r^2}{a^2}\right)^{-1}\left(\frac{dr}{d\phi}\right)^2 - r^2\left(\frac{d\theta}{d\phi}\right)^2 - r^2 \sin^2\theta = 0.$$

In the plane $\theta = \pi/2$ we have $d\theta/d\phi = 0$ and $dt/d\phi = t'/\phi' = r^2/cb$ where b is a constant. Hence

$$\left(\frac{dr}{d\phi}\right)^2 = \left(1 - \frac{r^2}{a^2}\right)\left(c^2\frac{t'^2}{\phi'^2} - r^2\right) = r^2\left(1 - \frac{r^2}{a^2}\right)\left(\frac{r^2}{b^2} - 1\right).$$

Putting $r^2 = u^{-1}$ we get $2r \, dr/d\phi = -u^{-2} du/d\phi$ and so

$$\left(\frac{du}{d\phi}\right)^2 = 4\left(u - \frac{1}{a^2}\right)\left(\frac{1}{b^2} - u\right).$$

To simplify this we set $\alpha = a^{-2}, \beta = b^{-2}$ and then integration gives

$$\phi + \epsilon = \frac{1}{2}\int \frac{du}{\sqrt{(u-\alpha)(\beta-u)}} = \frac{1}{2}\sin^{-1}\frac{u - (\alpha+\beta)/2}{(\alpha-\beta)/2}$$

where ϵ is an arbitrary constant. Hence

$$u = \frac{1}{r^2} = \frac{\alpha+\beta}{2} + \frac{\alpha-\beta}{2}\sin 2(\phi + \epsilon)$$

and taking $\epsilon = \pi/4$ this can be written as

$$r^{-2} = (\alpha+\beta)/2 + [(\alpha-\beta)/2]\cos 2\phi = \alpha\cos^2\phi + \beta\sin^2\phi.$$

Putting $x = r\cos\phi$ and $y = r\sin\phi$ we obtain the required ellipse $x^2/a^2 + y^2/b^2 = 1$.

Now we have

$$\frac{d\phi}{dt} = \frac{cb}{r^2} = cb(\alpha\cos^2\phi + \beta\sin^2\phi)$$

and so the time T taken by the photon to make a complete circuit of the ellipse is

$$T = \int_0^T dt = \frac{1}{cb}\int_0^{2\pi}\frac{d\phi}{\alpha\cos^2\phi + \beta\sin^2\phi}.$$

To do this integration we set $\xi = \tan\phi$ and then we get, as required,

$$T = \frac{4}{cb}\int_0^\infty \frac{d\xi}{\alpha + \beta\xi^2} = \frac{4a}{c}\int_0^\infty \frac{d\eta}{1 + \eta^2} = \frac{2\pi a}{c}.$$

15
QUANTUM THEORY

It was proposed by Max Planck (1858-1947) in 1900 that electromagnetic radiation is composed of quanta called *photons* or discrete packets of energy $h\nu$, where ν is the frequency of the radiation and h is *Planck's constant*. This explained the observed distribution of radiation emitted by black bodies as a function of frequency.

By performing experiments on the scattering of α-particles (helium nuclei) by metal foils, Rutherford was able to show in 1913 that atoms are composed of a heavy nucleus which is positively charged surrounded by orbiting electrons having negative charge and which are light, rather similar to a planetary system.

In 1885 Balmer showed that the light emitted by atomic hydrogen in the visible region has discrete frequencies (spectral lines) given by

$$\nu = R\left(\frac{1}{2^2} - \frac{1}{n^2}\right)$$

where R is called the Rydberg constant and n is a positive integer which is ≥ 3 for the Balmer series.

This led Niels Bohr (1885-1962) in 1913 to develop the *Old Quantum Theory* in which the angular momentum of the electron in a hydrogenic atom is quantized by the discrete amounts $n\hbar$, where $\hbar = h/2\pi$, leading to the quantization of the energy according to the formula

$$E_n = -\frac{me^4 Z^2}{2\hbar^2 n^2}$$

where $-e$ and m are the charge and the mass of the electron and Ze is the charge of the nucleus, giving $R = 2\pi^2 m e^4 / h^3$ for the Rydberg constant.

The radius of the $n = 1$ orbit for atomic hydrogen ($Z = 1$) is given by the *Bohr radius* $a_0 = \hbar^2 / me^2$.

In 1924 Louis de Broglie (1892-1987) suggested that electrons have a *wave-particle duality*. For light waves the wave length λ, frequency ν, energy E and momentum p are related by $\lambda\nu = c, h\nu = E, p = E/c$ giving $p = h/\lambda$. De Broglie proposed that the wave length λ associated with an electron is also related to its momentum p by $\lambda = h/p$ called the *de Broglie wave length*.

Debye pointed out in a colloquium in 1925 that the amplitude ψ of an electron wave must satisfy a *wave equation*.

Such an equation was obtained by Erwin Schrödinger (1887-1961) in 1926. This led to *wave mechanics* and *quantum theory* with the work in 1925 of Werner Heisenberg (1901-1976)) and in 1930 of Paul Dirac (1902-1984).

We can summarize quantum theory and wave mechanics by means of a set of *postulates:*

1. A quantum mechanical system is completely described by a state *wave function* ψ.

2. Every *observable* is represented by a linear Hermitian operator A with a complete set of orthonormal *eigenfunctions* ϕ_i.

3. The only possible result of a precise measurement of the observable represented by the operator A is one of its *eigenvalues* a_i.

4. The probability of a measurement being a_i is $|(\phi_i, \psi)|^2$ where $(\phi_i, \psi) = \int \phi_i^* \psi \, d\tau$ is the scalar product or inner product of ψ and ϕ_i and is known as the *probability amplitude*.

The operator representing linear momentum is $\mathbf{p} = -i\hbar\nabla$ and the operator representing energy E is the Hamiltonian H which gives the time-independent *Schrödinger equation* $H\psi = E\psi$. For a particle of mass m in a field of force with potential V, the Schrödinger equation becomes

$$-\frac{\hbar^2}{2m}\nabla^2\psi + V\psi = E\psi. \tag{15.1}$$

The time-dependent *Schrödinger equation* has the form

$$-\frac{\hbar^2}{2m}\nabla^2\psi + V\psi = i\hbar\frac{\partial\psi}{\partial t}, \tag{15.2}$$

which reduces to (15.1) if we set $\psi(\mathbf{r}, t) = \phi(\mathbf{r})\exp(-iEt/\hbar)$.

15-1 Exactly soluble problems

15.01 Use $< \mathbf{p} >= -i\hbar \int \psi^*\nabla\psi \, d\tau$ to show that the expectation value of the momentum $< \mathbf{p} >$ is real for a wave packet.

• *Solution*

The *expectation value* of a quantity represented by an operator Ω is given by

$$< \Omega >= \int \psi^*\Omega\psi \, d\tau \tag{15.3}$$

and so for the momentum $\mathbf{p} = -i\hbar\nabla$ we have $< \mathbf{p} >= -i\hbar \int \psi^*\nabla\psi \, d\tau$. Then

$$< \mathbf{p} > - < \mathbf{p} >^* = -i\hbar \int (\psi^*\nabla\psi + \psi\nabla\psi^*) \, d\tau = -i\hbar \int \nabla(\psi^*\psi) \, d\tau.$$

Now from (4.4) derived in problem 4.08, we have $\int_R \nabla\phi \, d\tau = \int_S \phi\hat{\mathbf{n}} \, dS$ where S is the boundary surface of the region R. Hence

$$< \mathbf{p} > - < \mathbf{p} >^* = -i\hbar \int \psi^*\psi\hat{\mathbf{n}} \, dS.$$

Since the wave function ψ is a wave packet it is confined to a finite region of space and thus gives a vanishing contribution to the surface integral as it recedes to infinity to make the volume integral cover all space.

Hence $< \mathbf{p} > - < \mathbf{p} >^* = 0$ and so $< \mathbf{p} >$ is real for a wave packet.

15.02 Show that for a 3-dimensional wave packet

$$m\frac{d < \mathbf{r}^2 >}{dt} =< \mathbf{r}.\mathbf{p} > + < \mathbf{p}.\mathbf{r} >$$

• *Solution*

We have

$$m\frac{d}{dt} < \mathbf{r}^2 >= m\frac{d}{dt} \int \psi^*\mathbf{r}^2\psi \, d\tau = m \int \frac{\partial\psi^*}{\partial t}\mathbf{r}^2\psi \, d\tau + m \int \psi^*\mathbf{r}^2\frac{\partial\psi}{\partial t} \, d\tau$$

and using the time-dependent Schrödinger equation (15.2) for ψ and its complex conjugate for ψ^* we obtain

$$m\frac{d}{dt} < \mathbf{r}^2 > = \frac{im}{\hbar} \int \mathbf{r}^2\psi \left(-\frac{\hbar^2}{2m}\nabla^2 + V\right)\psi^* \, d\tau - \frac{im}{\hbar} \int \mathbf{r}^2\psi^* \left(-\frac{\hbar^2}{2m}\nabla^2 + V\right)\psi \, d\tau$$

assuming that the potential V is real. Then we have

$$m\frac{d}{dt} < \mathbf{r}^2 > = \frac{1}{2}i\hbar \int (\mathbf{r}^2\psi^*\nabla^2\psi - \mathbf{r}^2\psi\nabla^2\psi^*) \, d\tau.$$

We now use *Green's theorem* which states that

$$\int_R (f\nabla^2 g - g\nabla^2 f) \, d\tau = \oint_S (f\nabla g - g\nabla f).\hat{\mathbf{n}} \, dS$$

where f and g are scalar functions of position with continuous second-order derivatives and $\hat{\mathbf{n}}$ is the unit vector in the outward normal direction to the boundary surface S enclosing the region R. We can apply this theorem to the functions $f = \mathbf{r}^2\psi$ and $g = \psi^*$ which enables us to write

$$m\frac{d}{dt} < \mathbf{r}^2 > = \frac{1}{2}i\hbar \int \psi^*[\mathbf{r}^2\nabla^2\psi - \nabla^2(\mathbf{r}^2\psi)] \, d\tau$$

since the contribution from the surface integral vanishes as we let the surface S recede to infinity because we have a wave packet.

Now $\nabla^2(\mathbf{r}^2\psi) = \nabla.(\mathbf{r}^2\nabla\psi + \psi\nabla\mathbf{r}^2) = \mathbf{r}^2\nabla^2\psi + 2\mathbf{r}.\nabla\psi + 2\nabla.(\mathbf{r}\psi)$ since $\mathbf{r}^2 = r^2$ and $\nabla r^2 = 2\mathbf{r}$, and so

$$m\frac{d}{dt} < \mathbf{r}^2 > = -i\hbar \int [\psi^*\mathbf{r}.\nabla\psi + \psi^*\nabla.(\mathbf{r}\psi)] \, d\tau = < \mathbf{r}.\mathbf{p} > + < \mathbf{p}.\mathbf{r} >$$

remembering that $\mathbf{p} = -i\hbar\nabla$.

15.03 Obtain the wave packet corresponding to the linear combination of one-dimensional waves given by

$$\psi(x) = \frac{1}{2\pi} \int_{-\infty}^{\infty} \phi(k)\exp(ikx) \, dk$$

where

$$\phi(k) = \begin{cases} 1 & (-\frac{1}{2}\Delta k + k_0 \leq k \leq \frac{1}{2}\Delta k + k_0) \\ 0 & (|k - k_0| > \frac{1}{2}\Delta k). \end{cases}$$

• *Solution*

We have, performing an elementary integration, that

$$\psi(x) = \frac{1}{2\pi} \int_{-\frac{1}{2}\Delta k + k_0}^{\frac{1}{2}\Delta k + k_0} \exp(ikx) \, dk = a(x)\exp(ik_0 x)$$

where

$$a(x) = \frac{\sin\left(\frac{1}{2}\Delta k\, x\right)}{\pi x}$$

is the amplitude of the wave packet.

We see that $a(x) = 0$ when $x = \pm 2\pi n/\Delta k$ $(n = 1, 2, ...)$ and that $a(0) = \Delta k/2\pi$.

Also, as $\Delta k \to \infty$ we see that $a(x) \to \delta(x)$ which is the Dirac delta function satisfying $\delta(x) = 0$ for $x \neq 0$ and

$$\int_{-\infty}^{\infty} \delta(x)\, dx = 1.$$

15.04 At zero time a one-dimensional wave packet has the form

$$\psi(x,0) = \frac{1}{[2\pi(\Delta x)^2]^{1/4}} \exp\left[-\frac{x^2}{4(\Delta x)^2}\right]. \tag{15.4}$$

Show that (i) $\psi(x,0)$ is normalized to unity, (ii) $< x >= 0$,   (iii) $< p >= 0$,   and (iv) that it is a minimum packet satisfying

$$\Delta x\, \Delta p = \frac{\hbar}{2}.$$

• *Solution*

(i) Since $\int_{-\infty}^{\infty} \exp(-\alpha x^2)\, dx = (\pi/\alpha)^{1/2}$ we have

$$\int_{-\infty}^{\infty} [\psi(x,0)]^2\, dx = \frac{1}{[2\pi(\Delta x)^2]^{1/2}} \int_{-\infty}^{\infty} \exp\left[-\frac{x^2}{2(\Delta x)^2}\right] dx = 1.$$

(ii) We see that

$$< x >= \int_{-\infty}^{\infty} [\psi(x,0)]^2 x\, dx = \frac{1}{[2\pi(\Delta x)^2]^{1/2}} \int_{-\infty}^{\infty} \exp\left[-\frac{x^2}{2(\Delta x)^2}\right] x\, dx = 0$$

since the integrand is an odd function of x.

(iii) We have

$$< p >= -i\hbar \int_{-\infty}^{\infty} \psi(x,0) \frac{d}{dx}\psi(x,0)\, dx = \frac{i\hbar}{2(\Delta x)^2} \int_{-\infty}^{\infty} [\psi(x,0)]^2 x\, dx = \frac{i\hbar < x >}{2(\Delta x)^2} = 0.$$

(iv) The *root-mean-square deviation* ΔA of A about its expectation value $< A >$ is given by

$$(\Delta A)^2 =< (A- < A >)^2 >$$

and so, using (iii), we have

$$(\Delta p)^2 =< p^2 >= -\hbar^2 \int_{-\infty}^{\infty} \psi(x,0) \frac{d^2}{dx^2}\psi(x,0)\, dx = -\hbar^2 \int_{-\infty}^{\infty} \psi(x,0)\left[-\frac{\psi(x,0)}{2(\Delta x)^2} + \frac{x^2\psi(x,0)}{4(\Delta x)^4}\right] dx.$$

Now $(\Delta x)^2 =< x^2 >$ and so, using (i), we obtain

$$(\Delta p)^2 = \frac{\hbar^2}{4(\Delta x)^2}$$

from which it follows that $\Delta x \Delta p = \hbar/2$ which defines a minimum packet since the *Heisenberg uncertainty principle* shows that $\Delta x \Delta p \geq \hbar/2$.

15.05 By expanding the one-dimensional wave packet $\psi(x,t)$ in terms of momentum eigenfunctions

$$u_k(x) = (2\pi)^{-\frac{1}{2}} \exp(ikx),$$

having energy eigenvalues $E_k = \hbar^2 k^2/2m$, in the form

$$\psi(x,t) = \int_{-\infty}^{\infty} a_k u_k(x) \exp(-iE_k t/\hbar)\, dk$$

where $\psi(x,0)$ is the minimum packet (15.4) defined in the previous problem, show that

$$\psi(x,t) = (2\pi)^{-\frac{1}{4}} \left(\Delta x + \frac{i\hbar t}{2m\Delta x}\right)^{-\frac{1}{2}} \exp\left[-\frac{x^2}{4(\Delta x)^2 + 2i\hbar t/m}\right].$$

Hence establish that the position probability density $|\psi(x,t)|^2$ has the same form as $|\psi(x,0)|^2$ with $(\Delta x)^2$ replaced by $(\Delta x)^2 + (\Delta p)^2 t^2/m^2$.

- *Solution*

The momentum eigenfunctions form the orthonormal set of solutions of the eigenvalue equation

$$-i\hbar \frac{\partial}{\partial x} u_k(x) = p u_k(x) \qquad (15.5)$$

where $p = \hbar k$.

Expanding in the form $\psi(x,t) = \int_{-\infty}^{\infty} a_k u_k(x) \exp(-iE_k t/\hbar)\, dk$ we see that

$$a_k = \int_{-\infty}^{\infty} u_k^*(x)\psi(x,0)\, dx = (2\pi)^{-1/2}[2\pi(\Delta x)^2]^{-1/4} \int_{-\infty}^{\infty} \exp[-x^2/4(\Delta x)^2] \exp(-ikx)\, dx.$$

Now

$$\int_{-\infty}^{\infty} \exp(-\alpha x^2) \exp(\pm i\beta x)\, dx = \int_{-\infty}^{\infty} \exp(-\alpha x^2) \cos\beta x\, dx = \sqrt{\pi/\alpha}\, \exp(-\beta^2/4\alpha) \qquad (15.6)$$

and so

$$a_k = \left[\frac{2(\Delta x)^2}{\pi}\right]^{1/4} \exp[-k^2(\Delta x)^2].$$

Hence

$$\psi(x,t) = (2\pi)^{-1/2} \left[\frac{2(\Delta x)^2}{\pi}\right]^{1/4} \int_{-\infty}^{\infty} \exp[-k^2(\Delta x)^2] \exp(-iE_k t/\hbar + ikx)\, dk$$

and so we find

$$\psi(x,t) = (2\pi)^{-1/2} \left[\frac{2(\Delta x)^2}{\pi}\right]^{1/4} \int_{-\infty}^{\infty} \exp\{-k^2[(\Delta x)^2 + i\hbar t/2m]\} \exp(ikx)\, dk.$$

Now using (15.6) again we get

$$\psi(x,t) = (2\pi)^{-1/2} \left[\frac{2(\Delta x)^2}{\pi}\right]^{1/4} \left\{\frac{\pi}{[(\Delta x)^2 + i\hbar t/2m]}\right\}^{1/2} \exp\left\{-\frac{x^2}{4[(\Delta x)^2 + i\hbar t/2m]}\right\}$$

which gives the required formula for $\psi(x,t)$.

Hence

$$|\psi(x,t)|^2 = (2\pi)^{-1/2} \left[(\Delta x)^2 + \frac{\hbar^2 t^2}{4m^2(\Delta x)^2}\right]^{-1/2} \exp\left\{-\frac{(\Delta x)^2 x^2}{2[(\Delta x)^4 + \hbar^2 t^2/4m^2]}\right\}$$

and since $(\Delta p)^2 = \hbar^2/4(\Delta x)^2$ from part (iv) of the previous problem, we see that

$$|\psi(x,t)|^2 = (2\pi)^{-1/2} \left[(\Delta x)^2 + \frac{(\Delta p)^2 t^2}{m^2}\right]^{-1/2} \exp\left\{-\frac{x^2}{2[(\Delta x)^2 + (\Delta p)^2 t^2/m^2]}\right\}$$

which has the same form as $|\psi(x,0)|^2$ with $(\Delta x)^2$ replaced by $(\Delta x)^2 + (\Delta p)^2 t^2/m^2$.

15.06 Let $u_1(\mathbf{r})$ and $u_2(\mathbf{r})$ be energy eigenfunctions for the same Hamiltonian corresponding to equal energy eigenvalues. Prove that

$$\int [u_1^* \mathbf{r}.\mathbf{p}u_2 + u_1^* \mathbf{p}.(\mathbf{r}u_2)]\, d\tau = 0$$

where $\mathbf{p} = -i\hbar\nabla$ is the linear momentum operator.

• *Solution*

We have

$$-\frac{\hbar^2}{2m}\nabla^2 u_1 + V u_1 = E u_1, \qquad -\frac{\hbar^2}{2m}\nabla^2 u_2 + V u_2 = E u_2$$

and so

$$\int (\mathbf{r}^2 u_1^* \nabla^2 u_2 - \mathbf{r}^2 u_2 \nabla^2 u_1^*)\, d\tau = 0.$$

Now using Green's theorem and the fact that the bound state eigenfunctions give a vanishing contribution to the surface integral as the boundary surface recedes to infinity, we get

$$\int [\mathbf{r}^2 u_1^* \nabla^2 u_2 - u_1^* \nabla^2 (\mathbf{r}^2 u_2)]\, d\tau = 0.$$

But, as in problem 15.02, we have $\nabla^2(\mathbf{r}^2 u_2) = \mathbf{r}^2\nabla^2 u_2 + 2\mathbf{r}.\nabla u_2 + 2\nabla.(\mathbf{r}u_2)$, and so the required result follows.

15.07 Obtain the energy eigenfunctions and eigenvalues for the one-dimensional potential

$$V(x) = \begin{cases} \infty & (x < 0) \\ \frac{1}{2}m\omega^2 x^2 & (x \geq 0) \end{cases}$$

from those for the simple harmonic oscillator by employing considerations of parity.

• *Solution*

The Schrödinger equation for this problem is

$$-\frac{\hbar^2}{2m}\frac{d^2 u}{dx^2} + \frac{1}{2}m\omega^2 x^2 u = E u \quad (0 \leq x < \infty)$$

with the boundary condition $u(0) = 0$.

This is the same as the Schrödinger equation for the simple harmonic oscillator but with the range confined to $x \geq 0$. The boundary condition at $x = 0$ demands only those solutions of the simple harmonic oscillator which have odd parity since these vanish at the origin whilst the even parity solutions do not.

The energy eigenfunctions of the simple harmonic oscillator with odd parity are

$$u_n(x) = \exp\left(-\frac{1}{2}\xi^2\right) H_{2n+1}(\xi) \quad (n = 0, 1, 2...)$$

with energy eigenvalues

$$E_n = \left(2n + 1 + \frac{1}{2}\right)\hbar\omega = \left(2n + \frac{3}{2}\right)\hbar\omega \quad (n = 0, 1, 2...)$$

where $\xi = (m\omega/\hbar)^{1/2}x$ and $H_{2n+1}(\xi)$ is a Hermite polynomial having the parity of $2n+1$ which is odd.

15.08 Show that the three-dimensional harmonic oscillator having the potential

$$V(x, y, z) = \frac{1}{2}m\omega^2(x^2 + y^2 + z^2)$$

has energy eigenvalues

$$E_n = \left(n + \frac{3}{2}\right)\hbar\omega \quad (n = 0, 1, 2, ...)$$

with degeneracy $(n + 1)(n + 2)/2$.

• *Solution*

The Schrödinger equation is

$$\left[-\frac{\hbar^2}{2m}\left(\frac{\partial^2}{\partial x^2} + \frac{\partial^2}{\partial y^2} + \frac{\partial^2}{\partial z^2}\right) + \frac{1}{2}m\omega^2(x^2 + y^2 + z^2)\right]\psi(x, y, z) = E\psi(x, y, z).$$

Separating the variables by writing $\psi(x, y, z) = u_{n_1}(x)u_{n_2}(y)u_{n_3}(z)$ we obtain

$$-\frac{\hbar^2}{2m}\frac{d^2 u_{n_1}}{dx^2} + \frac{1}{2}m\omega^2 x^2 u_{n_1} = E_{n_1}u_{n_1},$$

with two similar equations for u_{n_2} and u_{n_3}, where the energy eigenvalue $E = E_{n_1} + E_{n_2} + E_{n_3}$ and

$$\begin{array}{lll} E_{n_1} = & (n_1 + \frac{1}{2})\hbar\omega & (n_1 = 0, 1, 2, ...), \\ E_{n_2} = & (n_2 + \frac{1}{2})\hbar\omega & (n_2 = 0, 1, 2, ...), \\ E_{n_3} = & (n_3 + \frac{1}{2})\hbar\omega & (n_3 = 0, 1, 2, ...). \end{array}$$

Hence

$$E_n = \left(n + \frac{3}{2}\right)\hbar\omega \quad (n = 0, 1, 2, ...)$$

where $n = n_1 + n_2 + n_3$.

The number of states for the energy level n, that is the *degeneracy* of this level, is

$$\sum_{i=1}^{n+1} i = \frac{1}{2}(n + 1)(n + 2).$$

15.09 A particle in the square well potential with rigid walls

$$V(x) = \begin{cases} 0 & (0 \leq x \leq a) \\ \infty & (x < 0, \ x > a) \end{cases}$$

is described by the wave function

$$\psi(x) = Nx(a - x)$$

where N is a normalization constant.

Find the probability amplitudes for the different energy eigenvalues and also the expectation value of the energy.

• *Solution*

The Schrödinger equation is

$$\left[-\frac{\hbar^2}{2m}\frac{d^2}{dx^2} + V(x)\right]u(x) = Eu(x). \tag{15.7}$$

For $0 \leq x \leq a$ we have

$$\frac{d^2u}{dx^2} + k^2u = 0$$

where $k = (2mE/\hbar^2)^{1/2}$, and the general solution is $u(x) = A\sin kx + B\cos kx$. Now $u(0) = u(a) = 0$ and so $B = 0$ and $\sin ka = 0$ giving $ka = n\pi$ $(n = 1, 2, ...)$. Hence the eigenfunctions are $u_n(x) = A\sin(n\pi x/a)$ and the corresponding energy eigenvalues are

$$E_n = \frac{\pi^2\hbar^2}{2ma^2}n^2 \quad (n = 1, 2, ...).$$

To normalize the eigenfunctions to unity we set

$$\int_0^a [u_n(x)]^2\, dx = 1$$

which gives $A = \sqrt{2/a}$. Thus the normalized eigenfunctions are

$$u_n(x) = \sqrt{\frac{2}{a}}\sin\frac{n\pi x}{a} \quad (n = 1, 2, ...).$$

We first normalize the wave function $\psi(x) = Nx(a - x)$ to unity by setting $\int_0^a |\psi(x)|^2\, dx = 1$. Now $\int_0^a x^2(a - x)^2\, dx = a^5/30$ and so $N^2 = 30/a^5$.

Next we expand $\psi(x)$ in terms of the orthonormal set $u_n(x)$. Thus we write

$$\psi(x) = \sum_{n=1}^{\infty} c_n u_n(x)$$

where the probability amplitudes c_n are given by

$$c_n = (u_n, \psi) = \int_0^a u_n^*(x)\psi(x)\, dx = N\sqrt{\frac{2}{a}}\int_0^a x(a - x)\sin\frac{n\pi x}{a}\, dx.$$

Now, referring to the vibrating string problem 11.01, we have $\int_0^a x(a-x)\sin(n\pi x/a)\, dx = 2(a/n\pi)^3[1 - (-1)^n]$ and so we get for the probability amplitudes

$$c_n = \frac{4\sqrt{15}}{\pi^3}\frac{[1 - (-1)^n]}{n^3}.$$

The expectation value of the energy is

$$< H >= \int_0^a \psi^*(x)H\psi(x)\, dx = -\frac{\hbar^2}{2m}\int_0^a \psi^*(x)\frac{d^2}{dx^2}\psi(x)\, dx = \frac{\hbar^2 N^2}{m}\int_0^a x(a-x)\, dx = \frac{5\hbar^2}{ma^2}.$$

where $H = -(\hbar^2/2m)d^2/dx^2$ is the Hamiltonian for $0 \leq x \leq a$.

Alternatively, since $|c_n|^2$ is the probability of finding the particle in the state n with energy E_n we have, by our fourth quantum theory postulate stated in the preliminary section of this chapter, that

$$< H >= \sum_{n=1}^{\infty} E_n |c_n|^2 = \frac{480\hbar^2}{ma^2\pi^4}\sum_{n=0}^{\infty}\frac{1}{(2n + 1)^4} = \frac{5\hbar^2}{ma^2}$$

since $\sum_{n=0}^{\infty}(2n+1)^{-4} = \pi^4/96$, referring to formula (9.8) given in problem 9.05.

15.10 A particle of mass m is bound in the one-dimensional square-well potential

$$V(x) = \begin{cases} -\lambda^2 & (-a \leq x \leq a) \\ 0 & (|x| > a). \end{cases}$$

Derive an equation for the energy eigenvalues for the case when the parity is even.
Show that in the limit $\lambda^2 \to \infty$ and $a \to 0$ with $\lambda^2 a = \mu$, μ being a given constant, there is just one bound state of even parity. Find the energy eigenvalue and the normalized eigenfunction for this state. Discuss the case $\mu = \frac{1}{2}$.

- *Solution*

We have

$$\frac{d^2u}{dx^2} + \frac{2m}{\hbar^2}(\lambda^2 + E)u = 0 \quad (-a \leq x \leq a),$$

$$\frac{d^2u}{dx^2} = -\frac{2m}{\hbar^2}Eu \quad (|x| > a)$$

where $-\lambda^2 < E < 0$ for bound states.
Setting $k_1 = \sqrt{2m(\lambda^2 + E)}/\hbar$ and $k = \sqrt{-2mE}/\hbar$ we get

$$\frac{d^2u}{dx^2} + k_1^2 u = 0 \quad (-a \leq x \leq a), \qquad \frac{d^2u}{dx^2} = k^2 u \quad (|x| > a).$$

For even parity we see that

$$u(x) = \begin{cases} A\cos k_1 x & (|x| \leq a) \\ B\exp(-k|x|) & (|x| > a) \end{cases}$$

where

$$k_1^2 + k^2 = \frac{2m\lambda^2}{\hbar^2}.$$

The continuity of u and du/dx at $x = a$ gives $A\cos k_1 a = B\exp(-ka)$ and $k_1 A \sin k_1 a = kB\exp(-ka)$ so that

$$k_1 \tan k_1 a = k$$

which determines the energy eigenvalues.
As $\lambda \to \infty$ and $a \to 0$ we see that $\tan k_1 a \to k_1 a$ giving $k_1^2 a = k$ which leads to the single energy eigenvalue

$$E = -\frac{2m}{\hbar^2}\mu^2$$

since $\lambda^2 a = \mu$.
As $a \to 0$, the eigenfunction becomes $u(x) = B\exp(-k|x|)$ for all values of x. Normalization to unity requires $2B^2 \int_0^\infty \exp(-2kx)\,dx = 1$ giving $B^2 = k = \sqrt{-2mE}/\hbar = 2m\mu/\hbar^2$. Thus the normalized eigenfunction is

$$u(x) = \sqrt{\frac{2m\mu}{\hbar^2}}\exp\left(-\frac{2m\mu}{\hbar^2}|x|\right) \quad (-\infty < x < \infty)$$

whose first derivative is discontinuous at the origin.

Taking $\mu = \lambda^2 a = D/2$ we get $V(x) \to -D\delta(x)$ as $\lambda \to \infty$ and $a \to 0$, where $\delta(x)$ is the Dirac delta function. Then $E = -mD^2/2\hbar^2$ and $u(x) = \sqrt{mD/\hbar^2}\exp(-mD\,|x|\,/\hbar^2)$ for all x. If we put $D = 1$ we get $\mu = \frac{1}{2}$ which is the case we were asked to discuss.

15.11 A molecule is composed of two atoms having masses m_1 and m_2 which are rigidly joined together at a constant distance R apart. The molecule rotates about an axis through its centre of mass perpendicular to the internuclear line. Show that the rotational energy eigenvalues for this rigid rotator are given by

$$E_j = \frac{\hbar^2}{2I}j(j+1) \quad (j = 0, 1, 2, ...)$$

where $I = \mu R^2$ is the moment of inertia about the axis of rotation, $\mu = m_1 m_2/(m_1 + m_2)$ being the reduced mass.

- *Solution*

The moment of inertia of the molecule about an axis through its centre of mass G perpendicular to the internuclear line is given by $I = m_1 r_1^2 + m_2 r_2^2$ where $r_1 = m_2 R/(m_1 + m_2)$ and $r_2 = m_1 R/(m_1 + m_2)$ are the distances of the masses m_1 and m_2 from G respectively. Hence

$$I = \frac{m_1 m_2}{m_1 + m_2}R^2 = \mu R^2.$$

The square of the total angular momentum operator in spherical polar coordinates θ, ϕ

$$\mathbf{M}^2 = -\hbar^2 \left[\frac{1}{\sin\theta}\frac{\partial}{\partial\theta}\left(\sin\theta\frac{\partial}{\partial\theta}\right) + \frac{1}{\sin^2\theta}\frac{\partial^2}{\partial\phi^2} \right]$$

has eigenfunctions which are the spherical harmonics $Y_j^m(\theta, \phi)$ and eigenvalues $j(j+1)\hbar^2$ $(j = 0, 1, 2, ...)$.

Hence the eigenvalues of the Hamiltonian operator

$$H = \frac{1}{2I}\mathbf{M}^2$$

are

$$E_j = \frac{\hbar^2}{2I}j(j+1) \quad (j = 0, 1, 2, ...)$$

and these are the energy eigenvalues for the rigid rotator representing the rotational motion of the molecule about an axis through its centre of mass perpendicular to the internuclear line.

15.12 A particle of mass m is bound with energy E inside a three-dimensional square well potential

$$V(r) = \begin{cases} 0 & (r \le a) \\ \infty & (r > a). \end{cases}$$

Show that the radial wave function for the particle is $R(r) = Aj_l(kr)$ where $k = (2mE/\hbar^2)^{\frac{1}{2}}$ and j_l is a spherical Bessel function.

Further show that when $l = 0$ we have $R(r) = Aj_0(kr)$ where $j_0(\rho) = \sin\rho/\rho$. Deduce that $ka = n\pi$ $(n = 1, 2, ...)$ and show that the energy eigenvalues are $E_n = n^2\pi^2\hbar^2/2ma^2$ for zero angular momentum.

● *Solution*

The radial equation for a particle of mass m moving in a spherically symmetrical potential $V(r)$ is

$$\frac{1}{r^2}\frac{d}{dr}\left(r^2\frac{dR}{dr}\right) + \left\{\frac{2m}{\hbar^2}[E - V(r)] - \frac{l(l+1)}{r^2}\right\}R = 0 \tag{15.8}$$

where $R(r)$ is the radial wave function and r is the radial distance from the origin.

In the present problem, for $r \leq a$, this may be written

$$\frac{d^2R}{d\rho^2} + \frac{2}{\rho}\frac{dR}{d\rho} + \left[1 - \frac{l(l+1)}{\rho^2}\right]R = 0$$

where $\rho = kr$.

If we now make the substitution $R = \rho^{-1/2}J(\rho)$ we obtain

$$\frac{d^2J}{d\rho^2} + \frac{1}{\rho}\frac{dJ}{d\rho} + \left[1 - \frac{l(l+1) + \frac{1}{4}}{\rho^2}\right]J = 0$$

which is a special case of Bessel's equation

$$\frac{d^2y}{dv^2} + \frac{1}{v}\frac{dy}{dv} + \left(1 - \frac{n^2}{v^2}\right)y = 0$$

with $n = l + \frac{1}{2}$.

Now $R(r)$ must be finite at the origin $r = 0$ and so $J(\rho)$ must vanish at $\rho = 0$. Hence we have $R(r) = Aj_l(kr)$ where

$$j_l(\rho) = \sqrt{\frac{\pi}{2\rho}}J_{l+\frac{1}{2}}(\rho)$$

is a *spherical Bessel function*.

For $l = 0$ the radial equation becomes

$$\frac{d^2R}{d\rho^2} + \frac{2}{\rho}\frac{dR}{d\rho} + R = 0.$$

Putting $R(\rho) = \rho^{-1}y(\rho)$ we get

$$\frac{d^2y}{dx^2} + y(\rho) = 0$$

and so, since $y(\rho)$ must vanish at the origin, we obtain $y(\rho) = A\sin\rho$. Hence

$$R(r) = A\frac{\sin kr}{kr} = Aj_0(kr).$$

Since $R(r)$ must vanish when $r = a$ it follows that $\sin ka = 0$ and hence $ka = n\pi$ $(n = 1, 2, ...)$ which gives for the energy eigenvalues $E_n = \hbar^2k^2/2m = \hbar^2\pi^2n^2/2ma^2$, as required.

15.13 If $[\nabla^2 + V(r)]\psi_1 = E_1\psi_1$ and $[\nabla^2 + V(r)]\psi_2 = E_2\psi_2$ where E_1, E_2 are constants and $r\psi_i$, $r^2\partial\psi_i/\partial r$ $(i = 1, 2)$ are bounded as $r \to \infty$, show that

$$2\int \psi_1\nabla\psi_2\, d\tau = (E_1 - E_2)\int \mathbf{r}\psi_1\psi_2\, d\tau$$

the integrations being over all space.

- *Solution*

We have

$$\int (r\psi_2 \nabla^2 \psi_1 - \mathbf{r}\psi_1 \nabla^2 \psi_2)\, d\tau = (E_1 - E_2) \int \mathbf{r}\psi_1 \psi_2\, d\tau$$

and using Green's theorem we obtain

$$\int \psi_1[\nabla^2(\mathbf{r}\psi_2) - \mathbf{r}\nabla^2 \psi_2]\, d\tau = (E_1 - E_2) \int \mathbf{r}\psi_1 \psi_2\, d\tau$$

since the surface integral vanishes in the limit as the boundary surface recedes to infinity.

Now $\nabla^2(\mathbf{r}\psi_2) = \mathbf{r}\nabla^2 \psi_2 + 2\nabla\psi_2.\nabla\mathbf{r} = \mathbf{r}\nabla^2\psi_2 + 2\nabla\psi_2$ since $\mathbf{a}.\nabla\mathbf{r} = \mathbf{a}$, shown in problem 4.01, and so the required result follows.

15.14 The momentum distribution for a state with normalized wave function $u(r)$ is given by

$$g(\mathbf{p}) = \int u_{\mathbf{p}}^*(\mathbf{r}) u(\mathbf{r})\, d\tau$$

where

$$u_{\mathbf{p}}(\mathbf{r}) = (2\pi\hbar)^{-3/2} \exp(i\mathbf{p}.\mathbf{r}/\hbar)$$

is the normalized wave function for linear momentum $\mathbf{p}$.

Show that the momentum distribution of an electron in the 1s ground state of a hydrogenic atom is

$$\frac{2}{\pi} \left(\frac{2Z^5\hbar^5}{a_0^5} \right)^{\frac{1}{2}} \left[\left(\frac{Z\hbar}{a_0} \right)^2 + p^2 \right]^{-2}$$

where Ze is the nuclear charge and $a_0 = \hbar^2/me^2$ is the Bohr radius.

- *Solution*

The 1s ground state wave function for a hydrogenic atom is

$$u_{1s}(\mathbf{r}) = \left(\frac{Z^3}{\pi a_0^3} \right)^{\frac{1}{2}} \exp\left(-\frac{Zr}{a_0} \right).$$

Hence

$$g_{1s}(\mathbf{p}) = \frac{1}{(2\pi\hbar)^{3/2}} \left(\frac{Z^3}{\pi a_0^3} \right)^{\frac{1}{2}} \int \exp(-i\mathbf{p}.\mathbf{r}/\hbar)\exp(-Zr/a_0)\, d\tau.$$

Taking the polar axis in the direction of $\mathbf{p}$, integrating over the azimuthal angle ϕ and putting $\mu = \cos\theta$ we get

$$g_{1s}(\mathbf{p}) = \frac{2\pi}{(2\pi\hbar)^{3/2}} \left(\frac{Z^3}{\pi a_0^3} \right)^{\frac{1}{2}} \int_0^\infty \exp(-Zr/a_0)\, r^2\, dr \int_{-1}^1 \exp(-ipr\mu/\hbar)\, d\mu$$

and this gives

$$g_{1s}(\mathbf{p}) = \frac{2\hbar}{\pi p} \left(\frac{Z^3}{2\hbar^3 a_0^3} \right)^{\frac{1}{2}} \int_0^\infty \exp(-Zr/a_0)\, r \sin(pr/\hbar)\, dr.$$

Now $\int_0^\infty \exp(-\alpha r)\, r \sin(kr)\, dr = 2\alpha k[\alpha^2 + k^2]^{-2}$ and so the required result follows.

15.15 A particle of mass m and charge q is in one-dimensional motion under the action of a uniform electric field F. By solving the time-independent Schrödinger equation in the momentum representation and then transforming the resulting wave function to the coordinate representation, show that the wave function $\psi(x)$ for a particle with total energy E is proportional to the Airy function $\Phi(-y)$ where

$$\Phi(y) = \frac{1}{\sqrt{\pi}} \int_0^\infty \cos\left(\frac{u^3}{3} + uy\right) du$$

and $y = (2mF/\hbar^2)^{1/3}(E/F + x)$.

• *Solution*

The potential corresponding to a uniform electric field $F = -dV/dx$ is $V(x) = -Fx$ and so the Schrödinger equation in the coordinate representation is

$$\frac{-\hbar^2}{2m}\frac{d^2\psi}{dx^2} - Fx\psi = E\psi.$$

In the momentum representation the wave function is given by the Fourier transform

$$g(p) = \frac{1}{(2\pi\hbar)^{1/2}} \int_{-\infty}^{\infty} \psi(x)\exp(-ipx/\hbar)\,dx.$$

Then

$$i\hbar\frac{d}{dp}g(p) = \frac{1}{(2\pi\hbar)^{1/2}} \int_{-\infty}^{\infty} x\psi(x)\exp(-ipx/\hbar)\,dx$$

and so $i\hbar d/dp$ is the operator in momentum space which represents the coordinate x.

Further, the kinetic energy $p^2/2m$ corresponds to the operator $-(\hbar^2/2m)d^2/dx^2$, and so we may put

$$\frac{p^2}{2m}g(p) = \frac{1}{(2\pi\hbar)^{1/2}} \int_{-\infty}^{\infty} \left[-\frac{\hbar^2}{2m}\frac{d^2\psi(x)}{dx^2}\right]\exp(-ipx/\hbar)\,dx.$$

Hence in momentum space we have

$$\frac{p^2}{2m}g(p) - i\hbar F\frac{dg}{dp} = Eg(p)$$

and this can be readily integrated to give $i\hbar F \ln(g/C) = p^3/6m - Ep$ where C is a constant. Thus we obtain

$$g(p) = C\exp\left[\frac{1}{i\hbar F}\left(\frac{p^3}{6m} - Ep\right)\right].$$

Using the reciprocal Fourier transform we have

$$\psi(x) = \frac{1}{(2\pi\hbar)^{1/2}} \int_{-\infty}^{\infty} g(p)\exp(ipx/\hbar)\,dp$$

which gives

$$\psi(x) = \frac{C}{(2\pi\hbar)^{1/2}} \int_{-\infty}^{\infty} \exp\left\{-i\left[\frac{p^3}{6m\hbar F} - \frac{p}{\hbar}\left(\frac{E}{F} + x\right)\right]\right\}dp.$$

We now set $p = (2m\hbar F)^{1/3}u$ and since the imaginary part of the above integral is zero we see that

$$\psi(x) = \left(\frac{2}{\hbar}\right)^{1/2} (2m\hbar F)^{1/3} C\pi^{-1/2} \int_0^\infty \cos\left(\frac{u^3}{3} - yu\right) du = [(2/\hbar)^{1/2}(2m\hbar F)^{1/3}C]\Phi(-y)$$

where $\Phi(y)$ is the Airy integral, and so the required result is obtained.

15-2 Variational methods

The next few problems involve the use of *variational methods* together with *trial functions*. These could have been placed in the final chapter on *Variational principles* together with some other problems of the same kind but it is convenient to examine them here as they are essentially problems in quantum theory.

15.16 A particle of mass m is moving under the action of a force having the potential

$$V(r) = -Ar^{-1}\exp(-\lambda r).$$

Use the variational method for bound states with the trial function

$$\psi_t = \exp(-\alpha r),$$

α being an adjustable parameter, to find a least upper bound to the ground state energy eigenvalue.

• *Solution*

We use the variational method with a trial function ψ_t to obtain an upper bound to the ground state energy given by $E_t = I[\psi_t]$ where $I[\chi]$ is a functional defined by

$$I[\chi] = \frac{\int \chi^* H \chi \, d\tau}{\int \chi^* \chi \, d\tau} \tag{15.9}$$

which depends on the function χ, and

$$H = -\frac{\hbar^2}{2m}\nabla^2 + V(r)$$

is the Hamiltonian operator.

We have $\int \psi_t^* \psi_t \, d\tau = 4\pi \int_0^\infty \exp(-2\alpha r)\, r^2 \, dr = \pi/\alpha^3$ and

$$\int \psi_t^* V(r)\psi_t \, d\tau = -4\pi A \int_0^\infty \exp[-(2\alpha + \lambda)r]\, r \, dr = -\frac{4\pi A}{(2\alpha + \lambda)^2}.$$

Also

$$\int \psi_t^* \nabla^2 \psi_t \, d\tau = 4\pi \int_0^\infty \exp(-\alpha r)\frac{1}{r^2}\frac{d}{dr}\left[r^2 \frac{d}{dr}\exp(-\alpha r)\right]\, r^2 \, dr$$

so that we have

$$\int \psi_t^* \nabla^2 \psi_t \, d\tau = 4\pi \int_0^\infty (\alpha^2 r^2 - 2\alpha r)\exp(-2\alpha r)\, dr = -\frac{\pi}{\alpha}.$$

Hence, using (15.9) with χ replaced by ψ_t, we arrive at

$$E_t = \frac{\hbar^2}{2m}\alpha^2 - \frac{4A\alpha^3}{(2\alpha + \lambda)^2}.$$

To obtain the least upper bound to the ground state energy for the given trial function we set $\partial E_t/\partial \alpha = 0$. Now

$$\frac{\partial E_t}{\partial \alpha} = \frac{\hbar^2}{m}\alpha\left\{1 - B\left[\frac{3\alpha}{(2\alpha + \lambda)^2} - \frac{4\alpha^2}{(2\alpha + \lambda)^3}\right]\right\}$$

where $B = 4Am/\hbar^2$ and so we get

$$B = \frac{(2\alpha + \lambda)^3}{\alpha(2\alpha + 3\lambda)}$$

from which it follows that the least upper bound to the ground state energy is given by

$$E_t = -\frac{\hbar^2}{2m}\frac{\alpha^2(2\alpha - \lambda)}{2\alpha + 3\lambda}.$$

15.17 A particle of mass m is moving under the action of the attractive potential

$$V(r) = -\frac{Ze^2}{a_0}\exp(-\lambda r/a_0)$$

where $a_0 = \hbar^2/me^2$ is the Bohr radius. Using the variational method for bound states with the trial function

$$\psi_t(r) = \exp(-\alpha r/a_0),$$

show that an upper bound to the ground state energy eigenvalue is given by

$$-\frac{e^2}{6a_0}\frac{\alpha^2(4\alpha - \lambda)}{\lambda}$$

where α satisfies

$$24Z\alpha\lambda = (2\alpha + \lambda)^4.$$

Deduce that there exists a bound state of the particle with energy $E_0 < 0$ if $Z > 27\lambda^2/32$.

• *Solution*

In this problem we have $\int \psi_t^* \psi_t \, d\tau = 4\pi \int_0^\infty \exp(-2\alpha r/a_0)\, r^2\, dr = \pi a_0^3/\alpha^3$ and

$$\int \psi_t^* H \psi_t \, d\tau = -\frac{\hbar^2}{2m}\int \psi_t^*\left[\nabla^2 + \frac{2Z}{a_0^2}\exp(-\lambda r/a_0)\right]\psi_t \, d\tau$$

which gives, on doing the differentiations,

$$\int \psi_t^* H \psi_t \, d\tau = -4\pi\frac{e^2}{2a_0}\int_0^\infty \left[\alpha^2 r^2 - 2\alpha a_0 r + 2Zr^2\exp(-\lambda r/a_0)\right]\exp(-2\alpha r/a_0)\, dr.$$

Now carrying out the integrations we get

$$\int \psi_t^* H \psi_t \, d\tau = -4\pi\frac{e^2}{2a_0}a_0^3\left[-\frac{1}{4\alpha} + \frac{4Z}{(2\alpha + \lambda)^3}\right]$$

and so, using (15.9), we have

$$E_t = I[\psi_t] = -\frac{e^2}{2a_0}\left[-\alpha^2 + \frac{16Z\alpha^3}{(2\alpha + \lambda)^3}\right].$$

The optimum value of α is given by $\partial I/\partial\alpha = 0$ so that we get $-(2\alpha + \lambda)^4 + 24Z\alpha(2\alpha + \lambda) - 48Z\alpha^2 = 0$ and hence $24Z\alpha\lambda = (2\alpha + \lambda)^4$ as required. It follows that the least upper bound to the ground state energy is

$$E_t = -\frac{e^2}{2a_0}\left[-\alpha^2 + \frac{2\alpha^2(2\alpha + \lambda)}{3\lambda}\right] = -\frac{e^2}{6a_0}\frac{\alpha^2(4\alpha - \lambda)}{\lambda}.$$

From this we see that $I[\psi_t] = 0$ for $\alpha = \lambda/4$ and then we have $Z = 27\lambda^2/32$. Now taking $\alpha = \lambda/4$ we obtain for arbitrary Z

$$I[\psi_t] = -\frac{e^2}{2a_0}\left[-\frac{\lambda^2}{16} + \frac{2Z}{27}\right]$$

and hence $I[\psi_t] < 0$ if $Z > 27\lambda^2/32$. Thus, if this is satisfied, there exists a bound state since $I[\psi_t]$ is an upper bound to the ground state energy E_0.

15.18 A particle of mass m is bound by an attractive potential

$$V(r) = -\frac{Ze^2}{a_0}\frac{\exp(-\lambda r/a_0)}{1 - \exp(-\lambda r/a_0)}.$$

Using the variational method for bound states with the trial function

$$\psi_t(r) = r^{-1}\exp(-\alpha r/a_0)[1 - \exp(-\lambda r/a_0)]$$

show that

$$-\frac{e^2}{2a_0}\left\{\alpha^2 + \frac{\alpha}{\lambda}[2Z + \alpha^2 - (\alpha + \lambda)^2]\right\}$$

is an upper bound to the ground state energy eigenvalue.

Show further that the optimum value of the parameter α is given by $2Z = \lambda^2 + 2\alpha\lambda$ and that for this value of α the trial function is an exact solution of the Schrödinger equation with the energy eigenvalue $-\alpha^2 e^2/2a_0$.

- *Solution*

We have

$$\int \psi_t^* \psi_t \, d\tau = 4\pi\int_0^\infty \exp(-2\alpha r/a_0)\left[1 - \exp(-\lambda r/a_0)\right]^2 \, dr = 4\pi a_0\left[\frac{1}{2\alpha} - \frac{2}{2\alpha + \lambda} + \frac{1}{2(\alpha + \lambda)}\right]$$

and

$$\int \psi_t^*(r)H\psi_t(r)\, d\tau = -\frac{\hbar^2}{2m}4\pi\int_0^\infty \psi_t^*\left[\frac{1}{r^2}\frac{d}{dr}\left(r^2\frac{d\psi_t}{dr}\right) + \frac{2Z}{a_0^2}\frac{\exp(-\lambda r/a_0)\psi_t}{1 - \exp(-\lambda r/a_0)}\right] r^2\, dr$$

which, on doing the differentiations and integrations, yields

$$\int \psi_t^*(r)H\psi_t(r)\, d\tau = -\frac{\hbar^2}{2m}\frac{4\pi}{a_0}\left\{\alpha^2\left(\frac{1}{2\alpha} - \frac{1}{2\alpha + \lambda}\right) + \left[2Z - (\alpha + \lambda)^2\right]\left(\frac{1}{2\alpha + \lambda} - \frac{1}{2\alpha + 2\lambda}\right)\right\}$$

and thus, using (15.9), we have

$$I[\psi_t] = -\frac{e^2}{2a_0}\left\{\alpha^2 + \frac{\alpha}{\lambda}\left[2Z - (\alpha + \lambda)^2 + \alpha^2\right]\right\}$$

and this is an upper bound to the ground state energy eigenvalue.

The optimum value of the parameter α is given by $\partial I[\psi_t]/\partial\alpha = 0$ which leads to the required formula $2Z = \lambda^2 + 2\alpha\lambda$.

Now

$$H\psi_t = -\frac{\hbar^2}{2m}\frac{1}{r}\left\{\frac{\alpha^2}{a_0^2}\exp\left(-\frac{\alpha r}{a_0}\right)\left[1 - \exp\left(-\frac{\lambda r}{a_0}\right)\right] + \left[\frac{2Z}{a_0^2} - \frac{(\alpha + \lambda)^2}{a_0^2} + \frac{\alpha^2}{a_0^2}\right]\exp\left[-\frac{(\alpha + \lambda)r}{a_0}\right]\right\}$$

and hence, if $2Z = (\alpha + \lambda)^2 - \alpha^2 = \lambda^2 + 2\alpha\lambda$ we see that $H\psi_t = -(e^2/2a_0)\alpha^2\psi_t$ and so ψ_t is an energy eigenfunction with energy eigenvalue

$$-\frac{e^2}{2a_0}\alpha^2 = -\frac{e^2}{2a_0}\left(\frac{Z}{\lambda} - \frac{\lambda}{2}\right)^2.$$

15.19 A particle of mass m is in one-dimensional motion under the action of the potential $V(x) = kx^4$. Using the variational method with the trial function

$$u_t(x) = (1 + a^2x^2)^{-2}$$

where a is an adjustable parameter, find a least upper bound to the energy of the ground state.

• *Solution*

It can be readily shown by integrating by parts that the integral

$$I_n = \int_{-\infty}^{\infty} \frac{dz}{(1 + z^2)^n}$$

satisfies $2(n-1)I_n = (2n-3)I_{n-1}$ and so, since $I_1 = \pi$, it follows that

$$I_n = \frac{(2n-3)...3.1}{(n-1)! \, 2^{n-1}}\pi \quad (n = 2, 3, ...). \tag{15.10}$$

Hence we have

$$\int_{-\infty}^{\infty} u_t^* u_t \, dx = \int_{-\infty}^{\infty} \frac{dx}{(1 + a^2x^2)^4} = \frac{I_4}{a} = \frac{5\pi}{16a}$$

and

$$\int_{-\infty}^{\infty} u_t^* H u_t \, dx = -\frac{\hbar^2}{2m}\int_{-\infty}^{\infty} \frac{1}{(1 + a^2x^2)^2} \frac{d^2}{dx^2}\left[\frac{1}{(1 + a^2x^2)^2}\right] dx + k\int_{-\infty}^{\infty} \frac{x^4}{(1 + a^2x^2)^4} \, dx.$$

Carrying out the differentiations and integrations using (15.10) we find that

$$\int_{-\infty}^{\infty} u_t^* H u_t \, dx = \frac{\hbar^2}{2m}\frac{7a\pi}{16} + \frac{k}{a^4}\frac{\pi}{16a}$$

which gives, using the one-dimensional form of (15.9),

$$E_t = I[u_t] = \frac{1}{5}\left(\frac{\hbar^2}{2m}7a^2 + \frac{k}{a^4}\right).$$

The optimum value of a is obtained by taking $\partial E_t/\partial a = 0$ which yields $a^6 = 4km/7\hbar^2$ from which it follows that the least upper bound to the ground state energy for the given trial function is

$$E_t = \frac{21\hbar^2 a^2}{20m} = \frac{21}{20}\left(\frac{4\hbar^4 k}{7m^2}\right)^{1/3}.$$

15.20 By applying the variational method to a system with Hamiltonian operator H using the trial function

$$\psi_t = c_0\psi_0 + c_1\psi_1$$

where ψ_0 and ψ_1 are real orthonormal functions and c_0 and c_1 are arbitrary parameters, show that the optimum values of

$$E_t = \frac{\int \psi_t H \psi_t d\tau}{\int (\psi_t)^2 d\tau}$$

are given by

$$E_t = \frac{1}{2}(H_{00} + H_{11}) \pm \frac{1}{2}[(H_{00} - H_{11})^2 + 4H_{01}^2]^{\frac{1}{2}}$$

where $H_{ij} = \int \psi_i H \psi_j d\tau$.

• *Solution*

Using (15.9) we see that

$$E_t = I[\psi_t] = \frac{c_0^2 H_{00} + c_1^2 H_{11} + 2c_0 c_1 H_{01}}{c_0^2 + c_1^2}$$

and so $(E_t - H_{00})c_0^2 - 2H_{01}c_0 c_1 + (E_t - H_{11})c_1^2 = 0$. We optimize by taking $\partial E_t / c_0 = 0$ and $\partial E_t / c_1 = 0$. These give

$$
\begin{aligned}
c_0(E_t - H_{00}) \quad &-c_1 H_{01} \quad &= 0 \\
-c_0 H_{01} \quad &+c_1(E_t - H_{11}) \quad &= 0
\end{aligned}
$$

which are consistent if

$$
\begin{vmatrix}
E_t - H_{00} & -H_{01} \\
-H_{01} & E_t - H_{11}
\end{vmatrix} = 0.
$$

This secular equation can be written $(E_t - H_{00})(E_t - H_{11}) - H_{01}^2 = 0$, that is

$$E_t^2 - (H_{00} + H_{11})E_t + H_{00}H_{11} - H_{01}^2 = 0$$

whose solutions are the required optimum values of the energy.

15.21 A particle of mass m and charge e moves as a linear harmonic oscillator with classical angular frequency ω. By using the variational method with a trial function composed of a linear combination of the $n = 0$ and $n = 1$ unperturbed eigenfunctions of the harmonic oscillator, show that an electric field F acting on the particle in the $n = 0$ ground state causes the energy to change by $-\frac{1}{2}\alpha_p F^2$ to the second order in F where $\alpha_p = e^2/m\omega^2$ is the polarizability.

• *Solution*

The Hamiltonian for this problem is $H = H_0 + H'$ where

$$H_0 = -\frac{\hbar^2}{2m}\frac{d^2}{dx^2} + \frac{1}{2}m\omega^2 x^2 \tag{15.11}$$

is the unperturbed Hamiltonian and $H' = -eFx$ is the perturbation.

We take the trial function to be $u_t = c_0 u_0 + c_1 u_1$ where $u_0(\xi) = (m\omega/\hbar\pi)^{1/4}\exp(-\xi^2/2)$ and $u_1(\xi) = (m\omega/4\hbar\pi)^{1/4}2\xi\exp(-\xi^2/2)$, with $\xi = (m\omega/\hbar)^{1/2}x$, are the $n = 0$ and $n = 1$ unperturbed normalized eigenfunctions of the harmonic oscillator with energy eigenvalues $E_0 = \hbar\omega/2$ and $E_1 = 3\hbar\omega/2$ respectively.

Then
$$\int_{-\infty}^{\infty} u_t^* u_t \, dx = c_0^2 + c_1^2$$

and
$$\int_{-\infty}^{\infty} u_t^* H u_t \, dx = c_0^2 (E_0 + H'_{00}) + c_1^2 (E_1 + H'_{11}) + 2c_0 c_1 H'_{01}$$

where $H'_{ij} = \int u_i H' u_j \, dx$. Hence, using the one-dimensional form of (15.9), we have

$$E_t = \frac{c_0^2 (E_0 + H'_{00}) + c_1^2 (E_1 + H'_{11}) + 2c_0 c_1 H'_{01}}{c_0^2 + c_1^2}$$

and so $c_0^2 (E_t - E_0 - H'_{00}) + c_1^2 (E_t - E_1 - H'_{11}) - 2c_0 c_1 H'_{01} = 0$.

Optimizing with respect to c_0 and c_1 by taking $\partial E_t / c_0 = 0$ and $\partial E_t / c_1 = 0$ we obtain

$$\begin{aligned}
c_0(E_t - E_0 - H'_{00}) \quad -c_1 H'_{01} &= 0, \\
-c_0 H'_{01} \qquad\quad +c_1(E_t - E_1 - H'_{11}) &= 0
\end{aligned}$$

which yields the secular equation

$$\begin{vmatrix} E_t - E_0 - H'_{00} & -H'_{01} \\ -H'_{01} & E_t - E_1 - H'_{11} \end{vmatrix} = 0,$$

that is $(E_t - E_0 - H'_{00})(E_t - E_1 - H'_{11}) - (H'_{01})^2 = 0$.

Now $H'_{00} = -eF \int_{-\infty}^{\infty} u_0^2 \, x \, dx = 0$ and $H'_{11} = -eF \int_{-\infty}^{\infty} u_1^2 \, x \, dx = 0$ since the integrands are both odd functions of x.

Hence the secular equation becomes $(E_t - E_0)(E_t - E_1) - (H'_{01})^2 = 0$ which gives $E_t^2 - (E_0 + E_1)E_t + E_0 E_1 - (H'_{01})^2 = 0$ and so we get

$$E_t = \frac{1}{2} \left\{ E_0 + E_1 \pm [(E_1 - E_0)^2 + 4(H'_{01})^2]^{1/2} \right\}$$

where $H'_{01} = \int_{-\infty}^{\infty} u_0 H' u_1 \, dx = -eF(\hbar/m\omega)(2m\omega/\hbar\pi)^{1/2} \int_{-\infty}^{\infty} \xi^2 \exp(-\xi^2) \, d\xi$ which gives $H'_{01} = -eF(\hbar/2m\omega)^{1/2}$. Hence

$$E_t = \frac{1}{2} \left\{ E_0 + E_1 \pm (E_1 - E_0)[1 + 2\alpha_p F^2 / \hbar\omega]^{1/2} \right\}$$

and so, to the second-order in F, we get that $E_t \simeq E_0 - \frac{1}{2}\alpha_p F^2$ or $E_1 + \frac{1}{2}\alpha_p F^2$ where $\alpha_p = e^2 / m\omega^2$ is the polarizability.

Thus the energy of the ground $n = 0$ state is reduced by $-\frac{1}{2}\alpha_p F^2$ due to the presence of the electric field F.

15-3 Time-independent perturbation theory

15.22 A particle of mass m is confined within a one-dimensional box such that the potential is a square well with rigid walls given by

$$V(x) = \begin{cases} 0 & (0 \le x \le a \\ \infty & (x < 0, \ x > a). \end{cases}$$

If a small perturbing potential λx is applied to the particle in the box, show that the approximate increase in energy of the particle in the nth state is

$$\frac{1}{2}\lambda a + \frac{128m\lambda^2 a^4}{\pi^6 \hbar^2} \sum_{\substack{r \\ n-r \text{ odd}}} \frac{n^2 r^2}{(n^2 - r^2)^5}$$

where the summation is for $n - r$ odd.

• *Solution*

This is a problem on *time-independent perturbation theory* in which the energy is expanded in the form

$$E = E^{(0)} + \lambda E^{(1)} + \lambda^2 E^{(2)} + ... \tag{15.12}$$

and the eigenfunction is expanded as

$$\psi = \psi^{(0)} + \lambda \psi^{(1)} + \lambda^2 \psi^{(2)} + ... \tag{15.13}$$

From problem 15.09 we have that the unperturbed normalized energy eigenfunctions and eigenvalues are

$$\psi_n^{(0)}(x) = \sqrt{\frac{2}{a}} \sin \frac{n\pi x}{a}, \quad E_n^{(0)} = \frac{\pi^2 \hbar^2}{2ma^2} n^2 \quad (n = 1, 2, ...).$$

The first-order energy correction is

$$E_n^{(1)} = (\psi_n^{(0)}, \lambda x \psi_n^{(0)}) = \frac{2\lambda}{a} \int_0^a x \sin^2 \frac{n\pi x}{a} \, dx = \frac{1}{2}\lambda a.$$

The second-order energy correction is

$$E_n^{(2)} = -\sum_r{}' \frac{\left|(\psi_n^{(0)}, \lambda x \psi_r^{(0)})\right|^2}{E_r^{(0)} - E_n^{(0)}} \tag{15.14}$$

where the prime denotes that the sum excludes $r = n$.

We have

$$(\psi_n^{(0)}, \lambda x \psi_r^{(0)}) = \frac{2\lambda}{a} \int_0^a x \sin \frac{n\pi x}{a} \sin \frac{r\pi x}{a} \, dx = \frac{\lambda}{a} \int_0^a x \left[\cos \frac{(n-r)\pi x}{a} - \cos \frac{(n+r)\pi x}{a} \right] dx$$

and if $n - r$ is even this is zero, while if $n - r$ is odd we get

$$(\psi_n^{(0)}, \lambda x \psi_r^{(0)}) = -\frac{8\lambda a}{\pi^2} \frac{nr}{(n^2 - r^2)^2} \quad (n - r \text{ odd}).$$

Hence

$$E_n^{(2)} = \sum_{\substack{r \\ n-r \text{ odd}}} \frac{2ma^2}{\pi^2 \hbar^2 (n^2 - r^2)} \frac{64\lambda^2 a^2}{\pi^4} \frac{n^2 r^2}{(n^2 - r^2)^4} = \frac{128 m \lambda^2 a^4}{\pi^6 \hbar^2} \sum_{\substack{r \\ n-r \text{ odd}}} \frac{n^2 r^2}{(n^2 - r^2)^5}$$

and so the required formula for the approximate increase in the energy of the particle in the nth state is obtained.

15.23 A linear harmonic oscillator with classical angular frequency ω and mass m is in its ground state. The oscillator is subjected to the perturbation $H' = -eFx$. Show that $E^{(1)} = 0$ and find the differential equation satisfied by $\psi^{(1)}(x)$. Set $\psi^{(1)}(x) = f(x)\psi^{(0)}(x)$ and show that

$$\frac{d^2 f}{dx^2} - 2\alpha^2 x \frac{df}{dx} = -\frac{2meF}{\hbar^2} x \tag{15.15}$$

where $\alpha = (m\omega/\hbar)^{1/2}$. Hence obtain $\psi^{(1)}(x)$ and so evaluate $E^{(2)}$.

• *Solution*

The first order energy is given by $E^{(1)} = (\psi^{(0)}, H'\psi^{(0)})$ where $\psi^{(0)}(x) = (\alpha/\pi^{1/2})^{1/2}\exp(-\frac{1}{2}\alpha^2x^2)$ is the unperturbed ground $n = 0$ state wave function of the harmonic oscillator. Hence, as required, we get

$$E^{(1)} = -\frac{\alpha eF}{\pi^{1/2}}\int_{-\infty}^{\infty} x\exp(-\alpha^2x^2)\,dx = 0$$

since the integrand is an odd function of x.

The equation satisfied by $\psi^{(1)}(x)$ is

$$(H_0 - E^{(0)})\psi^{(1)} = (E^{(1)} - H')\psi^{(0)} \tag{15.16}$$

where the unperturbed Hamiltonian H_0 for the harmonic oscillator is given by (15.11) and the unperturbed energy of the $n = 0$ state is $E^{(0)} = \hbar\omega/2$. It follows that

$$\frac{d^2\psi^{(1)}}{dx^2} - \alpha^4x^2\psi^{(1)} + \alpha^2\psi^{(1)} = -\frac{2meF}{\hbar^2}x\psi^{(0)}$$

and putting $\psi^{(1)}(x) = f(x)\psi^{(0)}(x)$ we obtain equation (15.15) using $d^2\psi^{(0)}/dx^2 - \alpha^4x^2\psi^{(0)} + \alpha^2\psi^{(0)} = 0$.

A particular solution of (15.15) is $f(x) = eFx/\hbar\omega$ and so we get $\psi^{(1)}(x) = eFx\psi^{(0)}(x)/\hbar\omega$.

If the perturbation series (15.13) for the wave function is chosen so that $(\psi^{(0)}, \psi^{(r)}) = 0$ then

$$E^{(r)} = (\psi^{(0)}, H'\psi^{(r-1)}) \tag{15.17}$$

follows from the perturbation series (15.12) and (15.13), and so we have $E^{(2)} = (\psi^{(0)}, H'\psi^{(1)})$. Hence

$$E^{(2)} = -\frac{e^2F^2}{\hbar\omega}\frac{\alpha}{\pi^{1/2}}\int_{-\infty}^{\infty} x^2\exp(-\alpha^2x^2)\,dx$$

and since $\int_{-\infty}^{\infty} x^2\exp(-\alpha^2x^2)\,dx = \frac{1}{2}\pi^{1/2}/\alpha^3$ we obtain $E^{(2)} = -e^2F^2/(2m\omega^2)$ which is in accordance with the solution to problem (15.21).

15.24 A linear harmonic oscillator having classical angular frequency ω is perturbed by the potential λx^3. Show that to order λ^2 the eigenenergies are given by

$$E_n = \left(n + \frac{1}{2}\right)\hbar\omega - \frac{15\lambda^2\hbar^2}{4m^3\omega^4}\left(n^2 + n + \frac{11}{30}\right).$$

• *Solution*

In order to solve this problem we need to evaluate the matrix elements of x^3 between the states of the linear harmonic oscillator. The orthonormal eigenfunctions for the linear harmonic oscillator are given by

$$u_n(x) = N_n\exp(-\xi^2/2)H_n(\xi) \tag{15.18}$$

where $\xi = \alpha x, \alpha = (m\omega/\hbar)^{1/2}$ and the normalization factor is $N_n = [\alpha/(\pi^{1/2}2^nn!)]^{1/2}$.

To find the matrix elements we use the recurrence relation for the Hermite polynomials $H_n(\xi)$ given by

$$2\xi H_n = H_{n+1} + 2nH_{n-1}. \tag{15.19}$$

Applying this three times gives

$$(2\xi)^3H_n = H_{n+3} + 6(n + 1)H_{n+1} + 12n^2H_{n-1} + 8n(n - 1)(n - 2)H_{n-3}.$$

Using the orthonormality of the eigenfunctions we see that

$$\left|(u_{n+3}, x^3 u_n)\right|^2 = \frac{1}{(2\alpha)^6}\left|\frac{N_n}{N_{n+3}}\right|^2 = \frac{1}{(2\alpha)^6}\frac{2^{n+3}(n+3)!}{2^n n!} = \frac{1}{8\alpha^6}(n+3)(n+2)(n+1),$$

$$\left|(u_{n+1}, x^3 u_n)\right|^2 = \frac{1}{(2\alpha)^6}\left|6(n+1)\frac{N_n}{N_{n+1}}\right|^2 = \frac{9}{8\alpha^6}(n+1)^3,$$

and so

$$\left|(u_{n-1}, x^3 u_n)\right|^2 = \frac{9}{8\alpha^6}n^3, \quad \left|(u_{n-3}, x^3 u_n)\right|^2 = \frac{1}{8\alpha^6}n(n-1)(n-2).$$

The first-order correction to the energy is $E_n^{(1)} = (\psi_n^{(0)}, \lambda x^3 \psi_n^{(0)})$ where $\psi_n^{(0)}(x) = u_n(x)$ and so

$$E_n^{(1)} = \int_{-\infty}^{\infty} N_n^2 [H_n(\alpha x)]^2 \exp(-\alpha^2 x^2)\, \lambda x^3 \, dx = 0$$

because the integrand is odd.

The second order contribution to the energy is

$$E_n^{(2)} = -\sum_r {}' \frac{\left|(\psi_n^{(0)}, \lambda x^3 \psi_r^{(0)})\right|^2}{E_r^{(0)} - E_n^{(0)}} = -\lambda^2 \sum_r {}' \frac{\left|(u_r, x^3 u_n)\right|^2}{E_r^{(0)} - E_n^{(0)}}$$

where $E_n^{(0)} = \left(n + \frac{1}{2}\right)\hbar\omega$. Now using the matrix elements derived above we get

$$E_n^{(2)} = -\frac{\lambda^2}{8\alpha^6 \hbar\omega}\left[\frac{1}{3}(n+3)(n+2)(n+1) + 9(n+1)^3 - 9n^3 - \frac{1}{3}n(n-1)(n-2)\right]$$

which simplifies down to the required result.

15-4 Semi-classical approximation

15.25 Use the semi-classical approximation to show that the energy eigenvalues for a particle of mass m moving in the one-dimensional potential

$$V(x) = \lambda |x|^s \quad (-\infty < x < \infty),$$

where λ is a positive constant and $s > -2$, are given by

$$E_n = \left[\frac{(n + \frac{1}{2})\hbar\lambda^{1/s}}{4(2m)^{1/2} J_s}\right]^{2s/(s+2)}$$

with

$$J_s = \int_0^1 (1 - y^s)^{1/2}\, dy.$$

Hence show that the semi-classical approximation gives the correct eigenvalues $E_n = \left(n + \frac{1}{2}\right)\hbar\omega$ for the linear harmonic oscillator with classical angular frequency ω.

- *Solution*

The *Bohr-Sommerfeld quantization rule* for the energy levels E_n of a particle of mass m moving in a potential well $V(x)$ with classical turning points at $x = x_1$ and $x = x_2$ is

$$2 \int_{x_1}^{x_2} \{2m[E_n - V(x)]\}^{1/2} \, dx = \left(n + \frac{1}{2}\right) h \qquad (15.20)$$

where we have $V(x_1) = V(x_2) = E_n$. Now in this problem we have $V(x) = \lambda |x|^s$ where $\lambda > 0$ and so $|x_1| = |x_2| = (|E_n|/\lambda)^{1/s}$.

Hence for bound states with $E_n > 0$ we see that

$$4 \int_0^{(E_n/\lambda)^{1/s}} [2m(E_n - \lambda x^s)]^{1/2} \, dx = \left(n + \frac{1}{2}\right) h.$$

Now putting $y = (\lambda/E_n)^{1/s}x$ we obtain

$$4(2m)^{1/2}\lambda^{-1/s} E_n^{(s+2)/2s} \int_0^1 (1 - y^s)^{1/2} \, dy = \left(n + \frac{1}{2}\right) h$$

and so we get

$$E_n = \left[\frac{(n + \frac{1}{2})h\lambda^{1/s}}{4(2m)^{1/2}J_s}\right]^{2s/(s+2)}.$$

For the case of the linear harmonic oscillator we have $s = 2$ and $\lambda = \frac{1}{2}m\omega^2$. Then we find that

$$E_n = \frac{1}{8J_2}\left(n + \frac{1}{2}\right) h\omega$$

and since $J_2 = \int_0^1 (1 - y^2)^{1/2} \, dy = \pi/4$ we get the exact formula $E_n = \left(n + \frac{1}{2}\right)\hbar\omega$.

15.26 Use the semi-classical approximation to obtain the energy eigenvalues for the bound states of a particle of mass m in the potential

$$V(x) = \begin{cases} -V_0(1 - x/a) & (0 \le x \le a) \\ -V_0(1 + x/a) & (-a \le x \le 0) \\ 0 & (|x| > a). \end{cases}$$

• *Solution*

We again use the Bohr-Sommerfeld quantization rule given in the previous problem. Since the potential is symmetrical about the origin we see that the turning points satisfy $|x_1| = |x_2|$ and $E_n + V_0 (1 - x_1/a) = 0$.

Hence we have

$$4(2m)^{1/2} \int_0^{x_1} \left[E_n + V_0\left(1 - \frac{x}{a}\right)\right]^{1/2} \, dx = \left(n + \frac{1}{2}\right) h$$

which can be rewritten

$$\frac{4(2m)^{1/2}a}{V_0} \int_0^b y^{1/2} \, dy = \left(n + \frac{1}{2}\right) h$$

where $b = E_n + V_0$ and we have put $y = b - V_0 x/a$. It follows that the classical approximation to the energy eigenvalues yields

$$E_n = \left[\frac{3\left(n + \frac{1}{2}\right)hV_0}{8(2m)^{1/2}a}\right]^{2/3} - V_0.$$

15-5 Spin and orbital angular momenta

15.27 Show that the following are simultaneous 2-electron spin eigenvectors of the 2-electron spin operators $S_{1z} + S_{2z}$ and $(\mathbf{S_1} + \mathbf{S_2})^2$:

$$\alpha_1\alpha_2, \quad 2^{-\frac{1}{2}}(\alpha_1\beta_2 + \beta_1\alpha_2), \quad \beta_1\beta_2; \quad 2^{-\frac{1}{2}}(\alpha_1\beta_2 - \beta_1\alpha_2)$$

where α and β are the spin up and spin down eigenvectors respectively of a single electron.

- *Solution*

The *spin operator* $\mathbf{S} = \frac{1}{2}\hbar\sigma$ is defined in terms of σ whose components are the *Pauli spin matrices* given by

$$\sigma_x = \begin{pmatrix} 0 & 1 \\ 1 & 0 \end{pmatrix}, \quad \sigma_y = \begin{pmatrix} 0 & -i \\ i & 0 \end{pmatrix}, \quad \sigma_z = \begin{pmatrix} 1 & 0 \\ 0 & -1 \end{pmatrix} \tag{15.21}$$

and the eigenvectors are represented by

$$\alpha = \begin{pmatrix} 1 \\ 0 \end{pmatrix}, \quad \beta = \begin{pmatrix} 0 \\ 1 \end{pmatrix}. \tag{15.22}$$

Then we have

$$\begin{aligned} \sigma_x\alpha = \beta \quad \sigma_y\alpha = i\beta \quad \sigma_z\alpha = \alpha, \\ \sigma_x\beta = \alpha \quad \sigma_y\beta = -i\alpha \quad \sigma_z\beta = -\beta. \end{aligned} \tag{15.23}$$

We see that

$$(\sigma_{1z} + \sigma_{2z})\alpha_1\alpha_2 = 2\alpha_1\alpha_2, \quad (\sigma_{1z} + \sigma_{2z})2^{-\frac{1}{2}}(\alpha_1\beta_2 + \beta_1\alpha_2) = 0, \quad (\sigma_{1z} + \sigma_{2z})\beta_1\beta_2 = -2\beta_1\beta_2$$
$$(\sigma_{1z} + \sigma_{2z})2^{-\frac{1}{2}}(\alpha_1\beta_2 - \beta_1\alpha_2) = 0$$

and so $\alpha_1\alpha_2$, $2^{-\frac{1}{2}}(\alpha_1\beta_2 + \beta_1\alpha_2)$, $\beta_1\beta_2$; $2^{-\frac{1}{2}}(\alpha_1\beta_2 - \beta_1\alpha_2)$ are eigenvectors of $S_z = S_{1z} + S_{2z}$ with eigenvalues $\hbar, 0, -\hbar, 0$ respectively.

Also, using $(\sigma_1 + \sigma_2)^2 = (\sigma_{1x} + \sigma_{2x})^2 + (\sigma_{1y} + \sigma_{2y})^2 + (\sigma_{1z} + \sigma_{2z})^2$, we find that

$$(\sigma_1 + \sigma_2)^2\alpha_1\alpha_2 = 8\alpha_1\alpha_2, \quad (\sigma_1 + \sigma_2)^2 2^{-\frac{1}{2}}(\alpha_1\beta_2 + \beta_1\alpha_2) = 8[2^{-\frac{1}{2}}(\alpha_1\beta_2 + \beta_1\alpha_2)],$$
$$(\sigma_1 + \sigma_2)^2\beta_1\beta_2 = 8\beta_1\beta_2, \quad (\sigma_1 + \sigma_2)^2 2^{-\frac{1}{2}}(\alpha_1\beta_2 - \beta_1\alpha_2) = 0$$

and so $\alpha_1\alpha_2$, $2^{-\frac{1}{2}}(\alpha_1\beta_2 + \beta_1\alpha_2)$, $\beta_1\beta_2$; $2^{-\frac{1}{2}}(\alpha_1\beta_2 - \beta_1\alpha_2)$ are eigenvectors of $(\mathbf{S_1} + \mathbf{S_2})^2$ with eigenvalues $2\hbar^2, 2\hbar^2, 2\hbar^2, 0$ respectively.

15.28 Show that for $m \geq 0$

$$\begin{aligned} (L_x + iL_y)Y_l^m = -\hbar[(l + m + 1)(l - m)]^{\frac{1}{2}}Y_l^{m+1}, \\ (L_x - iL_y)Y_l^m = -\hbar[(l + m)(l - m + 1)]^{\frac{1}{2}}Y_l^{m-1} \end{aligned} \tag{15.24}$$

where L_x and L_y are the x and y components of the orbital angular momentum vector $\mathbf{L}$ and the $Y_l^m(\theta, \phi)$ are spherical harmonics.

- *Solution*

We have

$$L_x = i\hbar \left(\sin\phi \frac{\partial}{\partial\theta} + \cot\theta \cos\phi \frac{\partial}{\partial\phi} \right), \quad L_y = i\hbar \left(-\cos\phi \frac{\partial}{\partial\theta} + \cot\theta \sin\phi \frac{\partial}{\partial\phi} \right)$$

and so

$$L_x + iL_y = \hbar \exp(i\phi) \left(\frac{\partial}{\partial\theta} + i\cot\theta \frac{\partial}{\partial\phi} \right).$$

Now $Y_l^m(\theta,\phi) = N_l^m P_l^m(\cos\theta) \exp(im\phi)$ and hence

$$(L_x + iL_y)Y_l^m = \hbar N_l^m \exp[i(m+1)\phi] \left[\frac{\partial}{\partial\theta} P_l^m(\cos\theta) - m\cot\theta \, P_l^m(\cos\theta) \right].$$

But for $m \geq 0$ we have

$$\frac{\partial}{\partial\theta} P_l^m(\cos\theta) = m\cot\theta \, P_l^m - P_l^{m+1}$$

and so

$$(L_x + iL_y)Y_l^m = -\hbar N_l^m \exp[i(m+1)\phi] P_l^{m+1} = -\hbar Y_l^{m+1} \frac{N_l^m}{N_l^{m+1}}$$

where

$$N_l^m = \left[\frac{2l+1}{4\pi} \frac{(l-m)!}{(l+m)!} \right]^{\frac{1}{2}}.$$

Hence $N_l^m / N_l^{m+1} = [(l+m+1)(l-m)]^{\frac{1}{2}}$ and thus, as required, we obtain

$$(L_x + iL_y)Y_l^m = -\hbar[(l+m+1)(l-m)]^{\frac{1}{2}} Y_l^{m+1}.$$

Now $(L_x - iL_y)(L_x + iL_y) = L_x^2 + L_y^2 + i(L_xL_y - L_yL_x)$ and since $L_xL_y - L_yL_x = i\hbar L_z$ we see that $(L_x - iL_y)(L_x + iL_y) = \mathbf{L}^2 - L_z^2 - \hbar L_z$. But $\mathbf{L}^2 Y_l^m = \hbar^2 l(l+1)Y_l^m$ and $L_z Y_l^m = \hbar m Y_l^m$ from which it follows that

$$(L_x - iL_y)(L_x + iL_y)Y_l^m = [l(l+1) - m(m+1)]\hbar^2 Y_l^m.$$

Hence, using the first of equations (15.24), we get the second equation of (15.24):

$$(L_x - iL_y)Y_l^m = -\hbar \frac{[l(l+1) - m(m-1)]}{[(l+m)(l-m+1)]^{\frac{1}{2}}} Y_l^{m-1} = -\hbar[(l+m)(l-m+1)]^{\frac{1}{2}} Y_l^{m-1}.$$

$L_+ = L_x + iL_y$ is called the *positive shift operator* and $L_- = L_x - iL_y$ is called the *negative shift operator*.

15.29 A wave function ψ has the form

$$\psi = N \begin{pmatrix} \left(l + \frac{1}{2} - m \right)^{\frac{1}{2}} R(r) Y_l^{m-\frac{1}{2}}(\theta,\phi) \\ \left(l + \frac{1}{2} + m \right)^{\frac{1}{2}} R(r) Y_l^{m+\frac{1}{2}}(\theta,\phi) \end{pmatrix}$$

where $R(r)$ is a normalized radial function and $Y_l^{m-\frac{1}{2}}$, $Y_l^{m+\frac{1}{2}}$ are normalized spherical harmonics with $m \geq \frac{1}{2}$. Find the value of N for which ψ is normalized to unity. Show that ψ is an eigenfunction of the z component J_z of the total angular momentum $\mathbf{J}$, and the operator $\mathbf{J}^2$.

- *Solution*

We have $(\psi, \psi) = N^2 \left[\left(l + \frac{1}{2} - m \right) + \left(l + \frac{1}{2} + m \right) \right] = N^2 (2l + 1)$ and thus $N = (2l + 1)^{-\frac{1}{2}}$ if $(\psi, \psi) = 1$ so that ψ is normalized to unity.

Now the total angular momentum $\mathbf{J} = \mathbf{L} + \mathbf{S}$ where $\mathbf{L}$ is the orbital angular momentum and $\mathbf{S}$ is the spin angular momentum. We have

$$L_z Y_l^{m-\frac{1}{2}} = \hbar \left(m - \frac{1}{2} \right) Y_l^{m-\frac{1}{2}}, \quad L_z Y_l^{m+\frac{1}{2}} = \hbar \left(m + \frac{1}{2} \right) Y_l^{m+\frac{1}{2}}$$

and, using (15.23),

$$S_z \begin{pmatrix} 1 \\ 0 \end{pmatrix} = \frac{1}{2} \hbar \begin{pmatrix} 1 \\ 0 \end{pmatrix}, \quad S_z \begin{pmatrix} 0 \\ 1 \end{pmatrix} = -\frac{1}{2} \hbar \begin{pmatrix} 0 \\ 1 \end{pmatrix}$$

from which it follows that

$$J_z \psi = N \begin{pmatrix} (l + \frac{1}{2} - m)^{\frac{1}{2}} R(r)[\hbar(m - \frac{1}{2}) + \frac{1}{2}\hbar] Y_l^{m-\frac{1}{2}} \\ (l + \frac{1}{2} + m)^{\frac{1}{2}} R(r)[\hbar(m + \frac{1}{2}) - \frac{1}{2}\hbar] Y_l^{m+\frac{1}{2}} \end{pmatrix} = m\hbar\psi$$

and so we see that ψ is an eigenfunction of J_z with eigenvalue $m\hbar$.

Now $(S_x + iS_y) \begin{pmatrix} 1 \\ 0 \end{pmatrix} = 0$ and $(S_x + iS_y) \begin{pmatrix} 0 \\ 1 \end{pmatrix} = \hbar \begin{pmatrix} 1 \\ 0 \end{pmatrix}$ using (15.23). Also from the previous problem we see that

$$(L_x + iL_y)\psi = -\hbar N R(r) \begin{pmatrix} (l + \frac{1}{2} + m)^{\frac{1}{2}} (l - m + \frac{1}{2}) Y_l^{m+\frac{1}{2}} \\ [(l + m + \frac{3}{2})(l - m - \frac{1}{2})(l + m + \frac{1}{2})]^{\frac{1}{2}} Y_l^{m+\frac{3}{2}} \end{pmatrix}$$

and so we find that

$$(J_x + iJ_y)\psi = -\hbar N R(r) \begin{pmatrix} (l - \frac{1}{2} - m)(l + m + \frac{1}{2})^{\frac{1}{2}} Y_l^{m+\frac{1}{2}} \\ [(l + m + \frac{3}{2})(l - m - \frac{1}{2})(l + m + \frac{1}{2})]^{\frac{1}{2}} Y_l^{m+\frac{3}{2}} \end{pmatrix}.$$

It follows from the second equation of (15.24) that

$$(L_x - iL_y)(J_x + iJ_y)\psi = \hbar^2 N R(r) \begin{pmatrix} (l + m + \frac{1}{2})(l - \frac{1}{2} - m)(l - m + \frac{1}{2})^{\frac{1}{2}} Y_l^{m-\frac{1}{2}} \\ (l + m + \frac{3}{2})(l - m - \frac{1}{2})(l + m + \frac{1}{2})^{\frac{1}{2}} Y_l^{m+\frac{1}{2}} \end{pmatrix}$$

and since $(S_x - iS_y) \begin{pmatrix} 1 \\ 0 \end{pmatrix} = \hbar \begin{pmatrix} 0 \\ 1 \end{pmatrix}$ and $(S_x - iS_y) \begin{pmatrix} 0 \\ 1 \end{pmatrix} = 0$ we see that

$$(J_x - iJ_y)(J_x + iJ_y)\psi = \hbar^2 N R(r)[l^2 - (m + \frac{1}{2})^2] \begin{pmatrix} (l + \frac{1}{2} - m)^{\frac{1}{2}} Y_l^{m-\frac{1}{2}} \\ (l + \frac{1}{2} + m)^{\frac{1}{2}} Y_l^{m+\frac{1}{2}} \end{pmatrix},$$

that is

$$(J_x - iJ_y)(J_x + iJ_y)\psi = \hbar^2 [l^2 - (m + \frac{1}{2})^2]\psi.$$

Now $(J_x - iJ_y)(J_x + iJ_y) = J_x^2 + J_y^2 + i(J_x J_y - J_y J_x) = J_x^2 + J_y^2 - \hbar J_z$ since $J_x J_y - J_y J_x = i\hbar J_z$ and so

$$\mathbf{J}^2 \psi = \hbar^2 [l^2 - (m + \frac{1}{2})^2 + m + m^2]\psi = \hbar^2 (l^2 - \frac{1}{4})\psi = \hbar^2 j(j + 1)\psi$$

where $j = l - \frac{1}{2}$ is the total angular momentum quantum number.

15.30 If a particle is in a spin state given by the vector $\psi = \begin{pmatrix} \psi_1 \\ \psi_2 \end{pmatrix}$ with $\psi^\dagger \psi = 1$,

show that

$$(\Delta \sigma_x)^2 = 1 - 4[Re(\psi_1^* \psi_2)]^2,$$
$$(\Delta \sigma_y)^2 = 1 - 4[Im(\psi_1^* \psi_2)]^2$$

and $< \sigma_z > = \psi_1^* \psi_1 - \psi_2^* \psi_2$, where $< \sigma > = \psi^\dagger \sigma \psi$ is the expectation value of σ and $(\Delta \sigma)^2 = < (\sigma - < \sigma >)^2 > = < \sigma^2 > - < \sigma >^2$ is the mean square deviation of σ.

Hence verify that the uncertainty principle for spin matrices $\Delta \sigma_x \Delta \sigma_y \geq |< \sigma_z >|$ is satisfied.

• *Solution*

We have

$$< \sigma_x > = \psi^\dagger \sigma_x \psi = \begin{pmatrix} \psi_1^* & \psi_2^* \end{pmatrix} \begin{pmatrix} 0 & 1 \\ 1 & 0 \end{pmatrix} \begin{pmatrix} \psi_1 \\ \psi_2 \end{pmatrix} = \begin{pmatrix} \psi_1^* & \psi_2^* \end{pmatrix} \begin{pmatrix} \psi_2 \\ \psi_1 \end{pmatrix}$$

and so $< \sigma_x > = \psi_1^* \psi_2 + \psi_1 \psi_2^* = 2Re(\psi_1^* \psi_2)$. Also

$$< \sigma_x^2 > = \psi^\dagger \sigma_x^2 \psi = \psi^\dagger \psi = \psi_1^* \psi_1 + \psi_2^* \psi_2 = 1$$

giving for the mean square deviation of σ_x

$$(\Delta \sigma_x)^2 = 1 - 4[Re(\psi_1^* \psi_2)]^2.$$

Similarly

$$< \sigma_y > = \psi^\dagger \sigma_y \psi = \begin{pmatrix} \psi_1^* & \psi_2^* \end{pmatrix} \begin{pmatrix} 0 & -i \\ i & 0 \end{pmatrix} \begin{pmatrix} \psi_1 \\ \psi_2 \end{pmatrix} = i \begin{pmatrix} \psi_1^* & \psi_2^* \end{pmatrix} \begin{pmatrix} -\psi_2 \\ \psi_1 \end{pmatrix}$$

and so $< \sigma_y > = -i(\psi_1^* \psi_2 - \psi_1 \psi_2^*) = 2Im(\psi_1^* \psi_2)$. Also $< \sigma_y^2 > = 1$ giving for the mean square deviation of σ_y

$$(\Delta \sigma_y)^2 = 1 - 4[Im(\psi_1^* \psi_2)]^2.$$

Further

$$< \sigma_z > = \psi^\dagger \sigma_z \psi = \begin{pmatrix} \psi_1^* & \psi_2^* \end{pmatrix} \begin{pmatrix} 1 & 0 \\ 0 & -1 \end{pmatrix} \begin{pmatrix} \psi_1 \\ \psi_2 \end{pmatrix} = \psi_1^* \psi_1 - \psi_2^* \psi_2.$$

Hence

$$(\Delta \sigma_x)^2 (\Delta \sigma_y)^2 = \{1 - 4[Re(\psi_1^* \psi_2)]^2\} \{1 - 4[Im(\psi_1^* \psi_2)]^2\}$$
$$= 1 - 4 \{[Re(\psi_1^* \psi_2)]^2 + [Im(\psi_1^* \psi_2)]^2\} + 16[Re(\psi_1^* \psi_2)]^2 [Im(\psi_1^* \psi_2)]^2$$

and so we see that

$$(\Delta \sigma_x)^2 (\Delta \sigma_y)^2 \geq 1 - 4 |\psi_1^* \psi_2|^2 = (\psi_1^* \psi_1 + \psi_2^* \psi_2)^2 - 4\psi_1^* \psi_2 \psi_1 \psi_2^* = (\psi_1^* \psi_1 - \psi_2^* \psi_2)^2 = < \sigma_z >^2.$$

Hence

$$\Delta \sigma_x \Delta \sigma_y \geq |< \sigma_z >|$$

which is the uncertainty principle for spin operators.

The following is a problem on the *Feshbach operators* P and Q.

15.31 Let P and Q be operators satisfying

$$PQ = 0, \quad P + Q = I$$

where I is the identity operator. Show that

$$P^2 = P, \quad QP = 0, \quad Q^2 = Q.$$

If Ψ is the solution of the Schrödinger equation $(H - E)\Psi = 0$ where H is the Hamiltonian operator and E is the energy, show that

$$P(H + U - E)P\Psi = 0$$

where

$$U = -PHQ[Q(H - E)Q]^{-1}QHP.$$

- *Solution*

We have $P = PI = P(P + Q) = P^2$ since $P + Q = I$ and $PQ = 0$. Now we also have that $P = (P + Q)P = P^2 + QP$ and since we have shown that $P = P^2$ it follows that $QP = 0$. Further $Q = Q(Q + P) = Q^2 + QP$ and since we have proved that $QP = 0$ we deduce that $Q = Q^2$ as required.

Putting $\Psi = (P + Q)\Psi$ in the Schrödinger equation $(H - E)\Psi = 0$ and operating with P and Q respectively on the left we obtain

$$P(H - E)P\Psi + PHQ\Psi = 0, \quad Q(H - E)Q\Psi + QHP\Psi = 0$$

since $PQ = QP = 0$.

Hence $\Psi = -[Q(H - E)Q]^{-1}QHP\Psi$ from the second equation, and so substituting into the first equation we get

$$P(H - E)P\Psi - PHQ[Q(H - E)Q]^{-1}QHP\Psi = 0$$

which yields the required result since $P = P^2$.

15-6 Scattering theory

Our next few problems are on the *scattering of particles* by various potentials.

15.32 A particle of mass m and energy E is incident on a one-dimensional square potential barrier

$$V(x) = \begin{cases} D & (0 \leq x \leq a) \\ 0 & (x < 0, \ x > a). \end{cases}$$

It is represented by an incident wave Ae^{ikx} and a reflected wave Be^{-ikx} for $x < 0$ and a transmitted wave Ce^{ikx} for $x > a$ where $k = (2mE/\hbar^2)^{\frac{1}{2}}$. Show that for $E > D$, the reflection and transmission coefficients are given by

$$R = \left|\frac{B}{A}\right|^2 = \left[1 + \frac{4E(E - D)}{D^2 \sin^2 k_1 a}\right]^{-1},$$

$$T = \left|\frac{C}{A}\right|^2 = \left[1 + \frac{D^2 \sin^2 k_1 a}{4E(E - D)}\right]^{-1}$$

where $k_1 = \left[2m(E - D)/\hbar^2\right]^{\frac{1}{2}}$.

Further show that $|B/A|^2 + |C/A|^2 = 1$ and that the transmission coefficient T tends to $\left(1 + mDa^2/2\hbar^2\right)^{-1}$ as $E \to D$.

If $0 < E < D$ show that the transmission coefficient is approximately $[16E(D - E)/D^2]e^{-2\gamma a}$ for $\gamma a \gg 1$ where $k_1 = i\gamma$.

- *Solution*

This problem is analogous to problems 11.10 and 11.11 on the scattering of waves on an infinite string and problem 11.21 on the scattering of sound waves in a tube.

Here the one-dimensional wave equations are

$$\frac{d^2u}{dx^2} + k^2u = 0 \quad (x < 0, \ x > a), \qquad \frac{d^2u}{dx^2} + k_1^2u = 0 \quad (0 \le x \le a)$$

whose solutions may be written in the form

$$\begin{aligned}
u(x) &= Ae^{ikx} + Be^{-ikx} & (x < 0), \\
u(x) &= Ce^{ikx} & (x > a), \\
u(x) &= A_1e^{ik_1x} + B_1e^{-ik_1x} & (0 \le x \le a).
\end{aligned}$$

By the continuity of u and du/dx at $x = 0$ and $x = a$ we obtain the four equations

$$\begin{aligned}
A + B &= A_1 + B_1, \\
k(A - B) &= k_1(A_1 - B_1), \\
A_1e^{ik_1a} + B_1e^{-ik_1a} &= Ce^{ika}, \\
k_1(A_1e^{ik_1a} - B_1e^{-ik_1a}) &= kCe^{ika}.
\end{aligned}$$

Hence

$$\begin{aligned}
(k_1 + k)A + (k_1 - k)B &= 2k_1A_1, \\
2k_1A_1e^{ik_1a} &= (k_1 + k)Ce^{ika}, \\
(k_1 - k)A + (k_1 + k)B &= 2k_1B_1, \\
2k_1B_1e^{-ik_1a} &= (k_1 - k)Ce^{ika}
\end{aligned}$$

and so

$$\begin{aligned}
(k_1 + k)A + (k_1 - k)B &= (k_1 + k)Ce^{i(k-k_1)a}, \\
(k_1 - k)A + (k_1 + k)B &= (k_1 - k)Ce^{i(k+k_1)a}
\end{aligned}$$

which give

$$\frac{C}{A} = \frac{4kk_1e^{i(k_1-k)a}}{(k_1 + k)^2 - (k_1 - k)^2e^{2ik_1a}}.$$

Hence

$$\frac{B}{A} = \frac{k_1 + k}{k_1 - k}\left[\frac{C}{A}e^{i(k-k_1)a} - 1\right] = \frac{(k^2 - k_1^2)(1 - e^{2ik_1a})}{(k_1 + k)^2 - (k_1 - k)^2e^{2ik_1a}}.$$

It follows that

$$T = \left|\frac{C}{A}\right|^2 = \frac{(4kk_1)^2}{(k_1 + k)^4 + (k_1 - k)^4 - 2(k_1^2 - k^2)^2\cos 2k_1a} = \left[1 + \frac{(k^2 - k_1^2)^2\sin^2k_1a}{4k^2k_1^2}\right]^{-1}$$

which gives the required formula for the transmission coefficient, and

$$R = \left|\frac{B}{A}\right|^2 = \frac{4(k^2 - k_1^2)^2\sin^2k_1a}{(4kk_1)^2 + 4(k_1^2 - k^2)^2\sin^2k_1a} = \left[1 + \frac{4k^2k_1^2}{(k^2 - k_1^2)^2\sin^2k_1a}\right]^{-1}$$

which leads to the required expression for the reflection coefficient.

Hence

$$R + T = \left|\frac{B}{A}\right|^2 + \left|\frac{C}{A}\right|^2 = \frac{1}{1+\alpha} + \frac{1}{1+\alpha^{-1}} = 1$$

where $\alpha = 4E(E-D)/(D^2 \sin^2 k_1 a)$.

Now $\sin^2 k_1 a \to 2m(E-D)a^2/\hbar^2$ as $E \to D$ from above and so

$$\left|\frac{C}{A}\right|^2 = \left[1 + \frac{D^2 \sin^2 k_1 a}{4E(E-D)}\right]^{-1} \to \left[1 + \frac{mDa^2}{2\hbar^2}\right]^{-1}$$

in this limit.

If $0 < E < D$ we may set $k_1 = i\gamma$ where $\gamma = [2m(D-E)/\hbar^2]^{1/2}$, and then we obtain

$$\left|\frac{C}{A}\right|^2 = \left[1 + \frac{D^2 \sinh^2 \gamma a}{4E(D-E)}\right]^{-1}$$

and for $\gamma a \gg 1$ we have $\sinh^2 \gamma a \simeq e^{2\gamma a}/4$ so that, as required, we get

$$\left|\frac{C}{A}\right|^2 \simeq \frac{16E(D-E)}{D^2} e^{-2\gamma a}.$$

15.33 A particle of mass m is moving in one-dimension in a potential $V(x) = -D\delta(x)$ where $\delta(x)$ is the Dirac delta function. Show that although the wave function is continuous at all points, the first derivative of the wave function is discontinuous at the origin.

Express the energy of the bound state solution for this potential in terms of D and obtain the first order correction to this energy due to a perturbing term in the Hamiltonian given by $H' = \lambda x^2$.

A beam of particles with energy $E > 0$ and of amplitude A is incident on this potential in the positive x direction. Obtain expressions for the amplitude of the reflected wave B and the amplitude of the transmitted wave C in terms of A, E and D, showing that $|B|^2 + |C|^2 = |A|^2$.

• *Solution*

From the one-dimensional Schrödinger equation

$$\left[-\frac{\hbar^2}{2m}\frac{d^2}{dx^2} + V(x)\right] u(x) = Eu(x)$$

we obtain

$$\lim_{\epsilon \to 0} \int_{-\epsilon}^{\epsilon} \left[-\frac{\hbar^2}{2m}\frac{d^2}{dx^2} + V(x) - E\right] u(x)\,dx = 0$$

and so

$$\lim_{\epsilon \to 0} \int_{-\epsilon}^{\epsilon} \frac{d^2 u}{dx^2}\,dx = \lim_{\epsilon \to 0} \frac{2m}{\hbar^2} \int_{-\epsilon}^{\epsilon} V(x)u(x)\,dx.$$

Hence

$$\lim_{\epsilon \to 0} \left[\frac{du}{dx}\right]_{-\epsilon}^{\epsilon} = -\frac{2mD}{\hbar^2} \lim_{\epsilon \to 0} \int_{-\epsilon}^{\epsilon} \delta(x)u(x)\,dx$$

and so we find that

$$u'_+(0) - u'_-(0) = -\frac{2mD}{\hbar^2}u(0) \qquad\qquad (15.25)$$

where the primes denote differentiations with respect to x, and the $+$ and $-$ denote taking the limit from $x > 0$ and $x < 0$ respectively. Thus the first derivative of the wave function $u(x)$ is discontinuous at the origin $x = 0$.

For a bound state with energy $E^{(0)} < 0$ we have for $x \neq 0$

$$\frac{d^2 u}{dx^2} - k_0^2 u(x) = 0$$

where $k_0 = (-2mE^{(0)}/\hbar^2)^{1/2}$ and so $u(x) = ae^{-k_0 x} + be^{k_0 x}$. For $x > 0$ we must have $b = 0$ and for $x < 0$ we must have $a = 0$. Also $u(x)$ is continuous at $x = 0$ and so

$$\begin{aligned} u(x) &= ae^{-k_0 x} \quad (x > 0), \\ u(x) &= ae^{k_0 x} \quad (x < 0). \end{aligned}$$

Hence

$$u'_+(0) - u'_-(0) = -2k_0 a$$

and so, using (15.25), we get $k_0 = mD/\hbar^2$ which gives for the energy of the bound state

$$E^{(0)} = -\frac{mD^2}{2\hbar^2},$$

in agreement with the result obtained at the end of problem 15.10.

The first order correction to the energy due to the perturbation $H' = \lambda x^2$ is

$$E^{(1)} = \frac{\int_{-\infty}^{\infty} u H' u \, dx}{\int_{-\infty}^{\infty} u^2 \, dx}.$$

We have $\int_{-\infty}^{\infty} u^2 \, dx = 2a^2 \int_0^{\infty} e^{-2k_0 x} \, dx = a^2/k_0$ and $\int_{-\infty}^{\infty} u H' u \, dx = \int_{-\infty}^{\infty} \lambda x^2 u^2 dx = 2\lambda a^2 \int_0^{\infty} x^2 e^{-2k_0 x} \, dx = \lambda a^2/2k_0^3$ and so we find that the first order correction to the energy is

$$E^{(1)} = \frac{1}{2} \frac{\hbar^4 \lambda}{m^2 D^2}.$$

For the scattering of a beam of particles having energy E the incident wave is given by Ae^{ikx} where $k = (2mE/\hbar^2)^{1/2}$. Then we have for the wave function

$$\begin{aligned} u(x) &= Ae^{ikx} + Be^{-ikx} \quad (x < 0), \\ u(x) &= Ce^{ikx} \quad (x > 0). \end{aligned}$$

Continuity of $u(x)$ at $x = 0$ gives $A + B = C$. Also $u'_+(0) - u'_-(0) = ikC - ik(A - B)$ and so, using (15.25), we get $ikC - ik(A - B) = -2mDC/\hbar^2$ from which it follows that

$$C = \frac{ikA}{ik + mD/\hbar^2}$$

and

$$B = -\frac{mDA/\hbar^2}{ik + mD/\hbar^2}.$$

We see at once that $|B|^2 + |C|^2 = |A|^2$.

15.34 Use the Born approximation to obtain the scattering amplitude $f_B(K)$, where $K = 2k \sin \theta/2$, and the differential cross section $|f_B(\theta)|^2$, for the elastic scattering through an angle θ of electrons having wave number k by the static potential of atomic hydrogen in the ground state given by

$$V(r) = -e^2 \left(\frac{1}{a_0} + \frac{1}{r} \right) \exp(-2r/a_0). \qquad (15.26)$$

Further show that if $a_0 k \sin \theta / 2 \gg 1$, the differential cross section approaches that for the elastic scattering of electrons by the Coulomb potential $-e^2/r$ produced by a proton.

• *Solution*

Consider a hydrogen atom with the proton, charge e, situated at the origin and the electron, charge $-e$, at the point with position vector $\mathbf{r}'$. Then the potential of an external electron moving in the electrostatic field produced by the hydrogen atom at a point with position vector $\mathbf{r}$ is

$$V(\mathbf{r}, \mathbf{r}') = \frac{e^2}{|\mathbf{r} - \mathbf{r}'|} - \frac{e^2}{r}.$$

The static potential of the hydrogen atom at a point distance r from the nucleus is given by the expectation value of $V(\mathbf{r}, \mathbf{r}')$, that is

$$V(r) = \int u_{1s}^*(\mathbf{r}') \left(\frac{e^2}{|\mathbf{r} - \mathbf{r}'|} - \frac{e^2}{r} \right) u_{1s}(\mathbf{r}') \, d\tau'$$

where $u_{1s}(\mathbf{r}) = (\pi a_0^3)^{-1/2} \exp(-r/a_0)$ is the wave function of the ground $1s$ state of atomic hydrogen and a_0 is the Bohr radius.

Now

$$\frac{1}{|\mathbf{r} - \mathbf{r}'|} = \sum_{l=0}^{\infty} \gamma_l(r, r') P_l(\cos \Theta)$$

where Θ is the angle between $\mathbf{r}$ and $\mathbf{r}'$, given by $\cos \Theta = \cos \theta \cos \theta' + \sin \theta \sin \theta' \cos(\phi - \phi')$ in spherical polar coordinates, and $\gamma_l(r, r') = r_<^l / r_>^{l+1}$ where $r_<$ and $r_>$ are the lesser and the greater of r and r'. Hence, using the orthogonality property of the Legendre polynomials, we have

$$V(r) = \frac{e^2}{\pi a_0^3} 4\pi \int_0^\infty \exp(-2r'/a_0) \left[\gamma_0(r, r') - \frac{1}{r} \right] r'^2 \, dr'$$

and so

$$V(r) = \frac{4e^2}{a_0^3} \int_r^\infty \exp(-2r'/a_0) \left(\frac{1}{r'} - \frac{1}{r} \right) r'^2 \, dr'.$$

Now

$$\int_r^\infty \exp(-\alpha r') r' \, dr' = \frac{1}{\alpha^2} (\alpha r + 1) \exp(-\alpha r)$$

and

$$\int_r^\infty \exp(-\alpha r') r'^2 \, dr' = \frac{1}{\alpha^3} (\alpha^2 r^2 + 2\alpha r + 2) \exp(-\alpha r)$$

so that we have

$$\int_r^\infty \exp(-\alpha r') \left(r' - \frac{r'^2}{r} \right) \, dr' = -\frac{1}{\alpha^3} \left(\alpha + \frac{2}{r} \right) \exp(-\alpha r)$$

and the required formula for the static potential of the hydrogen atom follows.

Consider the elastic scattering by a potential $V(r)$ of a particle of mass m incident with wave vector $\mathbf{k_i}$ and scattered with wave vector $\mathbf{k_s}$.

The integral equation for the scattering amplitude is

$$f(\mathbf{k_s}, \mathbf{k_i}) = -\frac{1}{4\pi}\frac{2m}{\hbar^2} \int \exp(-i\mathbf{k_i}.\mathbf{r})V(r)\psi(\mathbf{r})\,d\tau$$

where the wave function ψ representing the scattered particle has the asymptotic form

$$\psi(\mathbf{r}) \sim \exp(i\mathbf{k_i}.\mathbf{r}) + r^{-1}\exp(ikr)f(\mathbf{k_s}, \mathbf{k_i}).$$

Then the scattering amplitude given by the Born approximation is obtained by replacing ψ by its undisturbed plane wave form $\exp(i\mathbf{k_i}.\mathbf{r})$ which yields, for elastic scattering through the angle θ,

$$f_B(\theta) = -\frac{1}{4\pi} \int \exp(i\mathbf{K}.\mathbf{r})U(r)\,d\tau \tag{15.27}$$

where $U(r) = (2m/\hbar^2)V(r)$ and $\mathbf{K} = \mathbf{k_i} - \mathbf{k_s}$. Then for the present problem, using (15.26), we have

$$f_B(K) = \frac{1}{4\pi}\frac{2me^2}{\hbar^2} \int \left(\frac{1}{a_0} + \frac{1}{r}\right) \exp(-2r/a_0)\exp(i\mathbf{K}.\mathbf{r})\,d\tau.$$

To evaluate this integral we take the polar axis in the direction of $\mathbf{K}$ where $K^2 = 2k^2(1 - \cos\theta) = 4k^2\sin^2(\theta/2)$. Then we see that

$$f_B(K) = \frac{1}{a_0}\int_0^\infty \left(\frac{1}{a_0} + \frac{1}{r}\right)\exp(-2r/a_0)r^2\,dr \int_{-1}^1 \exp(iKr\mu)\,d\mu$$

where $\mu = \cos\theta$, and since $\int_{-1}^1 \exp(iKr\mu)\,d\mu = 2\sin Kr/Kr$ we find that

$$f_B(K) = \frac{2}{a_0 K}\int_0^\infty \left(\frac{r}{a_0} + 1\right)\exp(-2r/a_0)\sin Kr\,dr = 2a_0\frac{8 + (Ka_0)^2}{[4 + (Ka_0)^2]^2}.$$

Hence the differential cross section given by the Born approximation is

$$I_B(\theta) = |f_B(\theta)|^2 = \frac{a_0^2[2 + a_0^2k^2\sin^2(\theta/2)]^2}{4[1 + a_0^2k^2\sin^2(\theta/2)]^4}.$$

If $a_0 k \sin(\theta/2) \gg 1$ we obtain

$$I_B(\theta) \simeq \frac{a_0^2}{4[a_0 k \sin(\theta/2)]^4}.$$

Now $\hbar k = mv$ where v is the speed of the electron and so

$$I_B(\theta) \simeq \frac{1}{4}\left(\frac{e^2}{mv^2}\right)^2 \frac{1}{\sin^4(\theta/2)}$$

which is the differential cross section for the scattering of electrons by the Coulomb potential $-e^2/r$.

15.35 Show that the zero-angular momentum scattering phase shift η_0 for the scattering of particles of mass m and energy E by the potential well

$$V(r) = \begin{cases} -D & (0 \leq r \leq a) \\ 0 & (r > a) \end{cases}$$

is given by

$$\eta_0 = \tan^{-1}\left(\frac{k}{k'}\tan k'a\right) - ka$$

where $k^2 = 2mE/\hbar^2$, $\quad k'^2 = k^2 + k_0^2$, $\quad k_0^2 = 2mD/\hbar^2$.

Also find the zero energy limit of the $l = 0$ partial cross section

$$\sigma_0 = \frac{4\pi}{k^2}\sin^2\eta_0.$$

• *Solution*

The radial equation for the $l = 0$ wave is

$$\frac{d^2u}{dr^2} + \left[k^2 - \frac{2m}{\hbar^2}V(r)\right]u(r) = 0$$

and thus we have

$$\frac{d^2u}{dr^2} + k'^2u = 0 \quad (0 \le r \le a), \qquad \frac{d^2u}{dr^2} + k^2u = 0 \quad (r > a).$$

Now $u(0) = 0$ and so

$$\begin{aligned} u(r) &= A\sin k'r & (0 \le r \le a), \\ u(r) &= B\sin(kr + \eta_0) & (r > a). \end{aligned}$$

Since $u^{-1}du/dr$ is continuous at $r = a$ we have

$$\frac{k'\cos k'a}{\sin k'a} = \frac{k\cos(ka + \eta_0)}{\sin(ka + \eta_0)}$$

and this yields the required expression for η_0.

As $k \to 0$ we see that

$$\eta_0 \to k\left(\frac{\tan k_0 a}{k_0} - a\right)$$

and so we obtain the formula

$$\sigma_0 \to 4\pi a^2\left(\frac{\tan k_0 a}{k_0 a} - 1\right)^2$$

for the zero energy limiting value of the $l = 0$ differential cross section.

15.36 Show that the zero-angular momentum scattering phase shift η_0 for the scattering of particles of mass m and energy E by the potential barrier

$$V(r) = \begin{cases} D & (0 \le r \le a) \\ 0 & (r > a) \end{cases}$$

is given by

$$\eta_0 = \tan^{-1}\left(\frac{k}{k'}\tanh k'a\right) - ka \quad (k < k_0)$$

where $k^2 = 2mE/\hbar^2$, $\quad k'^2 = k_0^2 - k^2$, $\quad k_0^2 = 2mD/\hbar^2$; and by

$$\eta_0 = \tan^{-1}\left(\frac{k}{k'}\tan k'a\right) - ka \quad (k > k_0)$$

where now $k'^2 = k^2 - k_0^2$.

Further show that for the case of the impenetrable sphere given by $D \to \infty$, the $l = 0$ partial cross section tends to $4\pi a^2$ in the zero energy limit.

• *Solution*

If we suppose that $k > k_0$ then, as in the previous problem for the potential well, we find that

$$\eta_0 \to \tan^{-1}\left(\frac{k}{k'} \tan k'a\right) - ka$$

where $k'^2 = k^2 - k_0^2$.

If now we suppose that $k < k_0$ we have

$$\frac{d^2u}{dr^2} - k'^2u = 0 \quad (0 \le r \le a), \qquad \frac{d^2u}{dr^2} + k^2u = 0 \quad (r > a)$$

where $k'^2 = k_0^2 - k^2$.

But $u(0) = 0$ and so

$$u(r) = A \sinh k'r \qquad (0 \le r \le a),$$
$$u(r) = B \sin(kr + \eta_0) \quad (r > a).$$

Since $u^{-1}du/dr$ is continuous at $r = a$ we have

$$\frac{k' \cosh k'a}{\sinh k'a} = \frac{k \cos(ka + \eta_0)}{\sin(ka + \eta_0)}$$

and this yields the required expression for η_0 for $k < k_0$.

Consider the case $D \to \infty$ corresponding to an impenetrable sphere. Then

$$\eta_0 \to \tan^{-1}\left(\frac{k}{k_0} \tanh k_0a\right) - ka \sim ka\left(\frac{\tanh k_0a}{k_0a} - 1\right) \sim -ka$$

as $k_0 \to \infty$. Hence

$$u(r) = 0 \qquad\qquad (0 \le r \le a),$$
$$u(r) = B \sin k(r - a) \quad (r > a)$$

and we see that the radial wave function vanishes at $r = a$.

Further the $l = 0$ partial cross section

$$\sigma_0 = \frac{4\pi}{k^2} \sin^2 ka \to 4\pi a^2$$

as $k \to 0$ which is 4×classical cross section πa^2 for an impenetrable sphere of radius a.

15-7 Time-dependent problems

15.37 The equilibrium point of a linear harmonic oscillator with classical angular frequency ω is suddenly moved from the origin to the point $x = a$ at time $t = 0$. If the oscillator was initially in the ground state, find the probability amplitude for the state with quantum number n after the sudden movement of the equilibrium point has occurred.

• *Solution*

Since the linear harmonic oscillator is initially in the state $n = 0$ we have that for time $t \le 0$ the wave function is given by $\psi_-(x, t) = u_0(x)\exp(-iE_0t/\hbar)$ where $u_0(x) = (\alpha/\pi^{1/2})^{1/2}\exp(-\frac{1}{2}\alpha^2x^2)$ with $\alpha = (m\omega/\hbar)^{1/2}$ and $E_0 = \frac{1}{2}\hbar\omega$.

For time $t \geq 0$ we expand the wave function in the form

$$\psi_+(x,t) = \sum_n a_n u_n(x-a) \exp(-iE_n t/\hbar)$$

where u_n is the linear harmonic oscillator wave function corresponding to the energy eigenvalue $E_n = (n + \tfrac{1}{2})\hbar\omega$.

Since $\psi_- = \psi_+$ at time $t = 0$ we have $u_0(x) = \sum_n a_n u_n(x-a)$ and so

$$a_n = \int_{-\infty}^{\infty} u_n(x)u_0(x+a)\,dx$$

which gives for the probability amplitude for the nth state

$$a_n = \left(\frac{\alpha}{\pi^{1/2}}\right)^{1/2} \int_{-\infty}^{\infty} u_n(x) \exp[-\tfrac{1}{2}\alpha^2(x+a)^2]\,dx.$$

Now $u_n(x) = [\alpha/(\pi^{1/2}2^n n!)]^{1/2} H_n(\alpha x)\exp(-\tfrac{1}{2}\alpha^2 x^2)$ and so, putting $\xi = \alpha x$, we get

$$a_n = \frac{1}{(\pi 2^n n!)^{1/2}} \int_{-\infty}^{\infty} H_n(\xi) \exp\left(-\xi^2 - \alpha a \xi - \tfrac{1}{2}\alpha^2 a^2\right)\,d\xi.$$

But

$$H_n(\xi)\exp(-\xi^2) = (-1)^n \frac{\partial^n}{\partial \xi^n}\exp(-\xi^2)$$

and so, integrating by parts n times, we obtain for the probability amplitude

$$a_n = \frac{(\alpha a)^n}{(\pi 2^n n!)^{1/2}} \int_{-\infty}^{\infty} \exp\left(-\xi^2 - \alpha a \xi - \tfrac{1}{2}\alpha^2 a^2\right)\,d\xi = \frac{(\alpha a)^n \exp(-\alpha^2 a^2/4)}{(2^n n!)^{1/2}}$$

since $\int_{-\infty}^{\infty} \exp[-(\xi + \alpha a/2)^2]\,d\xi = \pi^{1/2}$.

The remaining problems in this chapter are concerned with *time-dependent perturbation theory*.

15.38 A linear harmonic oscillator of mass m, charge $-e$ and classical angular frequency ω is acted on by a weak time-dependent uniform electric field

$$F(t) = \frac{A}{\sqrt{\pi}} \frac{\exp(-t^2/\tau^2)}{\tau}$$

where A and τ are constants. If at time $t = -\infty$ the oscillator is in its ground state, find the first order probability that at time $t = \infty$ it will be in its first excited state.

• *Solution*

In time-dependent problems the wave function $\psi(x,t)$ satisfies the time-dependent Schrödinger equation

$$i\hbar\frac{\partial \psi}{\partial t} = [H_0 + V(x,t)]\psi(x,t). \tag{15.28}$$

where H_0 is the unperturbed Hamiltonian and $V(x,t)$ is the time-dependent perturbation.

We expand the wave function in the form

$$\psi(x,t) = \sum_n a_n(t)u_n(x)\exp(-iE_n t/\hbar) \tag{15.29}$$

where $u_n(x)$ is the wave function of the nth state of the harmonic oscillator with energy $E_n = (n + \frac{1}{2})\hbar\omega$ and we have $H_0 u_n(x) = E_n u_n(x)$. Substituting (15.29) into (15.28), multiplying across by $u_n^*(x)\exp(iE_n t/\hbar)$ and integrating with respect to x, then gives

$$i\hbar \frac{\partial a_n}{\partial t} = \int u_n^*(x)\exp(iE_n t/\hbar)V(x,t)\psi(x,t)\,dx.$$

Now taking the unperturbed wave function to be $u_0(x)\exp(-iE_0 t/\hbar)$ we obtain for the first order amplitude

$$a_n^{(1)}(t) = \frac{1}{i\hbar}\int_{-\infty}^t V_{n0}(t')\exp[i(E_n - E_0)t'/\hbar]\,dt' \tag{15.30}$$

where $V_{nm}(t) = \int u_n^*(x)V(x,t)u_m(x)\,dx$ and denoting the ground state by 0.
Since $V(t) = -exF(t)$ we have

$$V_{n0}(t) = -\frac{eA}{\sqrt{\pi}}\frac{\exp(-t^2/\tau^2)}{\tau}\int_{-\infty}^\infty u_n^*(x)u_0(x)x\,dx$$

and so

$$a_n^{(1)}(\infty) = \frac{ieA}{\hbar\tau\sqrt{\pi}}\int_{-\infty}^\infty u_n^*(x)u_0(x)x\,dx\int_{-\infty}^\infty \exp[-t^2/\tau^2 + in\omega t]\,dt.$$

Now $\int_{-\infty}^\infty u_1(x)u_0(x)x\,dx = 1/(\alpha\sqrt{2})$ where $\alpha = (m\omega/\hbar)^{1/2}$, using $u_0(x) = (\alpha/\pi^{1/2})^{1/2}\exp(-\frac{1}{2}\alpha^2 x^2)$ and $u_1(x) = (2\alpha^3/\pi^{1/2})^{1/2}x\exp(-\frac{1}{2}\alpha^2 x^2)$. Also

$$\int_{-\infty}^\infty \exp(-t^2/\tau^2 + i\omega t)\,dt = \int_{-\infty}^\infty \exp(-t^2/\tau^2)\cos\omega t\,dt = \tau\sqrt{\pi}\exp(-\omega^2\tau^2/4)$$

using (15.6), so that

$$a_1^{(1)}(\infty) = \frac{ieA}{\hbar\alpha\sqrt{2}}\exp(-\omega^2\tau^2/4).$$

Thus the first-order probability of finding the linear harmonic oscillator in the first excited state is

$$|a_1^{(1)}(\infty)|^2 = \frac{e^2 A^2}{2\hbar m\omega}\exp(-\omega^2\tau^2/2).$$

15.39 A hydrogen atom in its ground state is perturbed by a homogeneous electric field having the time dependence

$$F(t) = \begin{cases} 0 & (t < 0) \\ F_0\exp(-t/\tau) & (t \geq 0). \end{cases}$$

Find the first order probability that the atom is eventually left in (i) the 2s state, (ii) the 2p states.

• *Solution*

The perturbation acting on the hydrogen atom is $V(\mathbf{r},t) = -eF(t)r\cos\theta$ choosing the polar axis in the direction of the electric field.
We use the same method as in the previous problem and expand the wave function in the form

$$\psi(\mathbf{r},t) = \sum_n a_n(t)\phi_n(\mathbf{r})\exp(-iE_n t/\hbar)$$

where $\phi_n(\mathbf{r})$ is the wave function of the nth state of the hydrogen atom with energy E_n. Then we obtain for the first order amplitude

$$a_n^{(1)} = \frac{1}{i\hbar} \int_0^t V_{n1}(t') \exp[i(E_n - E_1)t'/\hbar] \, dt'$$

where $V_{nm}(t) = \int \phi_n^*(\mathbf{r}) V(\mathbf{r}, t) \phi_m(\mathbf{r}) \, d\tau$ and the ground state is denoted by 1.
  For $t \geq 0$ we have

$$V_{n1}(t) = -eF_0 \exp(-t/\tau) \int \phi_n \phi_1 \, r \cos\theta \, d\tau.$$

Now $\int \phi_n \phi_1 \, r \cos\theta \, d\tau = 0$ for the $1s, 2s$ and $2p_{\pm 1}$ final states and so the first-order probability of exciting these states is zero . However for the $2p_0$ final state we get, using $\phi_{1s} = (\pi a_0^3)^{-1/2} \exp(-r/a_0)$ and $\phi_{2p_0} = (32\pi a_0^5)^{-1/2} r \cos\theta \exp(-r/2a_0)$,

$$\int \phi_{2p_0} \phi_{1s} r \cos\theta \, d\tau = \frac{1}{4\sqrt{2}\pi a_0^4} \int r^2 \exp(-3r/2a_0) \cos^2\theta \, d\tau$$

and so

$$V_{2p_0 1s} = -\frac{eF_0 \exp(-t/\tau)}{3\sqrt{2}a_0^4} \int_0^\infty r^4 \exp(-3r/2a_0) \, dr = -4\sqrt{2}\left(\frac{2}{3}\right)^5 a_0 eF_0 \exp(-t/\tau).$$

It follows that

$$a_{2p_0}^{(1)}(t) = \frac{ieF_0}{\hbar} 4\sqrt{2} \left(\frac{2}{3}\right)^5 a_0 \int_0^t \exp\left\{[-1/\tau + i(E_2 - E_1)/\hbar]t'\right\} dt'$$

and this gives

$$a_{2p_0}^{(1)}(t) = \frac{ieF_0 4\sqrt{2} \left(\frac{2}{3}\right)^5 a_0}{\hbar} \frac{\exp\left\{[-1/\tau + i(E_2 - E_1)/\hbar]t\right\} - 1}{i(E_2 - E_1)/\hbar - 1/\tau}.$$

In the limit as $t \to \infty$ we get

$$a_{2p_0}^{(1)}(\infty) = \frac{ieF_0 4\sqrt{2} \left(\frac{2}{3}\right)^5 a_0}{\hbar/\tau - i(E_2 - E_1)}$$

and since $E_2 - E_1 = (e^2/2a_0)(-1/4 + 1) = 3e^2/8a_0$ we arrive at the first-order probability for the excitation of the $2p_0$ state of hydrogen given by

$$\left|a_{2p_0}^{(1)}(\infty)\right|^2 = \frac{32\left(\frac{2}{3}\right)^{10} e^2 a_0^2 F_0^2}{9e^4/(64a_0^2) + \hbar^2/\tau^2}.$$

15.40 A hydrogen atom in its ground state is perturbed by a homogeneous electric field having the time dependence

$$F(t) = \frac{A\tau}{e\pi} \frac{1}{t^2 + \tau^2} \quad (-\infty < t < \infty)$$

and directed along the polar axis. Find the first order probability that the atom is eventually left in the $2p_0$ state.

• *Solution*

This is similar to the previous problem but now the relevant time integral is

$$\int_{-\infty}^{\infty} \frac{\exp[i(E_2 - E_1)t'/\hbar]}{t'^2 + \tau^2} \, dt'.$$

This can be evaluated by taking a semi-circular contour in the upper half plane. Since there is a pole at $i\tau$ the integral is given by

$$\frac{\pi}{\tau} \exp[-(E_2 - E_1)\tau/\hbar]$$

and so we obtain

$$\left| a_{2p0}^{(1)}(\infty) \right|^2 = 32 \left(\frac{2}{3} \right)^{10} \left(\frac{a_0 A}{\hbar} \right)^2 \exp[-2(E_2 - E_1)\tau/\hbar].$$

16
VARIATIONAL PRINCIPLES

Variational principles and *variational methods* provide powerful ways of solving problems in various branches of applied mathematics and theoretical physics. We have already used them in the chapters on *classical dynamics*, on the *theory of relativity*, and on *quantum theory* where they are of great importance. In this chapter we will solve a variety of problems including the determination of *geodesics* and the approximate evaluation of eigenvalues for *wave motions*.

In the *calculus of variations* we consider curves Γ joining two fixed end points corresponding to values x_1 and x_2 of the independent variable x, in the n-dimensional space where $y_1, y_2, ..., y_n$ are curvilinear coordinates.

The fundamental problem is to find the curve Γ_0 defined by $y_r = \phi_r(x)$ $(r = 1, ..., n)$ that gives a stationary value of the integral

$$I = \int_{x_1}^{x_2} F(x; y_1, ..., y_n; y_1', ..., y_n') \, dx \tag{16.1}$$

where the primes denote differentiations with respect to x.

The *Euler-Lagrange equations*

$$\frac{\partial F}{\partial y_r} - \frac{d}{dx}\left(\frac{\partial F}{\partial y_r'}\right) = 0 \quad (r = 1, ..., n) \tag{16.2}$$

are a necessary condition for the curve Γ_0 to provide a stationary value of I.

They are associated with Lagrange because they become *Lagrange's equations*

$$\frac{\partial L}{\partial q_r} - \frac{d}{dt}\left(\frac{\partial L}{\partial \dot{q}_r}\right) = 0$$

for a holonomic dynamical system characterized by n generalized coordinates $q_1(t), ..., q_n(t)$ if F is the *Lagrangian function* $L(t; q_1, ..., q_n; \dot{q}_1, ..., \dot{q}_n)$, x is the time t and the y_r are the q_r.

If F is independent of x, the Euler-Lagrange equations can be integrated to give

$$U \equiv F - \sum_{r=1}^{n} y_r' \frac{\partial F}{\partial y_r'} = C \tag{16.3}$$

where C is a constant. This is the energy equation in classical dynamics.

16-1 Geodesics

The first few problems are on *geodesics*, that is on problems concerned with finding the shortest path between two points on a given surface.

16.01 Find the geodesics on a sphere by minimizing the path integral $\int ds$ using spherical polar coordinates.

• *Solution*

On a sphere of radius a an element of path ds is given by the metric

$$ds^2 = a^2 d\theta^2 + a^2 \sin^2\theta \, d\phi^2$$

in spherical polar coordinates θ, ϕ.

Then we have

$$\int ds = a \int \sqrt{\left(\frac{d\theta}{d\phi}\right)^2 + \sin^2\theta} \, d\phi$$

and so we require to minimize

$$\int F(\theta, \theta') \, d\phi$$

where $F(\theta, \theta') = \sqrt{\theta'^2 + \sin^2\theta}$ and $\theta' = d\theta/d\phi$.

Since F is independent of ϕ the Euler-Lagrange equation can be integrated using (16.3) to give

$$U \equiv F - \theta' \frac{\partial F}{\partial \theta'} = \lambda$$

where λ is a constant. Hence

$$\sqrt{\theta'^2 + \sin^2\theta} - \frac{\theta'^2}{\sqrt{\theta'^2 + \sin^2\theta}} = \lambda,$$

that is

$$\frac{\sin^2\theta}{\sqrt{\theta'^2 + \sin^2\theta}} = \lambda.$$

Hence, rearranging and integrating we obtain

$$\phi = \lambda \int \frac{d\theta}{\sin\theta\sqrt{\sin^2\theta - \lambda^2}}.$$

Now setting $u = \lambda \cot\theta/\sqrt{1 - \lambda^2}$ we get

$$\phi = -\int \frac{du}{\sqrt{1 - u^2}}$$

which gives $\phi + \alpha = -\sin^{-1}u$ where α is a constant. It follows that

$$\sin\alpha \sin\theta \cos\phi + \cos\alpha \sin\theta \sin\phi + \tan\beta \cos\theta = 0$$

where $\tan\beta = \lambda/\sqrt{1 - \lambda^2}$. Then putting $x = a\sin\theta \cos\phi$, $y = a\sin\theta \sin\phi$, $z = a\cos\theta$ we get

$$x \sin\alpha \cos\beta + y \cos\alpha \cos\beta + z \sin\beta = 0$$

which is the equation of a plane passing through the centre of the sphere whose normal is in the direction of the unit vector $\hat{\mathbf{n}} = (\sin\alpha \cos\beta, \cos\alpha \cos\beta, \sin\beta)$. Thus the geodesics are great circles on the sphere.

16.02 Find the differential equations of a geodesic on any surface of revolution given by

$$x = r\cos\theta, \quad y = r\sin\theta, \quad z = f(r) \tag{16.4}$$

and show that $r^2 \, d\theta/ds =$ constant where s denotes arc length. Hence prove that the geodesics on a right circular cylinder cut its generators at a constant angle and are therefore helices.

VARIATIONAL PRINCIPLES
295

• *Solution*

If ds is an element of arc on the surface of revolution, we have

$$ds^2 = dx^2 + dy^2 + dz^2 = r^2 d\theta^2 + \left[1 + \left(\frac{df}{dr}\right)^2\right] dr^2.$$

Thus we have to minimize

$$\int ds = \int F(r, r')\, d\theta$$

where

$$F(r, r') = \left\{r^2 + \left[1 + \left(\frac{df}{dr}\right)^2\right] r'^2\right\}^{\frac{1}{2}}$$

and $r' = dr/d\theta$.

Since F does not depend on θ, the first integral of the Euler-Lagrange equation is

$$U \equiv F - r'\frac{\partial F}{\partial r'} = b$$

where b is a constant.

Hence we have

$$r^2 \left\{r^2 + \left[1 + \left(\frac{df}{dr}\right)^2\right] r'^2\right\}^{-\frac{1}{2}} = b, \qquad (16.5)$$

that is

$$r^2 \frac{d\theta}{ds} = b.$$

For a right circular cylinder $r = a$ where a is the constant radius of the cylinder, and so $a\,d\theta/ds = b/a$ which means that the geodesics cut the generators of the cylinder at a constant angle $\alpha = \sin^{-1}(b/a)$.

16.03 Find the geodesics on a right circular cone, verifying that when the cone is developed into a plane, the geodesics become straight lines.

• *Solution*

This is a special case of the previous problem for which $x = r\cos\theta$, $y = r\sin\theta$, $z = r\cot\alpha$ where α is the semi-angle at the vertex. Then $f(r) = r\cot\alpha$ and so from (16.5) we have

$$r^2[r^2 + (1 + \cot^2\alpha)r'^2]^{-1/2} = b$$

which gives

$$r'^2 = \frac{\sin^2\alpha}{b^2} r^2(r^2 - b^2).$$

To solve this equation we put $r = u^{-1}$. Then $r' = -u'u^{-2}$ and so

$$u'^2 = \sin^2\alpha\,(b^{-2} - u^2)$$

which yields $u = b^{-1}\sin(\theta\sin\alpha)$, that is

$$r = \frac{b}{\sin(\theta\sin\alpha)}.$$

The cone develops into a plane when $\alpha \to \pi/2$ giving the straight line $r \sin \theta = b$.

16.04 Find the differential equation defining the geodesics on the paraboloid of revolution

$$x = r \cos \theta, \quad y = r \sin \theta, \quad z = r^2/2a.$$

• *Solution*

This is another special case of problem 16.02 for which $f(r) = r^2/2a$ so that $df/dr = r/a$. Then we have the first integral (16.5) of the Euler-Lagrange equation

$$U \equiv r^2 \left\{ r^2 + \left[1 + \left(\frac{r}{a} \right)^2 \right] r'^2 \right\}^{-\frac{1}{2}} = b$$

where $r' = dr/d\theta$, which yields

$$\left(\frac{dr}{d\theta} \right)^2 = \frac{a^2 r^2 (r^2 - b^2)}{b^2 (r^2 + a^2)}.$$

16-2 Mechanics

The first problem in this section is on *statics*.

16.05 Using the principle that for stable equilibrium the potential energy of a system must be a minimum, find the shape of a uniform chain suspended from two fixed points.

• *Solution*

The potential energy of a chain with fixed end points at A and B is given by

$$V = \rho g \int_A^B y \, ds$$

where ρ is the mass per unit length of the chain.

Hence, if the fixed ends of the chain are at $x = a$ and $x = b$, we have

$$V = \rho g \int_a^b y \sqrt{1 + y'^2} \, dx \qquad (16.6)$$

where $y' = dy/dx$, and this must be made stationary subject to the condition

$$l = \int_A^B ds = \int_a^b \sqrt{1 + y'^2} \, dx \qquad (16.7)$$

where l is the length of the chain.

According to *Euler's rule* we require the extremals for

$$\int_a^b (y + \lambda) \sqrt{1 + y'^2} \, dx$$

where λ is an unknown constant.

Euler's equation has the first integral $U \equiv F - y' \partial F / \partial y' = \beta$ and so we get $y + \lambda = \beta \sqrt{1 + y'^2}$ where β is a constant. Hence we have

$$\left(\frac{dy}{dx} \right)^2 = \frac{(y + \lambda)^2}{\beta^2} - 1$$

and thus we obtain the equation of a catenary

$$y + \lambda = \beta \cosh \frac{x - \alpha}{\beta}$$

where α is a constant.

Now $ds/dx = \sqrt{1 + y'^2} = \cosh[(x - \alpha)/\beta]$. Hence the arc length is given by

$$s = \beta \sinh \frac{x - a}{\beta}$$

taking $\alpha = a$ so that $s = 0$ at the first point of the chain $x = a$. Since $s = l$ when $x = b$ we see that β is determined by $l = \beta \sinh[(b - a)/\beta]$.

16.06 Use Jacobi's form of the principle of least action to find the orbits of a particle moving in a plane under the attractive forces (i) μ/r^2, (ii) $m\omega^2 r$, (iii)$2\mu/r^3$ towards a fixed point O, where r is the distance from O.

• *Solution*

The *Jacobi form* of the *principle of least action* requires the minimization of the *action integral*

$$A = \int \sqrt{2m(E - V)}\, ds \tag{16.8}$$

where E is the total energy and V is the potential so that $\sqrt{2m(E - V)} = p$ is the momentum of the particle. Since the potential is a function of r only and $ds/d\theta = \sqrt{r'^2 + r^2}$ we have to minimize

$$\int \sqrt{2m[E - V(r)]}\sqrt{r'^2 + r^2}\, d\theta$$

where r, θ are the polar angles and $r' = dr/d\theta$.

Now $F(r, r') = \sqrt{2m[E - V(r)]}\sqrt{r'^2 + r^2}$ is independent of θ and so $U \equiv F - r'\partial F/\partial r' = \lambda$ where λ is a positive constant. It follows that

$$\frac{r^2\sqrt{2[E - V(r)]}}{\sqrt{r'^2 + r^2}} = \lambda$$

and this gives

$$r'^2 = \frac{2[E - V(r)]}{\lambda^2}r^4 - r^2. \tag{16.9}$$

(i) $V(r) = -\mu/r$ where $\mu > 0$.

To solve the differential equation (16.9) we put $r = u^{-1}$. Then $r' = -u^{-2}u'$ and so

$$u'^2 = \frac{2E}{\lambda^2} + \frac{\mu^2}{\lambda^4} - \left(u - \frac{\mu}{\lambda^2}\right)^2$$

which gives

$$u = \frac{\mu}{\lambda^2} + A\sin(\theta + \epsilon)$$

where $A = \mu\lambda^{-2}\sqrt{1 + 2E\lambda^2/\mu^2}$. Thus we have the conic

$$\frac{l}{r} = 1 + e\sin(\theta + \epsilon)$$

where the semi-latus rectum $l = \lambda^2/\mu$ and the eccentricity $e = \sqrt{1 + 2E\lambda^2/\mu^2}$.

If $E \begin{cases} < 0 \\ = 0 \\ > 0 \end{cases}$, the path is an $\begin{cases} \text{ellipse} \\ \text{parabola} \\ \text{hyperbola} \end{cases}$ since $e \begin{cases} < 1 \\ = 1 \\ > 1 \end{cases}$ respectively.

(ii) $V(r) = \frac{1}{2}m\omega^2 r^2$.

Here we put $u = r^{-2}$ in (16.9). Then $u' = -2r'r^{-3}$ and so

$$u'^2 = \frac{4E^2}{\lambda^4} - \frac{4m\omega^2}{\lambda^2} - 4\left(u - \frac{E}{\lambda^2}\right)^2.$$

Hence

$$u = \frac{E}{\lambda^2} + A\cos 2(\theta + \epsilon)$$

where $A = \lambda^{-1}\sqrt{E^2/\lambda^2 - m\omega^2}$. It follows that we get the ellipse

$$\frac{x^2}{a^2} + \frac{y^2}{b^2} = 1$$

where $x = r\cos(\theta + \epsilon)$, $y = r\sin(\theta + \epsilon)$ and $a^{-2} = E\lambda^{-2} + A$, $b^{-2} = E\lambda^{-2} - A$ since $E\lambda^{-2} > A$.

(iii) $V(r) = -\mu/r^2$ where $\mu > 0$.

We put $u = r^{-1}$ in (16.9). Then $u' = -r'r^{-2}$ and so

$$u'^2 = \frac{2E}{\lambda^2} + \left(\frac{2\mu}{\lambda^2} - 1\right)u^2$$

which gives

$$\frac{d^2u}{d\theta^2} + \left(1 - \frac{2\mu}{\lambda^2}\right)u = 0.$$

Hence, if $2\mu < \lambda^2$, we get

$$u = \frac{1}{r} = A\sin\left(\sqrt{1 - \frac{2\mu}{\lambda^2}}\,\theta + \epsilon\right)$$

where $A = \sqrt{2E/(\lambda^2 - 2\mu)}$ and $E > 0$.

However if $2\mu > \lambda^2$ we obtain

$$u = \frac{1}{r} = B\sinh\left(\sqrt{\frac{2\mu}{\lambda^2} - 1}\,\theta + \epsilon\right)$$

where $B = \sqrt{2E/(2\mu - \lambda^2}$ and $E > 0$.

Also if $2\mu = \lambda^2$ we get $r^{-1} = \sqrt{2E/\lambda^2}\,\theta + \epsilon$ and $E > 0$.

16.07 Show that the principle of least action can be generalized to include dynamical systems for which the Hamiltonian

$$H = \sum_{r=1}^{n} \dot{q}_r \frac{\partial L}{\partial \dot{q}_r} - L \tag{16.10}$$

is not an integral of energy and thus H is not a constant, by showing that

$$\int \left(\sum_{r=1}^{n} \dot{q}_r \frac{\partial L}{\partial \dot{q}_r} + t\frac{dH}{dt}\right) dt$$

is stationary for variations such that the actual path and the neighbouring path have coincident terminal points, and $\Delta H = 0$ at the terminal points.

- *Solution*

Suppose that the arc AB in configuration space represents a part of a *trajectory*, that is an actual path of the dynamical system, and let $A'B'$ be an adjacent arc which need not be a trajectory.

Suppose further that t_1, t_2 and $t_1 + \Delta t_1$, $t_2 + \Delta t_2$ are the times corresponding to the end-points A, B and A', B' of the arcs, respectively.

If δ denotes a weak variation of a quantity between a point of the arc AB and a point of the arc $A'B'$ associated with the same time t, known as a *contemporaneous variation*, then

$$\int_{A'B'} L\,dt - \int_{AB} L\,dt = L_B \Delta t_2 - L_A \Delta t_1 + \int_{t_1}^{t_2} \delta L\,dt$$

where $L(t; q_r, \dot{q}_r)$ is the Lagrangian function of the system.

Since

$$\delta L = \sum_{r=1}^{n} \left(\frac{\partial L}{\partial \dot{q}_r} \delta \dot{q}_r + \frac{\partial L}{\partial q_r} \delta q_r \right)$$

and using Lagrange's equations

$$\frac{d}{dt}\left(\frac{\partial L}{\partial \dot{q}_r} \right) = \frac{\partial L}{\partial q_r}$$

we get

$$\int_{A'B'} L\,dt - \int_{AB} L\,dt = L_B \Delta t_2 - L_A \Delta t_1 + \int_{t_1}^{t_2} \left(\sum_{r=1}^{n} \frac{\partial L}{\partial \dot{q}_r} \delta \dot{q}_r + \frac{d}{dt}\left(\frac{\partial L}{\partial \dot{q}_r} \right) \delta q_r \right) dt.$$

Now $\delta \dot{q}_r = d(\delta q_r)/dt$ and so

$$\int_{A'B'} L\,dt - \int_{AB} L\,dt = \left[L\Delta t + \sum_{r=1}^{n} \frac{\partial L}{\partial \dot{q}_r} \delta q_r \right]_A^B.$$

Denoting the changes in q_r between A and A' and between B and B' by $(\Delta q_r)_A = (\delta q_r)_A + (\dot{q}_r)_A \,\Delta t_1$ and $(\Delta q_r)_B = (\delta q_r)_B + (\dot{q}_r)_B \,\Delta t_2$ respectively, we obtain

$$\int_{A'B'} L\,dt - \int_{AB} L\,dt = \left[\sum_{r=1}^{n} \frac{\partial L}{\partial \dot{q}_r} \Delta q_r - H\Delta t \right]_A^B$$

where H is the Hamiltonian function given by (16.10).

Now

$$\sum_{r=1}^{n} \dot{q}_r \frac{\partial L}{\partial \dot{q}_r} + t\frac{dH}{dt} = L + H + t\frac{dH}{dt} = L + \frac{d}{dt}(tH)$$

and hence

$$\int_{A'B'} \left(\sum_{r=1}^{n} \dot{q}_r \frac{\partial L}{\partial \dot{q}_r} + t\frac{dH}{dt} \right) dt - \int_{AB} \left(\sum_{r=1}^{n} \dot{q}_r \frac{\partial L}{\partial \dot{q}_r} + t\frac{dH}{dt} \right) dt = \left[\sum_{r=1}^{n} \frac{\partial L}{\partial \dot{q}_r} \Delta q_r - H\Delta t + \Delta(tH) \right]_A^B.$$

Thus we have

$$\int_{A'B'} \left(\sum_{r=1}^{n} \dot{q}_r \frac{\partial L}{\partial \dot{q}_r} + t\frac{dH}{dt} \right) dt - \int_{AB} \left(\sum_{r=1}^{n} \dot{q}_r \frac{\partial L}{\partial \dot{q}_r} + t\frac{dH}{dt} \right) dt = \left[\sum_{r=1}^{n} \frac{\partial L}{\partial \dot{q}_r} \Delta q_r + t\Delta H \right]_A^B.$$

and hence if $\Delta q_r = 0$ $(r = 1, ..., n)$ and $\Delta H = 0$ at A and B we see that

$$\int \left(\sum_{r=1}^{n} \dot{q}_r \frac{\partial L}{\partial \dot{q}_r} + t \frac{dH}{dt} \right) dt$$

is stationary as required.

16.08 A relativistic particle having rest mass m_0 and charge q is moving with four-velocity $(V_\nu) = (\gamma \mathbf{v}, ic\gamma)$ where $\gamma = (1 - v^2/c^2)^{-1/2}$ in an electromagnetic field with four-potential $(\psi_\mu) = (\mathbf{A}, iV)$. Using the variational principle

$$\delta \int_{\tau_1}^{\tau_2} L \, d\tau = 0$$

where τ is the proper time and the Lagrangian is given by

$$L = \frac{1}{2} m_0 \sum_{\nu=1}^{4} V_\nu^2 + \frac{q}{c} \sum_{\mu=1}^{4} V_\mu \psi_\mu,$$

show that the Euler-Lagrange equations yield the equations of motion of the particle in the form

$$\frac{d(m_0 V_\nu)}{d\tau} = F_\nu$$

and find an expression for the Minkowski four-force F_ν showing that its spatial part is given by $\gamma \mathbf{f}$ where $\mathbf{f}$ is the Lorentz force.

• *Solution*

The Euler-Lagrange equations are

$$\frac{d}{d\tau} \left(\frac{\partial L}{\partial V_\nu} \right) = \frac{\partial L}{\partial x_\nu} \quad (\nu = 1, 2, 3, 4)$$

where the components of the four-velocity are given by $V_\nu = dx_\nu/d\tau$.

Hence

$$\frac{d}{d\tau}(m_0 V_\nu) + \frac{d}{d\tau} \left(\frac{q}{c} \psi_\nu \right) = \frac{\partial}{\partial x_\nu} \left(\frac{q}{c} \sum_{\mu=1}^{4} V_\mu \psi_\mu \right),$$

that is

$$\frac{d(m_0 V_\nu)}{d\tau} = F_\nu$$

where

$$F_\nu = \frac{\partial}{\partial x_\nu} \left(\frac{q}{c} \sum_{\mu=1}^{4} V_\mu \psi_\mu \right) - \frac{d}{d\tau} \left(\frac{q}{c} \psi_\nu \right).$$

The spatial part of the Minkowski force F_ν is

$$\mathbf{F} = \nabla \left[\frac{q}{c} (\gamma \mathbf{A}.\mathbf{v} - c\gamma V) \right] - \gamma \frac{d}{dt} \left(\frac{q}{c} \mathbf{A} \right)$$

since $\gamma d\tau = dt$, that is $\mathbf{F} = \gamma \mathbf{f}$ where

$$\mathbf{f} = q \left[-\nabla V - \frac{1}{c} \frac{d\mathbf{A}}{dt} + \frac{1}{c} \nabla (\mathbf{A}.\mathbf{v}) \right]$$

is the Lorentz force.

This is the same as the definition of $\mathbf{f} = q(\mathbf{E} + c^{-1} \mathbf{v} \times \mathbf{H})$ given in problem 14.17 since $d\mathbf{A}/dt = \partial \mathbf{A}/\partial t + \mathbf{v}.\nabla \mathbf{A}$, $\mathbf{E} = -\nabla V - c^{-1} \partial \mathbf{A}/\partial t$ and $\mathbf{H} = \operatorname{curl} \mathbf{A}$ so that $\mathbf{v} \times \mathbf{H} = \mathbf{v} \times (\nabla \times \mathbf{A}) = \nabla(\mathbf{v}.\mathbf{A}) - \mathbf{v}.\nabla \mathbf{A}$.

16-3 Light rays

The next three problems are concerned with *Fermat's principle* which states that a light ray takes the least possible time in travelling between two fixed points.

16.09 Two homogeneous media in which the speeds of light are v_1 and v_2 are separated by a plane interface. A light ray passes from a point A in medium 1 to a point B in medium 2. If the angles α_1 and α_2 of incidence and refraction of the light ray at the interface satisfy *Snell's law*

$$\frac{\sin \alpha_1}{\sin \alpha_2} = \frac{v_1}{v_2}$$

show that the time taken by the light ray in travelling from A to B is the least possible.

• *Solution*

If the points A and B are perpendicular distances a and b from the plane interface respectively, the time taken by the light ray to travel from A to B is

$$T = \frac{a}{v_1 \cos \alpha_1} + \frac{b}{v_2 \cos \alpha_2}$$

and so

$$\frac{dT}{d\alpha_1} = \frac{a \sin \alpha_1}{v_1 \cos^2 \alpha_1} + \frac{b \sin \alpha_2}{v_2 \cos^2 \alpha_2} \frac{d\alpha_2}{d\alpha_1}.$$

Since A and B are fixed points we have

$$a \tan \alpha_1 + b \tan \alpha_2 = d$$

where d is a constant, and so we see that

$$a \sec^2 \alpha_1 + b \sec^2 \alpha_2 \frac{d\alpha_2}{d\alpha_1} = 0.$$

Hence

$$\frac{dT}{d\alpha_1} = \frac{a}{\cos^2 \alpha_1} \left(\frac{\sin \alpha_1}{v_1} - \frac{\sin \alpha_2}{v_2} \right) = 0$$

by Snell's law.

Also

$$\frac{d^2 T}{d\alpha_1^2} = \frac{a}{\cos^2 \alpha_1} \left(\frac{\cos \alpha_1}{v_1} + \frac{\cos \alpha_2}{v_2} \frac{a \cos^2 \alpha_2}{b \cos^2 \alpha_1} \right) > 0$$

and so the time taken is a minimum.

This is a special case of *Fermat's principle of least time*.

16.10 The refractive index μ in an optical medium is a function of y only, where $O(x, y)$ is a rectangular Cartesian frame of reference. Show from Fermat's principle of least time that, for a light ray in the $z = 0$ plane, $\mu \cos \psi$ is constant along the ray where ψ is the angle the ray makes with the x-axis. Hence find the dependence of μ on y if the light path is part of an ellipse

$$\frac{x^2}{a^2} + \frac{y^2}{b^2} = 1.$$

• *Solution*

The refractive index in an optical medium is defined by $\mu = c/v$ where c is the velocity of light *in vacuo* and v is the velocity of light in the medium. Then Fermat's principle requires that $\int \mu \, ds$ be a minimum where ds is an element of arc length. Thus we need to minimize $\int \mu(y) \sqrt{1 + y'^2} \, dx$ where $y' = dy/dx$.

Since the integrand $F(y, y') = \mu(y) \sqrt{1 + y'^2}$ is independent of x it follows that $U \equiv F - y' \partial F / \partial y' =$ constant and so we obtain $\mu(y)/\sqrt{1 + y'^2} =$ constant.

Putting $y' = \tan \psi$ we get $\sqrt{1 + y'^2} = 1/\cos \psi$ and hence we obtain $\mu(y) \cos \psi =$ constant along the ray, as required.

Now if the light ray is a part of the ellipse $x^2/a^2 + y^2/b^2 = 1$ we have $x/a^2 + yy'/b^2 = 0$ which gives $y'^2 = b^2(b^2 - y^2)/(a^2 y^2)$. It follows that

$$\mu(y) = \mu(b) \left[1 + \frac{b^2}{a^2} \frac{b^2 - y^2}{y^2} \right]^{\frac{1}{2}}.$$

16.11 Using Fermat's principle, find the path of a light ray in a medium whose refractive index is given by

$$\mu(r) = \left(1 + \frac{k^2}{r^2} \right)^{\frac{1}{2}},$$

assuming that the ray is at right angles to the radius vector when $r = a$.

• *Solution*

In this problem we have to minimize $\int \mu(r) \, ds = \int \mu(r) \sqrt{r'^2 + r^2} \, d\theta$ where r, θ are polar coordinates and $r' = dr/d\theta$. Since the integrand $F(r, r') = \mu(r) \sqrt{r'^2 + r^2}$ is independent of θ we have $U \equiv F - r' \partial F / \partial r' = \lambda$ where λ is a constant and so

$$\frac{r^2 \mu(r)}{\sqrt{r'^2 + r^2}} = \lambda.$$

If $\mu(r) = \sqrt{1 + k^2/r^2}$ then

$$r'^2 = \frac{r^4}{\lambda^2} \left(1 + \frac{k^2}{r^2} \right) - r^2.$$

Now supposing that the ray is at right angles to the radius vector when $r = a$ we have $r' = 0$ when $r = a$ and hence, since $\lambda = a\mu(a)$, we may write

$$r'^2 = \frac{r^2(r^2 - a^2)}{[a\mu(a)]^2}.$$

To solve this equation we put $r = u^{-1}$. Then $r' = -u'u^{-2}$ and so $u'^2 = (1 - a^2 u^2)/[a\mu(a)]^2$ which yields

$$\frac{d^2 u}{d\theta^2} + \frac{u}{[\mu(a)]^2} = 0.$$

Hence $u = A \cos[\theta/\mu(a) + \epsilon]$ where A and ϵ are constants, and if $r = a$ when $\theta = 0$ we get

$$r = a \sec \frac{\theta}{\mu(a)}.$$

16.12 According to the general theory of relativity the path of a light ray in the neighbourhood of the Sun minimizes

$$\int \sqrt{k^2 r'^2 + kr^2}\, d\phi$$

where

$$k(r) = \left(1 - \frac{m}{r}\right)^{-1}$$

and $m = 2GM/c^2$ is a constant, M being the mass of the Sun and G the gravitational constant. Show that the differential equation of the path is

$$\frac{d^2 u}{d\phi^2} + u = \frac{3}{2}mu^2$$

where $u = 1/r$.

• *Solution*

Here we have to minimize $\int F(r, r')\, d\phi$ where $r' = dr/d\phi$ and $F(r, r') = \sqrt{k^2 r'^2 + kr^2}$ is independent of ϕ. Hence $U \equiv F - r'\partial F/\partial r'$ is a constant and so

$$\frac{kr^2}{\sqrt{k^2 r'^2 + kr^2}} = \lambda$$

where λ is a constant.

Hence we have

$$r'^2 = \frac{r^4}{\lambda^2} - \frac{r^2}{k(r)}.$$

To solve this equation we put $r = u^{-1}$. Then $r' = -u^{-2}u'$ and so we get $u'^2 = \lambda^{-2} - u^2(1 - mu)$ and thus

$$\frac{d}{du}\left(\frac{1}{2}u'^2\right) + u = \frac{3}{2}mu^2.$$

Since $d(\frac{1}{2}u'^2)/du = d^2 u/d\phi^2$ the required equation for the path of the light ray follows.

16-4 Field equations

The next few problems are concerned with variational principles for field equations.

Consider a field whose n components are the real functions $\psi_\sigma(x_\mu)$ $(\sigma = 1, ..., n)$ where the x_μ $(\mu = 1, ..., m)$ are a set of m coordinates specifying location in space and time.

To derive the field equations we introduce a *Lagrangian density* $\mathcal{L}(x_\mu, \psi_\sigma, \partial\psi_\sigma/\partial x_\mu)$ and then define the integral

$$I = \int_T \mathcal{L}(x_\mu, \psi_\sigma, \partial\psi_\sigma/\partial x_\mu)\, d\tau \tag{16.11}$$

where the domain of integration T is a hypervolume in the m-dimensional space of the coordinates x_μ.

If the integral I is assumed to be stationary for weak variations $\delta\psi_\sigma$ which vanish over the boundary hypersurface of T, the Euler-Lagrange field equations

$$\frac{\partial\mathcal{L}}{\partial\psi_\sigma} - \sum_{\mu=1}^{m}\frac{\partial}{\partial x_\mu}\left[\frac{\partial\mathcal{L}}{\partial(\partial\psi_\sigma/\partial x_\mu)}\right] = 0 \quad (\sigma = 1, ..., n) \tag{16.12}$$

are obtained.

16.13 If the Lagrangian density for a field with m components ψ_ν ($\nu = 1, ..., m$) is given by

$$\mathcal{L} = \frac{1}{2} \sum_{\mu=1}^{m} \sum_{\nu=1}^{m} \left(\frac{\partial \psi_\nu}{\partial x_\mu} - \frac{\partial \psi_\mu}{\partial x_\nu} \right)^2 + \lambda^2 \sum_{\nu=1}^{m} \psi_\nu^2 \tag{16.13}$$

where λ is a non-zero constant, show that

$$\sum_{\nu=1}^{m} \frac{\partial \psi_\nu}{\partial x_\nu} = 0 \tag{16.14}$$

and

$$\sum_{\mu=1}^{m} \frac{\partial^2 \psi_\nu}{\partial x_\mu^2} = \lambda^2 \psi_\nu \quad (\nu = 1, ..., m). \tag{16.15}$$

• *Solution*

The Euler-Lagrange equation (16.12) for the field component ψ_ν gives

$$\sum_{\mu=1}^{m} \frac{\partial}{\partial x_\mu} \left(\frac{\partial \psi_\nu}{\partial x_\mu} - \frac{\partial \psi_\mu}{\partial x_\nu} \right) = \lambda^2 \psi_\nu \quad (\nu = 1, ..., m). \tag{16.16}$$

Differentiating with respect to x_ν and summing over ν yields

$$\sum_{\nu=1}^{m} \sum_{\mu=1}^{m} \frac{\partial^2}{\partial x_\nu \partial x_\mu} \left(\frac{\partial \psi_\nu}{\partial x_\mu} - \frac{\partial \psi_\mu}{\partial x_\nu} \right) = \lambda^2 \sum_{\nu=1}^{m} \frac{\partial \psi_\nu}{\partial x_\nu}.$$

The left hand side must vanish since if we interchange μ and ν it changes sign. Now $\lambda^2 \neq 0$ and so (16.14) follows as required. Equation (16.15) now comes at once from (16.16).

16.14 Show that the Lagrangian density for small longitudinal vibrations of a continuous elastic rod is

$$\mathcal{L} = \frac{1}{2} \left[\rho \left(\frac{\partial \xi}{\partial t} \right)^2 - \lambda \left(\frac{\partial \xi}{\partial x} \right)^2 \right] \tag{16.17}$$

where $\xi(x, t)$ is the longitudinal displacement of the rod at the point with coordinate x at time t, the mass per unit length is ρ, and λ is the modulus of elasticity. Hence derive the wave equation for longitudinal vibrations of the rod.

• *Solution*

The kinetic energy of an element of the rod between x and $x + \delta x$ is

$$\frac{1}{2} \rho \left(\frac{\partial \xi}{\partial t} \right)^2 \delta x.$$

The potential energy of this element is $\frac{1}{2} P \delta \xi$ where $P = \lambda \partial \xi / \partial x$ is the tension in the rod, and thus the potential energy of the element is

$$\frac{1}{2} \lambda \frac{\partial \xi}{\partial x} \delta \xi = \frac{1}{2} \lambda \left(\frac{\partial \xi}{\partial x} \right)^2 \delta x.$$

Since the Lagrangian function is given by $L = T - V$ where T is the kinetic energy and V is the potential energy it follows that

$$L = \int \mathcal{L} \, dx$$

where $\mathcal{L}$ is the Lagrangian density (16.17).

The Euler -Lagrange equation is

$$\frac{\partial}{\partial x} \left[\frac{\partial \mathcal{L}}{\partial(\partial \xi/\partial x)} \right] + \frac{\partial}{\partial t} \left[\frac{\partial \mathcal{L}}{\partial(\partial \xi/\partial t)} \right] = 0$$

which gives the wave equation for longitudinal vibrations of the rod to be

$$\rho \frac{\partial^2 \xi}{\partial t^2} = \lambda \frac{\partial^2 \xi}{\partial x^2}. \tag{16.18}$$

16.15 Find the Lagrangian density for a vibrating membrane and hence derive the wave equation using the Euler-Lagrange equation.

• *Solution*

Suppose the membrane has mass ρ per unit area and that it is stretched at tension P.

Let $z(x, y, t)$ be the displacement of the membrane at the point with coordinates x, y and at time t.

The kinetic energy of an element $\delta x \delta y$ of the membrane is

$$\frac{1}{2} \rho \left(\frac{\partial z}{\partial t} \right)^2 \delta x \delta y$$

and its potential energy is

$$\frac{1}{2} P \left[\left(\frac{\partial z}{\partial x} \right)^2 + \left(\frac{\partial z}{\partial y} \right)^2 \right] \delta x \delta y.$$

Hence the Lagrangian density is

$$\mathcal{L} = \frac{1}{2} \rho \left(\frac{\partial z}{\partial t} \right)^2 - \frac{1}{2} P \left[\left(\frac{\partial z}{\partial x} \right)^2 + \left(\frac{\partial z}{\partial y} \right)^2 \right]. \tag{16.19}$$

For the present problem the Euler-Lagrange equation is

$$\frac{\partial}{\partial x} \left[\frac{\partial \mathcal{L}}{\partial(\partial z/\partial x)} \right] + \frac{\partial}{\partial y} \left[\frac{\partial \mathcal{L}}{\partial(\partial z/\partial y)} \right] + \frac{\partial}{\partial t} \left[\frac{\partial \mathcal{L}}{\partial(\partial z/\partial t)} \right] = 0$$

which yields the wave equation for a vibrating membrane

$$\rho \frac{\partial^2 z}{\partial t^2} - P \left(\frac{\partial^2 z}{\partial x^2} + \frac{\partial^2 z}{\partial y^2} \right) = 0. \tag{16.20}$$

16.16 The Lagrangian density for a gravitational field having scalar potential V arising from a continuous distribution of matter with density ρ is

$$\mathcal{L} = \frac{1}{8\pi G}(\nabla V)^2 + \rho V \tag{16.21}$$

where G is the gravitational constant. Show that the Euler-Lagrange equation is Poisson's equation

$$\nabla^2 V = 4\pi G\rho.$$

- *Solution*

The Euler -Lagrange equation for the gravitational field V is

$$\frac{\partial \mathcal{L}}{\partial V} - \sum_{k=1}^{3} \frac{\partial}{\partial x_k} \left(\frac{\partial \mathcal{L}}{\partial (\partial V / \partial x_k)} \right) = 0$$

where

$$\mathcal{L} = \frac{1}{8\pi G} \sum_{k=1}^{3} \left(\frac{\partial V}{\partial x_k} \right)^2 + \rho V$$

and so

$$\rho - \frac{1}{4\pi G} \sum_{k=1}^{3} \frac{\partial^2 V}{\partial x_k^2} = 0$$

which is Poisson's equation $\nabla^2 V = 4\pi G\rho$.

16.17 Find the Euler- Lagrange equation for a field with a single component ψ and Lagrangian density having the form

$$\mathcal{L} = \frac{1}{2} \sum_{\mu} \sum_{\nu} g_{\mu\nu} \frac{\partial \psi}{\partial x_\mu} \frac{\partial \psi}{\partial x_\nu} + f\psi.$$

Hence obtain the Lagrangian density which yields the Maxwell equation

$$\text{div } \mathbf{D} = \sum_{i=1}^{3} \frac{\partial D_i}{\partial x_i} = 4\pi\rho$$

for an inhomogeneous anisotropic dielectric medium. Here ρ is the charge density, the displacement vector $\mathbf{D}$ has components

$$D_i = \sum_{j=1}^{3} K_{ij} E_j \quad (i = 1, 2, 3)$$

where the K_{ij} are the elements of a 3×3 symmetric tensor characterizing the dielectric properties of the medium, and the electric field vector is given in terms of a scalar potential V by $\mathbf{E} = -\nabla V$.

- *Solution*

Since the Euler-Lagrange equation for a field with the single component ψ has the form

$$\frac{\partial \mathcal{L}}{\partial \psi} - \sum_{\mu} \frac{\partial}{\partial x_\mu} \left[\frac{\partial \mathcal{L}}{\partial (\partial \psi / \partial x_\mu)} \right] = 0$$

we get

$$\sum_{\mu} \sum_{\nu} \frac{\partial}{\partial x_\mu} \left(g_{\mu\nu} \frac{\partial \psi}{\partial x_\nu} \right) = f.$$

Now

$$\sum_{i=1}^{3} \frac{\partial D_i}{\partial x_i} = 4\pi\rho$$

and

$$D_i = -\sum_{j=1}^{3} K_{ij} \frac{\partial V}{\partial x_j}.$$

Hence

$$\sum_{i=1}^{3} \sum_{j=1}^{3} \frac{\partial}{\partial x_i} \left(K_{ij} \frac{\partial V}{\partial x_j} \right) = -4\pi\rho$$

and so

$$\mathcal{L} = \frac{1}{2} \sum_{i=1}^{3} \sum_{j=1}^{3} K_{ij} \frac{\partial V}{\partial x_i} \frac{\partial V}{\partial x_j} - 4\pi\rho V.$$

16.18 If $\psi(\mathbf{r})$ vanishes on the boundary surface S of a region R, prove that the integral

$$I = \int_R [2\mu\psi - (\nabla\psi)^2 - \lambda^2\psi^2]\, d\tau,$$

where λ and μ are constants, is stationary for the function $\psi = \phi$ satisfying the equation

$$\nabla^2\phi - \lambda^2\phi + \mu = 0 \qquad\qquad (16.22)$$

throughout the region R.

If the stationary value of the integral I is J show that

$$J = \mu \int_R \phi\, d\tau$$

and that $I \leq J$ for all ψ which vanish over S.

• *Solution*

The Lagrangian density is

$$\mathcal{L} = 2\mu\psi - \left(\frac{\partial\psi}{\partial x}\right)^2 - \left(\frac{\partial\psi}{\partial y}\right)^2 - \left(\frac{\partial\psi}{\partial z}\right)^2 - \lambda^2\psi^2$$

and the Euler-Lagrange equation is

$$\frac{\partial\mathcal{L}}{\partial\psi} - \frac{\partial}{\partial x}\left[\frac{\partial\mathcal{L}}{\partial(\partial\psi/\partial x)}\right] - \frac{\partial}{\partial y}\left[\frac{\partial\mathcal{L}}{\partial(\partial\psi/\partial y)}\right] - \frac{\partial}{\partial z}\left[\frac{\partial\mathcal{L}}{\partial(\partial\psi/\partial z)}\right] = 0$$

which gives

$$\frac{\partial^2\psi}{\partial x^2} + \frac{\partial^2\psi}{\partial y^2} + \frac{\partial^2\psi}{\partial z^2} + \mu - \lambda^2\psi = 0,$$

that is

$$\nabla^2\psi - \lambda^2\psi + \mu = 0.$$

It follows that I is stationary for the function $\psi = \phi$ satisfying (16.22).
Now, by Green's theorem,

$$\int_R [\psi\nabla^2\psi + (\nabla\psi)^2]\, d\tau = \int_S \psi\frac{\partial\psi}{\partial n}\, dS = 0$$

since $\psi = 0$ over S and so we have

$$I = \int_R (2\mu\psi + \psi\nabla^2\psi - \lambda^2\psi^2)\, d\tau.$$

Hence

$$J = \int_R (2\mu\phi + \phi\nabla^2\phi - \lambda^2\phi^2)\,d\tau = \mu\int_R \phi\,d\tau$$

as required.

Let us set $\psi = \phi + \eta$ in I where η vanishes over the boundary surface S. Then we obtain, using (16.22) and $\int_R \phi\nabla^2\eta\,d\tau = \int_R \eta\nabla^2\phi\,d\tau$ by Green's theorem, that

$$I = \int_R \mu\phi\,d\tau - \int_R [(\nabla\eta)^2 + \lambda^2\eta^2]\,d\tau \le \int_R \mu\phi\,d\tau = J$$

and thus $I \le J$ for all ψ.

16-5 Eigenvalue problems

The next problem is concerned with obtaining an upper bound to the least angular frequency of vibration of a loaded string using Rayleigh's principle.

16.19 A light string of length $4a$ is fixed at its two ends and stretched at tension P. Three particles of equal masses m are attached to the string dividing it into four equal parts of lengths a. If y_1, y_2, y_3 are the transverse displacements of the particles from their equilibrium positions, in order along the length of the string, find expressions for the kinetic and potential energies. Hence use Rayleigh's principle to obtain an upper bound to the lowest angular frequency of vibration by taking $\mu y_1 = y_2 = \mu y_3$ and optimizing with respect to μ.

• *Solution*

The kinetic energy of the loaded string is

$$T = \frac{1}{2}m(\dot{y}_1^2 + \dot{y}_2^2 + \dot{y}_3^2)$$

and the potential energy of the string is

$$V = \frac{P}{2a}[y_1^2 + (y_2 - y_1)^2 + (y_3 - y_2)^2 + y_3^2].$$

Now we put $y_1 = y_3 = y$ and $y_2 = \mu y$ where $y = A\sin(\omega t + \epsilon)$. Then the mean values of the kinetic energy and the potential energy obtained by averaging over a long time interval are given by

$$\overline{T} = \frac{1}{4}mA^2\omega^2(1 + \mu^2 + 1),$$

$$\overline{V} = \frac{PA^2}{4a}[1 + (\mu - 1)^2 + (1 - \mu)^2 + 1].$$

Rayleigh's principle states that the angular frequency ω given by setting $\overline{T} = \overline{V}$ is an upper bound to the lowest angular frequency of vibration ω_1. This yields

$$\omega^2 = \frac{P}{ma}\frac{4 - 4\mu + 2\mu^2}{2 + \mu^2}.$$

We have

$$\frac{\partial\omega^2}{\partial\mu} = \frac{4P}{ma}\frac{\mu^2 - 2}{(2 + \mu^2)^2}$$

and this vanishes when $\mu = \sqrt{2}$. Then we obtain for the optimized upper bound to ω_1^2 :

$$\omega^2 = \frac{P}{ma}(2 - \sqrt{2}).$$

Actually this is the exact value of ω_1^2.

The variational method we shall use in the next two problems on the vibrations of a stretched membrane depends on the following analysis:

If we substitute $\Psi(\mathbf{r}, t) = \psi(\mathbf{r})\sin(\omega t + \epsilon)$ into the wave equation

$$\nabla^2 \Psi = \frac{1}{c^2}\frac{\partial^2 \Psi}{\partial t^2} \tag{16.23}$$

where c is the phase velocity of the wave motion, we obtain the *Helmholtz equation*

$$\nabla^2 \psi + \frac{\omega^2}{c^2}\psi = 0. \tag{16.24}$$

Multiplying across by ψ and integrating we obtain

$$\frac{\omega^2}{c^2} = -\frac{\int \psi \nabla^2 \psi \, d\tau}{\int \psi^2 \, d\tau}.$$

Now suppose that $\psi = 0$ over the boundary S of the region of integration. Then by Green's theorem we have

$$\int \psi \nabla^2 \psi \, d\tau + \int (\nabla \psi)^2 \, d\tau = \int_S \psi \nabla \psi . d\mathbf{S} = 0$$

and so

$$\frac{\omega^2}{c^2} = \frac{\int (\nabla \psi)^2 \, d\tau}{\int \psi^2 \, d\tau}.$$

Defining the functional

$$I[\chi] = \frac{\int (\nabla \chi)^2 \, d\tau}{\int \chi^2 \, d\tau} \tag{16.25}$$

we may write

$$I[\psi] = \frac{\omega^2}{c^2} \tag{16.26}$$

where ψ is a solution of the Helmholtz equation (16.24) satisfying the boundary condition $\psi = 0$.

Clearly $I[\chi] \geq 0$ for all functions χ and so the function ψ_1 which makes $I[\chi]$ attain its least possible value and satisfies the boundary condition $\psi_1 = 0$ is the exact solution of the Helmholtz equation corresponding to the least angular frequency ω_1. It follows that for all functions χ we have

$$I[\chi] \geq \frac{\omega_1^2}{c^2} \tag{16.27}$$

which may be used to obtain an upper bound to ω_1 as we shall see in the following problem.

16.20 A circular membrane of radius a, which is clamped at its boundary, has mass ρ per unit area and is stretched at tension P. Using the trial function

$$\chi(r) = (a^2 - r^2)^s$$

and treating s as a variable parameter, find an upper bound to the least angular frequency of vibration of the membrane.

● *Solution*

We have

$$I[\chi] = \frac{\int_0^a \left(\frac{\partial\chi}{\partial r}\right)^2 r\, dr}{\int_0^a \chi^2 r\, dr} = \frac{2s(2s+1)}{a^2(2s-1)}.$$

The least value of $I[\chi]$ occurs when $\partial I/\partial s = 0$ which gives $s = \frac{1}{2}(1+\sqrt{2})$. Then using (16.27) we get

$$I[\chi] = \frac{3+2\sqrt{2}}{a^2} = \frac{5.828}{a^2} \geq \frac{\omega_1^2}{c^2}$$

where the phase velocity is given by $c = \sqrt{P/\rho}$.
The exact value of $\omega_1^2 a^2/c^2$ is 5.783.

16.21 A membrane having the shape of a sector of radius a and angle $\pi/3$ is clamped at its boundary composed of the circular arc $r = a$ and the straight edges given by $\theta = 0$ and $\theta = \pi/3$ where r, θ are circular polar coordinates. Using the trial function

$$\chi(r,\theta) = r(a^2 - r^2)\sin 3\theta$$

find an upper bound to the least angular frequency of vibration of the membrane.

● *Solution*

We have

$$\int_S (\nabla_1\chi)^2 dS = \int_0^{\pi/3} d\theta \int_0^a (\nabla_1\chi)^2 r\, dr$$

where the circular polar components of $\nabla_1\chi$ are $\partial\chi/\partial r$ and $r^{-1}\partial\chi/\partial\theta$.
    Now

$$\frac{\partial\chi}{\partial r} = (a^2 - 3r^2)\sin 3\theta, \qquad \frac{1}{r}\frac{\partial\chi}{\partial\theta} = 3(a^2 - r^2)\cos 3\theta$$

and so

$$\int_S (\nabla_1\chi)^2 dS = \int_0^{\pi/3} d\theta \int_0^a [(a^2 - 3r^2)^2 \sin^2 3\theta + 9(a^2 - r^2)^2 \cos^2 3\theta] r\, dr = \frac{\pi a^6}{3}.$$

Also

$$\int_S \chi^2 dS = \int_0^{\pi/3} d\theta \int_0^a \chi^2 r\, dr = \int_0^{\pi/3} \sin^2 3\theta\, d\theta \int_0^a r^3(a^2 - r^2)^2\, dr = \frac{\pi a^8}{144}.$$

Hence

$$I[\chi] = \frac{\int_S (\nabla_1\chi)^2\, dS}{\int_S \chi^2\, dS} = \frac{48}{a^2}$$

and so using (16.27) we have that

$$\frac{\omega_1^2}{c^2} \leq \frac{48}{a^2}$$

where the phase velocity is given by $c = \sqrt{P/\rho}$.
    Hence $4\sqrt{3}\, c/a$ is an upper bound to the least angular frequency ω_1.

Finally we shall use the variational method to solve some eigenvalue problems in quantum theory. We have already considered a number of such problems, 15.16 to 15.21, in the previous chapter, but we shall complete this book with a few more.

The first one of these uses a trial function based on time-independent perturbation theory.

16.22 Use the variational method for bound states with the trial function

$$\psi_t = \psi^{(0)} + \lambda\psi^{(1)},$$

composed of the first two terms of the time-independent perturbation theory expansion of the wave function, to show that

$$E^{(0)} + \lambda E^{(1)} + \frac{\lambda^2[E^{(2)} + \lambda E^{(3)}]}{1 + \lambda^2(\psi^{(1)}, \psi^{(1)})}$$

is an upper bound to the ground state energy eigenvalue where $E^{(0)} + \lambda E^{(1)} + \lambda^2 E^{(2)} + \lambda^3 E^{(3)}$ is the energy of the ground state evaluated to third order using perturbation theory.

• *Solution*

The variational method for bound states gives

$$E_0 \leq \frac{(\psi_t, H\psi_t)}{(\psi_t, \psi_t)} \tag{16.28}$$

where E_0 is the ground state energy, $H = H_0 + \lambda H'$ is the Hamiltonian of the perturbed system and H_0 is the Hamiltonian of the unperturbed system.

Then we have $(\psi_t, \psi_t) = (\psi^{(0)}, \psi^{(0)}) + \lambda[(\psi^{(1)}, \psi^{(0)}) + (\psi^{(0)}, \psi^{(1)})] + \lambda^2(\psi^{(1)}, \psi^{(1)})$ and since $(\psi^{(0)}, \psi^{(r)}) = (\psi^{(r)}, \psi^{(0)}) = 0$ for $r \neq 0$ and $(\psi^{(0)}, \psi^{(0)}) = 1$ we get

$$(\psi_t, \psi_t) = 1 + \lambda^2(\psi^{(1)}, \psi^{(1)}). \tag{16.29}$$

Further

$$(\psi_t, H\psi_t) = \quad (\psi^{(0)}, H_0\psi^{(0)}) + \lambda[(\psi^{(1)}, H_0\psi^{(0)}) + (\psi^{(0)}, H_0\psi^{(1)})] + \lambda^2(\psi^{(1)}, H_0\psi^{(1)})$$
$$+ \lambda(\psi^{(0)}, H'\psi^{(0)}) + \lambda^2[(\psi^{(1)}, H'\psi^{(0)}) + (\psi^{(0)}, H'\psi^{(1)})] + \lambda^3(\psi^{(1)}, H'\psi^{(1)})$$

and so

$$(\psi_t, H\psi_t) = E^{(0)} + \lambda^2(\psi^{(1)}, H_0\psi^{(1)}) + \lambda E^{(1)} + 2\lambda^2 E^{(2)} + \lambda^3(\psi^{(1)}, H'\psi^{(1)}) \tag{16.30}$$

using $H_0\psi^{(0)} = E^{(0)}\psi^{(0)}$, $(\psi^{(0)}, H_0\psi^{(1)}) = (\psi^{(1)}, H_0\psi^{(0)}) = 0$ and $E^{(r)} = (\psi^{(0)}, H'\psi^{(r-1)})$.

Now from the time-independent perturbation theory we have also

$$(H_0 - E^{(0)})\psi^{(1)} = (E^{(1)} - H')\psi^{(0)} \tag{16.31}$$

so that $(\psi^{(1)}, H_0\psi^{(1)}) = E^{(0)}(\psi^{(1)}, \psi^{(1)}) - E^{(2)}$. Further we have

$$(H_0 - E^{(0)})\psi^{(2)} = (E^{(1)} - H')\psi^{(1)} + E^{(2)}\psi^{(0)} \tag{16.32}$$

and so $(\psi^{(1)}, H'\psi^{(1)}) = E^{(1)}(\psi^{(1)}, \psi^{(1)}) - (\psi^{(1)}, (H_0 - E^{(0)})\psi^{(2)})$. But from 16.31 we see that $(\psi^{(1)}, (H_0 - E^{(0)})\psi^{(2)}) = -E^{(3)}$ and hence (16.30) yields

$$(\psi_t, H\psi_t) = (E^{(0)} + \lambda E^{(1)})(1 + \lambda^2(\psi^{(1)}, \psi^{(1)}) + \lambda^2 E^{(2)} + \lambda^3 E^{(3)}. \tag{16.33}$$

Hence, using (16.28), (16.29) and (16.33), the desired upper bound to the ground state energy E_0 is obtained.

16.23 A one-dimensional oscillating system has the Hamiltonian

$$H = -\frac{\hbar^2}{2m}\frac{d^2}{dx^2} + \frac{1}{2}m\omega^2 x^2 + \mu x^4.$$

Use the variational method for bound states with the unnormalized trial function $u_t(x) = \exp(-\beta x^2/2)$ to obtain an equation for the parameter β which gives the least upper bound to the ground state energy of the system. Evaluate this least upper bound to first order in μ and verify that this agrees with first order perturbation theory.

- *Solution*

We have that

$$\int_{-\infty}^{\infty} u_t^2(x)\,dx = \int_{-\infty}^{\infty} \exp(-\beta x^2)\,dx = \sqrt{\frac{\pi}{\beta}}$$

and

$$\int_{-\infty}^{\infty} u_t H u_t\,dx = \int_{-\infty}^{\infty} \exp(-\beta x^2)\left[-\frac{\hbar^2}{2m}\left(\beta^2 x^2 - \beta\right) + \frac{1}{2}m\omega^2 x^2 + \mu x^4\right]dx.$$

Now

$$(-1)^n \frac{\partial^n}{\partial \beta^n}\int_{-\infty}^{\infty}\exp(-\beta x^2)\,dx = \int_{-\infty}^{\infty} x^{2n}\exp(-\beta x^2)\,dx$$

and so

$$\int_{-\infty}^{\infty} x^2 \exp(-\beta x^2)\,dx = \frac{1}{2\beta}\sqrt{\frac{\pi}{\beta}}, \quad \int_{-\infty}^{\infty} x^4 \exp(-\beta x^2)\,dx = \frac{3}{4\beta^2}\sqrt{\frac{\pi}{\beta}}. \tag{16.34}$$

Hence

$$I = \frac{\int_{-\infty}^{\infty} u_t H u_t\,dx}{\int_{-\infty}^{\infty} u_t^2(x)\,dx} = \frac{\hbar^2}{2m}\frac{\beta}{2} + \frac{1}{2}m\omega^2\frac{1}{2\beta} + \frac{3\mu}{4\beta^2} \tag{16.35}$$

and so the value of the parameter β which gives the least upper bound to the ground state energy of the system, given by $\partial I/\partial \beta = 0$, satisfies

$$\beta^2 = \alpha^4 + \frac{6m\mu}{\hbar^2 \beta}$$

where $\alpha = (m\omega/\hbar)^{1/2}$.

To zero-order in μ we get $\beta = \alpha^2 = m\omega/\hbar$ and $I = \frac{1}{2}\hbar\omega$ which corresponds to the exact eigenfunction and energy eigenvalue for the ground state of the simple harmonic oscillator.

To first-order in μ we find that $\beta = \alpha^2 + 3\mu/m\omega^2$ and

$$I = \frac{1}{2}\hbar\omega + \frac{3\mu\hbar^2}{4m^2\omega^2}. \tag{16.36}$$

The first-order energy given by perturbation theory is, using the second integral of (16.34),

$$E^{(1)} = \frac{\int_{-\infty}^{\infty}\mu x^4 \exp(-\alpha^2 x^2)\,dx}{\int_{-\infty}^{\infty}\exp(-\alpha^2 x^2)\,dx} = \frac{3\mu}{4\alpha^4}$$

and this, as required, is the same as the second term of (16.36).

16.24 If E_0 and $E_1(> E_0)$ are the two least energy eigenvalues for a Hamiltonian operator H, and ψ_t is a trial function such that $H\psi_t \neq E_0\psi_t$, show that

$$\frac{(H\psi_t, [H - E_0]\psi_t)}{(\psi_t, [H - E_0]\psi_t)} \geq E_1. \tag{16.37}$$

Hence, by taking $\psi_t = \exp(-\alpha r/a_0)$ and $E_0 = -e^2/2a_0$, show that $-e^2/10a_0$ is an upper bound to the energy of the first excited state of a hydrogen atom where $a_0 = \hbar^2/me^2$ is the Bohr radius.

- *Solution*

To verify the correctness of (16.37) we expand $\psi_t = \sum_n a_n\psi_n$ in terms of the complete set of orthonormal eigenfunctions ψ_n of the Hamiltonian H such that $H\psi_n = E_n\psi_n$. Then we find that

$$\frac{(H\psi_t, [H - E_0]\psi_t)}{(\psi_t, [H - E_0]\psi_t)} = \frac{\sum_n a_n \sum_m a_m E_n(E_m - E_0)(\psi_n, \psi_m)}{\sum_n a_n \sum_m a_m(E_m - E_0)(\psi_n, \psi_m)} = \frac{\sum_n a_n^2 E_n(E_n - E_0)}{\sum_n a_n^2(E_n - E_0)} \geq E_1$$

since $E_n \geq E_1$ for $n \geq 1$ and thus $\sum_n a_n^2 E_n(E_n - E_0) \geq E_1 \sum_n a_n^2(E_n - E_0)$.

Taking $\psi_t = \exp(-\alpha r/a_0)$ and $E_0 = -e^2/2a_0$ we obtain

$$(H - E_0)\psi_t = \left[-\frac{\hbar^2}{2m}\frac{1}{r^2}\frac{d}{dr}\left(r^2\frac{d}{dr}\right) - \frac{e^2}{r} + \frac{e^2}{2a_0} \right] \exp(-\alpha r/a_0)$$

so that

$$(H - E_0)\psi_t = \left[\frac{e^2}{2a_0}\left(1 - \alpha^2\right) - \frac{e^2}{r}(1 - \alpha) \right] \exp(-\alpha r/a_0)$$

which yields

$$(\psi_t, [H - E_0]\psi_t) = 8\pi\frac{e^2}{2a_0}\left(\frac{a_0}{2\alpha}\right)^3 (\alpha - 1)^2$$

using $\int_0^\infty r^n \exp(-\lambda r)\, dr = n!/\lambda^{n+1}$.

Also

$$(H\psi_t, [H - E_0]\psi_t) = 8\pi \left(\frac{e^2}{2a_0}\right)^2 \left(\frac{a_0}{2\alpha}\right)^3 \left(5\alpha^4 - 12\alpha^3 + 9\alpha^2 - 2\alpha\right)$$

and thus

$$I[\psi_t] = \frac{(H\psi_t, [H - E_0]\psi_t)}{(\psi_t, [H - E_0]\psi_t)} = \frac{e^2}{2a_0}\frac{5\alpha^4 - 12\alpha^3 + 9\alpha^2 - 2\alpha}{(\alpha - 1)^2}. \tag{16.38}$$

The optimum value of α is obtained by setting $dI/d\alpha = 0$ which gives $(5\alpha - 1)(\alpha - 1)^3 = 0$. Now $\alpha = 1$ yields the exact ground state wave function and energy so that $H\psi_t = E_0\psi_t$ which is disallowed. Hence we have the optimum value $\alpha = 1/5$ and then (16.38) gives $I[\psi_t] = -e^2/10a_0$ as an upper bound to the energy E_1 of the first excited state of atomic hydrogen which has the exact value $-e^2/8a_0$.

In this final chapter we have seen how to solve a wide variety of problems in applied mathematics or mathematical physics by using variational principles based on Lagrangian functions F which lead to Euler-Lagrange equations by making $I = \int F\, dx$ stationary.

In the case of geodesics we have to minimize the path length $\int ds$ on the surface being considered or in the space-time continuum in the general theory of relativity.

However in dynamics the independent variable x is the time t and the Lagrangian function F is taken to be $L = T - V$ where T is the kinetic energy and V is the potential energy of the dynamical system.

To obtain field equations, such as those for electromagnetism, the integral $I = \int \mathcal{L} \, d\tau$ is made stationary for weak variations of the field components where $\mathcal{L}$ is the appropriate Lagrangian density.

Eigenvalue problems are treated differently. For example, in the case of wave mechanics, we have to minimize the functional for the energy given by $I[\chi] = \int \chi^* H \chi \, d\tau / \int \chi^* \chi \, d\tau$ where $H = T + V$ is the Hamiltonian operator. This leads to an upper bound for the lowest energy eigenvalue and to upper bounds for higher energy eigenvalues as well.

Variational methods provide very powerful methods for solving problems in applied mathematics. Indeed in certain cases using a variational method offers the best possible method for obtaining an accurate approximate solution.

BIBLIOGRAPHY

Adler, R., Bazin, M. and Schiffer, M., *Introduction to General Relativity* , McGraw Hill, New York, 1975.

Arthurs, A. M., *Calculus of Variations,* Routledge and Kegan Paul, London, 1975.

Chorlton, F., *Textbook of Fluid Dynamics,* van Nostrand, London, 1967.

Churchill, R. V., *Fourier Series and Boundary Value Problems,* McGraw-Hill, New York, 1963.

Churchill, R. V., *Operational Mathematics,* McGraw-Hill, New York, 1958.

Clegg, J. C., *Calculus of Variations,* Oliver and Boyd, Edinburgh, 1968.

Coulson, C. A., *Electricity,* Oliver and Boyd, London, 1948.

Coulson, C. A., *Waves,* Oliver and Boyd, London, 1955.

Coulson, C. A. and Boyd, T.J.M., *Electricity,* Longman, London.

Coulson, C. A. and Jeffrey, A., *Waves,* Longman, London.

Goldstein, H., *Classical Mechanics,* Addison-Wesley, Cambridge, Mass., 1950.

Lawden, D. F., *Mathematical Principles of Quantum Mechanics,* Methuen, 1967.

Lawden, D. F., *Tensor Calculus and Relativity,* Methuen, 1967.

Lovitt, W. V., *Linear Integral Equations,* Dover, New York, 1950.

Moiseiwitsch, B. L., *Integral equations,* Longman, London, 1977; Dover, New York, 2005.

Moiseiwitsch, B. L., *Variational Principles,* Wiley, London, 1966; Dover, New York, 2004.

Morse, P. M. and Feshbach, H., *Methods of Theoretical Physics,* McGraw-Hill, New York, 1953.

Ramsey, A. S., *Hydrodynamics,* Bell, London, 1929.

Rindler, W., *Special Relativity,* Oliver and Boyd, London, 1966.

Schiff, L. I., *Quantum mechanics,* McGraw-Hill, New York, 1955.

Smythe, W. R., *Static and Dynamic Electricity,* McGraw-Hill, New York, 1939.

Weatherburn, C. E., *Elementary Vector Analysis,* Bell, London, 1946.

Weatherburn, C. E., *Advanced Vector Analysis,* Bell, London, 1947.

Whittaker, E. T., *Analytical Dynamics, 4th ed.,* Cambridge University Press, 1937.

Index

A CATALOG OF SELECTED

DOVER BOOKS

IN SCIENCE AND MATHEMATICS

Mathematics

FUNCTIONAL ANALYSIS (Second Corrected Edition), George Bachman and Lawrence Narici. Excellent treatment of subject geared toward students with background in linear algebra, advanced calculus, physics and engineering. Text covers introduction to inner-product spaces, normed, metric spaces, and topological spaces; complete orthonormal sets, the Hahn-Banach Theorem and its consequences, and many other related subjects. 1966 ed. 544pp. 6⅛ x 9¼.   0-486-40251-7

DIFFERENTIAL MANIFOLDS, Antoni A. Kosinski. Introductory text for advanced undergraduates and graduate students presents systematic study of the topological structure of smooth manifolds, starting with elements of theory and concluding with method of surgery. 1993 edition. 288pp. 5⅜ x 8½.   0-486-46244-7

VECTOR AND TENSOR ANALYSIS WITH APPLICATIONS, A. I. Borisenko and I. E. Tarapov. Concise introduction. Worked-out problems, solutions, exercises. 257pp. 5⅝ x 8¼.   0-486-63833-2

AN INTRODUCTION TO ORDINARY DIFFERENTIAL EQUATIONS, Earl A. Coddington. A thorough and systematic first course in elementary differential equations for undergraduates in mathematics and science, with many exercises and problems (with answers). Index. 304pp. 5⅜ x 8½.   0-486-65942-9

FOURIER SERIES AND ORTHOGONAL FUNCTIONS, Harry F. Davis. An incisive text combining theory and practical example to introduce Fourier series, orthogonal functions and applications of the Fourier method to boundary-value problems. 570 exercises. Answers and notes. 416pp. 5⅜ x 8½.   0-486-65973-9

COMPUTABILITY AND UNSOLVABILITY, Martin Davis. Classic graduate-level introduction to theory of computability, usually referred to as theory of recurrent functions. New preface and appendix. 288pp. 5⅜ x 8½.   0-486-61471-9

AN INTRODUCTION TO MATHEMATICAL ANALYSIS, Robert A. Rankin. Dealing chiefly with functions of a single real variable, this text by a distinguished educator introduces limits, continuity, differentiability, integration, convergence of infinite series, double series, and infinite products. 1963 edition. 624pp. 5⅜ x 8½.   0-486-46251-X

METHODS OF NUMERICAL INTEGRATION (SECOND EDITION), Philip J. Davis and Philip Rabinowitz. Requiring only a background in calculus, this text covers approximate integration over finite and infinite intervals, error analysis, approximate integration in two or more dimensions, and automatic integration. 1984 edition. 624pp. 5⅜ x 8½.   0-486-45339-1

INTRODUCTION TO LINEAR ALGEBRA AND DIFFERENTIAL EQUATIONS, John W. Dettman. Excellent text covers complex numbers, determinants, orthonormal bases, Laplace transforms, much more. Exercises with solutions. Undergraduate level. 416pp. 5⅜ x 8½.   0-486-65191-6

RIEMANN'S ZETA FUNCTION, H. M. Edwards. Superb, high-level study of landmark 1859 publication entitled "On the Number of Primes Less Than a Given Magnitude" traces developments in mathematical theory that it inspired. xiv+315pp. 5⅜ x 8½.
0-486-41740-9

CALCULUS OF VARIATIONS WITH APPLICATIONS, George M. Ewing. Applications-oriented introduction to variational theory develops insight and promotes understanding of specialized books, research papers. Suitable for advanced undergraduate/graduate students as primary, supplementary text. 352pp. 5⅜ x 8½.
0-486-64856-7

MATHEMATICIAN'S DELIGHT, W. W. Sawyer. "Recommended with confidence" by *The Times Literary Supplement,* this lively survey was written by a renowned teacher. It starts with arithmetic and algebra, gradually proceeding to trigonometry and calculus. 1943 edition. 240pp. 5⅜ x 8½.
0-486-46240-4

ADVANCED EUCLIDEAN GEOMETRY, Roger A. Johnson. This classic text explores the geometry of the triangle and the circle, concentrating on extensions of Euclidean theory, and examining in detail many relatively recent theorems. 1929 edition. 336pp. 5⅜ x 8½.
0-486-46237-4

COUNTEREXAMPLES IN ANALYSIS, Bernard R. Gelbaum and John M. H. Olmsted. These counterexamples deal mostly with the part of analysis known as "real variables." The first half covers the real number system, and the second half encompasses higher dimensions. 1962 edition. xxiv+198pp. 5⅜ x 8½.
0-486-42875-3

CATASTROPHE THEORY FOR SCIENTISTS AND ENGINEERS, Robert Gilmore. Advanced-level treatment describes mathematics of theory grounded in the work of Poincaré, R. Thom, other mathematicians. Also important applications to problems in mathematics, physics, chemistry and engineering. 1981 edition. References. 28 tables. 397 black-and-white illustrations. xvii + 666pp. 6⅛ x 9¼.
0-486-67539-4

COMPLEX VARIABLES: Second Edition, Robert B. Ash and W. P. Novinger. Suitable for advanced undergraduates and graduate students, this newly revised treatment covers Cauchy theorem and its applications, analytic functions, and the prime number theorem. Numerous problems and solutions. 2004 edition. 224pp. 6½ x 9¼.
0-486-46250-1

NUMERICAL METHODS FOR SCIENTISTS AND ENGINEERS, Richard Hamming. Classic text stresses frequency approach in coverage of algorithms, polynomial approximation, Fourier approximation, exponential approximation, other topics. Revised and enlarged 2nd edition. 721pp. 5⅜ x 8½.
0-486-65241-6

INTRODUCTION TO NUMERICAL ANALYSIS (2nd Edition), F. B. Hildebrand. Classic, fundamental treatment covers computation, approximation, interpolation, numerical differentiation and integration, other topics. 150 new problems. 669pp. 5⅜ x 8½.
0-486-65363-3

MARKOV PROCESSES AND POTENTIAL THEORY, Robert M. Blumental and Ronald K. Getoor. This graduate-level text explores the relationship between Markov processes and potential theory in terms of excessive functions, multiplicative functionals and subprocesses, additive functionals and their potentials, and dual processes. 1968 edition. 320pp. 5⅜ x 8½.
0-486-46263-3

ABSTRACT SETS AND FINITE ORDINALS: An Introduction to the Study of Set Theory, G. B. Keene. This text unites logical and philosophical aspects of set theory in a manner intelligible to mathematicians without training in formal logic and to logicians without a mathematical background. 1961 edition. 112pp. 5⅜ x 8½.   0-486-46249-8

INTRODUCTORY REAL ANALYSIS, A.N. Kolmogorov, S. V. Fomin. Translated by Richard A. Silverman. Self-contained, evenly paced introduction to real and functional analysis. Some 350 problems. 403pp. 5⅜ x 8½. 0-486-61226-0

APPLIED ANALYSIS, Cornelius Lanczos. Classic work on analysis and design of finite processes for approximating solution of analytical problems. Algebraic equations, matrices, harmonic analysis, quadrature methods, much more. 559pp. 5⅜ x 8½. 0-486-65656-X

AN INTRODUCTION TO ALGEBRAIC STRUCTURES, Joseph Landin. Superb self-contained text covers "abstract algebra": sets and numbers, theory of groups, theory of rings, much more. Numerous well-chosen examples, exercises. 247pp. 5⅜ x 8½.
0-486-65940-2

QUALITATIVE THEORY OF DIFFERENTIAL EQUATIONS, V. V. Nemytskii and V.V. Stepanov. Classic graduate-level text by two prominent Soviet mathematicians covers classical differential equations as well as topological dynamics and ergodic theory. Bibliographies. 523pp. 5⅜ x 8½. 0-486-65954-2

THEORY OF MATRICES, Sam Perlis. Outstanding text covering rank, nonsingularity and inverses in connection with the development of canonical matrices under the relation of equivalence, and without the intervention of determinants. Includes exercises. 237pp. 5⅜ x 8½. 0-486-66810-X

INTRODUCTION TO ANALYSIS, Maxwell Rosenlicht. Unusually clear, accessible coverage of set theory, real number system, metric spaces, continuous functions, Riemann integration, multiple integrals, more. Wide range of problems. Undergraduate level. Bibliography. 254pp. 5⅜ x 8½. 0-486-65038-3

MODERN NONLINEAR EQUATIONS, Thomas L. Saaty. Emphasizes practical solution of problems; covers seven types of equations. ". . . a welcome contribution to the existing literature. . . ."—*Math Reviews*. 490pp. 5⅜ x 8½. 0-486-64232-1

MATRICES AND LINEAR ALGEBRA, Hans Schneider and George Phillip Barker. Basic textbook covers theory of matrices and its applications to systems of linear equations and related topics such as determinants, eigenvalues and differential equations. Numerous exercises. 432pp. 5⅜ x 8½. 0-486-66014-1

LINEAR ALGEBRA, Georgi E. Shilov. Determinants, linear spaces, matrix algebras, similar topics. For advanced undergraduates, graduates. Silverman translation. 387pp. 5⅜ x 8½. 0-486-63518-X

MATHEMATICAL METHODS OF GAME AND ECONOMIC THEORY: Revised Edition, Jean-Pierre Aubin. This text begins with optimization theory and convex analysis, followed by topics in game theory and mathematical economics, and concluding with an introduction to nonlinear analysis and control theory. 1982 edition. 656pp. 6⅛ x 9¼.
0-486-46265-X

SET THEORY AND LOGIC, Robert R. Stoll. Lucid introduction to unified theory of mathematical concepts. Set theory and logic seen as tools for conceptual understanding of real number system. 496pp. 5⅜ x 8¼. 0-486-63829-4

Math—Decision Theory, Statistics, Probability

INTRODUCTION TO PROBABILITY, John E. Freund. Featured topics include permutations and factorials, probabilities and odds, frequency interpretation, mathematical expectation, decision-making, postulates of probability, rule of elimination, much more. Exercises with some solutions. Summary. 1973 edition. 247pp. 5⅜ x 8½.
0-486-67549-1

STATISTICAL AND INDUCTIVE PROBABILITIES, Hugues Leblanc. This treatment addresses a decades-old dispute among probability theorists, asserting that both statistical and inductive probabilities may be treated as sentence-theoretic measurements, and that the latter qualify as estimates of the former. 1962 edition. 160pp. 5⅜ x 8½.
0-486-44980-7

APPLIED MULTIVARIATE ANALYSIS: Using Bayesian and Frequentist Methods of Inference, Second Edition, S. James Press. This two-part treatment deals with foundations as well as models and applications. Topics include continuous multivariate distributions; regression and analysis of variance; factor analysis and latent structure analysis; and structuring multivariate populations. 1982 edition. 692pp. 5⅜ x 8½.   0-486-44236-5

LINEAR PROGRAMMING AND ECONOMIC ANALYSIS, Robert Dorfman, Paul A. Samuelson and Robert M. Solow. First comprehensive treatment of linear programming in standard economic analysis. Game theory, modern welfare economics, Leontief input-output, more. 525pp. 5⅜ x 8½.   0-486-65491-5

PROBABILITY: AN INTRODUCTION, Samuel Goldberg. Excellent basic text covers set theory, probability theory for finite sample spaces, binomial theorem, much more. 360 problems. Bibliographies. 322pp. 5⅜ x 8½.   0-486-65252-1

GAMES AND DECISIONS: INTRODUCTION AND CRITICAL SURVEY, R. Duncan Luce and Howard Raiffa. Superb nontechnical introduction to game theory, primarily applied to social sciences. Utility theory, zero-sum games, n-person games, decision-making, much more. Bibliography. 509pp. 5⅜ x 8½.   0-486-65943-7

INTRODUCTION TO THE THEORY OF GAMES, J. C. C. McKinsey. This comprehensive overview of the mathematical theory of games illustrates applications to situations involving conflicts of interest, including economic, social, political, and military contexts. Appropriate for advanced undergraduate and graduate courses; advanced calculus a prerequisite. 1952 ed. x+372pp. 5⅜ x 8½.   0-486-42811-7

FIFTY CHALLENGING PROBLEMS IN PROBABILITY WITH SOLUTIONS, Frederick Mosteller. Remarkable puzzlers, graded in difficulty, illustrate elementary and advanced aspects of probability. Detailed solutions. 88pp. 5⅜ x 8½.   0-486-65355-2

PROBABILITY THEORY: A CONCISE COURSE, Y. A. Rozanov. Highly readable, self-contained introduction covers combination of events, dependent events, Bernoulli trials, etc. 148pp. 5⅜ x 8¼.   0-486-63544-9

THE STATISTICAL ANALYSIS OF EXPERIMENTAL DATA, John Mandel. First half of book presents fundamental mathematical definitions, concepts and facts while remaining half deals with statistics primarily as an interpretive tool. Well-written text, numerous worked examples with step-by-step presentation. Includes 116 tables. 448pp. 5⅜ x 8½.   0-486-64666-1

Math—Geometry and Topology

ELEMENTARY CONCEPTS OF TOPOLOGY, Paul Alexandroff. Elegant, intuitive approach to topology from set-theoretic topology to Betti groups; how concepts of topology are useful in math and physics. 25 figures. 57pp. 5³/₈ x 8¹/₂.   0-486-60747-X

A LONG WAY FROM EUCLID, Constance Reid. Lively guide by a prominent historian focuses on the role of Euclid's Elements in subsequent mathematical developments. Elementary algebra and plane geometry are sole prerequisites. 80 drawings. 1963 edition. 304pp. 5³/₈ x 8¹/₂.   0-486-43613-6

EXPERIMENTS IN TOPOLOGY, Stephen Barr. Classic, lively explanation of one of the byways of mathematics. Klein bottles, Moebius strips, projective planes, map coloring, problem of the Koenigsberg bridges, much more, described with clarity and wit. 43 figures. 210pp. 5³/₈ x 8¹/₂.   0-486-25933-1

THE GEOMETRY OF RENÉ DESCARTES, René Descartes. The great work founded analytical geometry. Original French text, Descartes's own diagrams, together with definitive Smith-Latham translation. 244pp. 5³/₈ x 8¹/₂.   0-486-60068-8

EUCLIDEAN GEOMETRY AND TRANSFORMATIONS, Clayton W. Dodge. This introduction to Euclidean geometry emphasizes transformations, particularly isometries and similarities. Suitable for undergraduate courses, it includes numerous examples, many with detailed answers. 1972 ed. viii+296pp. 6¹/₈ x 9¹/₄.   0-486-43476-1

EXCURSIONS IN GEOMETRY, C. Stanley Ogilvy. A straightedge, compass, and a little thought are all that's needed to discover the intellectual excitement of geometry. Harmonic division and Apollonian circles, inversive geometry, hexlet, Golden Section, more. 132 illustrations. 192pp. 5³/₈ x 8¹/₂.   0-486-26530-7

THE THIRTEEN BOOKS OF EUCLID'S ELEMENTS, translated with introduction and commentary by Sir Thomas L. Heath. Definitive edition. Textual and linguistic notes, mathematical analysis. 2,500 years of critical commentary. Unabridged. 1,414pp. 5³/₈ x 8¹/₂. Three-vol. set.
Vol. I: 0-486-60088-2   Vol. II: 0-486-60089-0   Vol. III: 0-486-60090-4

SPACE AND GEOMETRY: IN THE LIGHT OF PHYSIOLOGICAL, PSYCHOLOGICAL AND PHYSICAL INQUIRY, Ernst Mach. Three essays by an eminent philosopher and scientist explore the nature, origin, and development of our concepts of space, with a distinctness and precision suitable for undergraduate students and other readers. 1906 ed. vi+148pp. 5³/₈ x 8¹/₂.   0-486-43909-7

GEOMETRY OF COMPLEX NUMBERS, Hans Schwerdtfeger. Illuminating, widely praised book on analytic geometry of circles, the Moebius transformation, and two-dimensional non-Euclidean geometries. 200pp. 5³/₈ x 8¹/₄.   0-486-63830-8

DIFFERENTIAL GEOMETRY, Heinrich W. Guggenheimer. Local differential geometry as an application of advanced calculus and linear algebra. Curvature, transformation groups, surfaces, more. Exercises. 62 figures. 378pp. 5³/₈ x 8¹/₂.   0-486-63433-7

A TREATISE ON ELECTRICITY AND MAGNETISM, James Clerk Maxwell. Important foundation work of modern physics. Brings to final form Maxwell's theory of electromagnetism and rigorously derives his general equations of field theory. 1,084pp. 5³/₈ x 8¹/₂. Two-vol. set.   Vol. I: 0-486-60636-8   Vol. II: 0-486-60637-6

MATHEMATICS FOR PHYSICISTS, Philippe Dennery and Andre Krzywicki. Superb text provides math needed to understand today's more advanced topics in physics and engineering. Theory of functions of a complex variable, linear vector spaces, much more. Problems. 1967 edition. 400pp. 6¹/₂ x 9¹/₄.   0-486-69193-4

INTRODUCTION TO QUANTUM MECHANICS WITH APPLICATIONS TO CHEMISTRY, Linus Pauling & E. Bright Wilson, Jr. Classic undergraduate text by Nobel Prize winner applies quantum mechanics to chemical and physical problems. Numerous tables and figures enhance the text. Chapter bibliographies. Appendices. Index. 468pp. 5³/₈ x 8¹/₂.   0-486-64871-0

METHODS OF THERMODYNAMICS, Howard Reiss. Outstanding text focuses on physical technique of thermodynamics, typical problem areas of understanding, and significance and use of thermodynamic potential. 1965 edition. 238pp. 5³/₈ x 8¹/₂.

0-486-69445-3

THE ELECTROMAGNETIC FIELD, Albert Shadowitz. Comprehensive under- graduate text covers basics of electric and magnetic fields, builds up to electromagnetic theory. Also related topics, including relativity. Over 900 problems. 768pp. 5³/₈ x 8¹/₄.

0-486-65660-8

GREAT EXPERIMENTS IN PHYSICS: FIRSTHAND ACCOUNTS FROM GALILEO TO EINSTEIN, Morris H. Shamos (ed.). 25 crucial discoveries: Newton's laws of motion, Chadwick's study of the neutron, Hertz on electromagnetic waves, more. Original accounts clearly annotated. 370pp. 5³/₈ x 8¹/₂.   0-486-25346-5

EINSTEIN'S LEGACY, Julian Schwinger. A Nobel Laureate relates fascinating story of Einstein and development of relativity theory in well-illustrated, nontechnical volume. Subjects include meaning of time, paradoxes of space travel, gravity and its effect on light, non-Euclidean geometry and curving of space-time, impact of radio astronomy and space-age discoveries, and more. 189 b/w illustrations. xiv+250pp. 8³/₈ x 9¹/₄.   0-486-41974-6

THE VARIATIONAL PRINCIPLES OF MECHANICS, Cornelius Lanczos. Philosophic, less formalistic approach to analytical mechanics offers model of clear, scholarly exposition at graduate level with coverage of basics, calculus of variations, principle of virtual work, equations of motion, more. 418pp. 5³/₈ x 8¹/₂.   0-486-65067-7

Paperbound unless otherwise indicated. Available at your book dealer, online at www.doverpublications.com, or by writing to Dept. GI, Dover Publications, Inc., 31 East 2nd Street, Mineola, NY 11501. For current price information or for free catalogues (please indicate field of interest), write to Dover Publications or log on to www.doverpublications.com and see every Dover book in print. Dover publishes more than 400 books each year on science, elementary and advanced mathematics, biology, music, art, literary history, social sciences, and other areas.